MODELS AND MODELERS OF
HYDROGEN

MODELS AND MODELERS OF
HYDROGEN

THALES THOMSON
RUTHERFORD BOHR SOMMERFELD
GOUDSMIT HEISENBERG SCHRÖDINGER
DIRAC SALLHOFER

Editor
Akhlesh Lakhtakia
Pennsylvania State University

World Scientific
Singapore • New Jersey • London • Hong Kong

Published by

World Scientific Publishing Co. Pte. Ltd.

P O Box 128, Farrer Road, Singapore 912805

USA office: Suite 1B, 1060 Main Street, River Edge, NJ 07661

UK office: 57 Shelton Street, Covent Garden, London WC2H 9HE

British Library Cataloguing-in-Publication Data
A catalogue record for this book is available from the British Library.

The publisher is grateful to the following for permission to reproduce the photographs found in this volume:
J.J. Thomson, *Courtesy:* AIP Emilio Segré Visual Archives (WF Meggers Collection); E. Rutherford, c. 1919, *Courtesy:* Cambridge University Library and AIP Emilio Segré Visual Archives; N. H. D. Bohr, *Courtesy:* Princeton University and AIP Emilio Segré Visual Archives; A. J. W. Sommerfeld, c. 1949, *Courtesy:* AIP Emilio Segré Visual Archives; S. A. Goudsmit, photo by Heka Photo, *Courtesy:* AIP Emilio Segré Visual Archives; W. K. Heisenberg, photo by A. Börtzells Truckeri, *Courtesy:* AIP Emilio Segré Visual Archives; E. Schrödinger, photo by Francis Simon, *Courtesy:* AIP Emilio Segré Visual Archives (Francis Simon Collection); P. A. M. Dirac, *Courtesy:* AIP Emilio Segré Visual Archives; H. H. Sallhofer, 1994, *Courtesy:* H. H. Sallhofer.

First published 1996
Reprinted 2001

MODELS AND MODELERS OF HYDROGEN

ISBN 981-02-2302-1

This book is printed on acid-free paper.

Printed in Singapore by Uto-Print

Preface

Patrick M. Lenahan

Professor of Engineering Science and Mechanics,
The Pennsylvania State University,
University Park, PA 16802-1401, USA

Over two millennia ago, Democritus concluded that matter was composed of indivisible atoms. The ideas of Democritus and other Greek atomists were interesting speculation, unsupported by experiment. Why should one believe such ideas? Early in the nineteenth century, John Dalton and other chemists began to investigate the world on an atomic scale, testing and refining their ideas with quantitative experiments. Nevertheless, just one hundred years ago relatively little was known about the microscopic world on the atomic scale and nothing was known of the inner workings of an individual atom. This state of profound ignorance changed dramatically at the turn of the century.

J. J. Thomson's discovery of the electron and Max Planck's development of a quantum theory of black body radiation initiated a thirty-year period of astonishing progress, culminating in P. A. M. Dirac's development of relativistically consistent quantum mechanics. The development of quantum mechanics — a theory which describes nature on atomic, subatomic and microscopic scales, to great precision — is one of the greatest intellectual achievements in human history. This achievement is primarily due to a few (mostly) very young men. This book tells the story of the development of quantum mechanics by telling us about their work and their lives.

Hydrogen, the simplest and most abundant atom in the universe, played a uniquely important role in the development of quantum mechanics. It provided a system from which experimentalists could secure comprehensible data and with which theoreticians could test models of increasing sophistication.

Niels Bohr realized that the optical emissions from hydrogen are consistent with the quantization of orbital angular momentum. From this realization grew his simple, seminal, not quite right, model; the model was later interpreted by de Broglie to imply the existence of electron standing waves. In turn, de Broglie's work inspired Schrödinger development of nonrelativistic quantum mechanics based upon his famous equation. Schrödinger's triumph became apparent when his equation accurately described the optical emissions from hydrogen.

This book ends with an interview with a practitioner of physics from the end of the twentieth century who doesn't agree with the conventional (Copenhagen) interpretation of quantum mechanics. Some of the essayists (as well as the author of this preface) do accept the Copenhagen interpretation. Is this bad? No. Real scientists argue, constantly. Are these arguments resolvable? Absolutely, eventually.

Scientific arguments are resolvable by experiment, by observation. Arguments between theorists and experimentalists are eventually resolved by experiment. This book is about theories and experiments. Experimental observations with hydrogen lead to theories which could be applied to hydrogen. The interplay between theory and observation resulted in a precise and detailed understanding of the world which continues to expand and deepen to this day.

To The Reader

Akhlesh Lakhtakia

Department of Engineering Science and Mechanics,
The Pennsylvania State University,
University Park, PA 16802-1401, USA

Hydrogen is primitive matter. The hydrogen atom is a simple planetary system: a negatively charged particle of minuscule mass orbitting a positively charged massive particle. The hydrogen atom is electrically neutral and the only stable two-body system possible. The sole orbitting electron can occupy one of many stable energy states which can be computed with very high precision. It may transit from a low to a high energy state by absorbing discrete units of externally supplied energy, and it may revert to a lower energy state on emitting discrete units of energy. Thus, atomic hydrogen displays a rich spectrum ranging from the microwave to the ultraviolet frequencies. Add to this its abundance in the universe, and you will come to realize the importance of hydrogen.

The hydrogen spectrum came to be appreciated long before the hydrogen atom was resolved as an electron orbitting a proton. The Swedish astronomer Anders J. Ångström identified the first resonance of the hydrogen atom in 1862. Shortly thereafter, Sir William Higgins identified several resonances in the visible spectra of stars. In 1885, the Basel schoolteacher Johann J. Balmer proposed a general formula for several resonances that are together said to comprise the Balmer series. This empirical formula was further generalized by the spectroscopist Johannes R. Rydberg at Lund in 1889, but only in 1913 — after the emergence of the quantum concept — came its first theoretical interpretation, given by the Danish giant Niels Bohr.

So there we have the discrete nature of electrical charge central to the hydrogen atom and much of twentieth-century science and technology. The road to Bohr's understanding of the hydrogen atom was tortuous. This simple atom continued to be a profound mystery till the Cambridge scholar Paul Dirac conjured up his now famous Dirac equation.

I do not want to convey the impression that Dirac's work completely elucidated the nature of matter. Indeed, it did not; but it set the tone for subsequent research. The hydrogen atom has come to be known very well. Its energetics are avidly studied by electrochemists and nuclear physicists alike. Its spectrum is sought in the light coming from the farthest corners of the universe. It is commonly used in metrology as well as in quantum electrodynamics. If you wish to acquaint yourself with some modern developments, I suggest a review paper by B. Cagniac, M. D. Plimmer, L. Julien and F. Biraben in the journal *Reports of Progress in Physics* **57** (1994) 853–893.

The book before you is not about these modern developments. Instead, you will find here an account of models of hydrogen atom culminating with that given by Dirac. The cut-off is deliberate. Much of what is described in this book has entered low-level physics courses and some of it can even be found printed on T-shirts. Gentle reader, I imagine you to be an undergraduate student in the USA, an advanced pre-university student in most of the rest of the world, a schoolteacher or the instructor of an undergraduate course with an intriguing title such as *The Poetry of Physics*, or a non-physicist with an abiding interest in the history of science. Therefore, you will not be overburdened with mathematical details often, though mathematical acumen and training shall be invaluable. If you are a student, you may well decide to study physics and other sciences, pure as well as applied, more seriously after skimming through the following pages. Or, you may be content with becoming an early member of the science-literate citizenry of the world that Sir Peter Medawar envisioned at the dawn of the age of lunar and planetary exploration.

And now, a few words about this book. It is divided into two parts. The first part contains nine essays on nine models and modelers of the hydrogen atom and an interview. Not much is known of the first, Thales, who represents the pre-Socratic natural philosophers of ancient Greece. The idea of the atom was not peculiar to the Greeks, it having formed a part of the Hindu view of life too. What is interesting is that the Greeks were aware of the wave-particle duality, as Dennis Rose, a Buddhist monk from the British Isles, will bring out in his essay.

The second essay is devoted to the plum pudding model of the hydrogen atom devised by Sir Joseph John Thomson. This was not a planetary model, but it managed to explain many of the experimental facts known by the end of the nineteenth century. Maximo Barón, a physics professor at the University of Buenos Aires, will explain that Thomson's model paved the way for a model due to Lord Ernest Rutherford.

Rutherford, a native of New Zealand, gave the first planetary model: a tiny nucleus about which scurries a tinier electron. In the third essay, Homer Tilton of the Pima Community College in Arizona will point out that not only did Rutherford come up with model, he even tested it by bombarding atoms with miniscule projectiles.

Unfortunately, an electron scurrying around the positively charged nucleus has no chance of survival according to classical electromagnetic theory. The electron shall be inexorably drawn to the nucleus and gobbled up. Thus the hydrogen atom would go *Poof*! sooner or later, a very unsatisfactory state of affairs for the Rutherford model. Enter Niels Bohr. In 1913, he applied the quantum concept of Max Planck to natural phenomena other than light, and — as Daniel Robertson (retired from the British Defense Research Agency) will inform you in the fourth essay — thereby conceptualized stable orbits for the electron.

An orbitting electron must move almost at the speed of light. Relativistic considerations cannot be ignored, according to my distinguished Penn State colleague William White. The great synthesizer, Arnold Sömmerfeld, therefore blended Bohr's stable orbits with the theory of relativity. Circular orbits were replaced by elliptical ones.

Advances came fast and furious during the 1920s. There was a "sweet confusion" then prevailing in the world of physics and Samuel Goudsmit was, as Tarun Biswas of the State University of New York wrote for the sixth essay, an oddball. In order to explain the exclusion principle of Wolfgang Pauli, Goudsmit dreamt up an oddball idea: every electron has a spin. Spin produces angular momentum, but it should not be confused with rotational motion. There was no theory of electron spin at that time, but one came shortly thereafter from the pen of Paul Dirac.

But before the electron spin could emerge naturally from a model, rather than decreed by fiat, quantum mechanics had to be formalized. The name of Werner Heisenberg rings loudly in the Quantum Mechanics Hall of Fame. James Edmonds, who teaches at the McNeese State University in Louisiana, will take you on an exciting ride, should you be willing to hold his hand. He will acquaint you the basics of quantum mechanics in the seventh essay.

When Heisenberg comes, can Erwin Schrödinger be far behind? (Some years ago, an artist friend loaned me a portrait of Schrödinger, who now looks down from the wall at my desk. I know all is well with what I am doing if I can detect him smiling.) Professor Owen de Lange of the University of Natal shall offer you the next ride — on wave mechanics and how Schrödinger applied wave mecahnics to the hydrogen atom. I suggest that you gloss over the seventh and the eighth essays in a first reading but return to them for details after reading the ninth essay on Paul Dirac.

Dirac was surely a magician. If you are a cricket fan, think of him as the Ranjitsinhji of physics; if you prefer baseball, he was the Babe Ruth of physics. The consummate grace with which he conjured up the Dirac equation — neatly wrapping up all known aspects of the hydrogen atom in it — is described in the ninth essay by Vilas Rajopadhye, who taught physics for many years at the Indian Institute of Technology, Kharagapur.

Scientific theories never die. They keep on being tinkered with, adjusted for new phenomena, refined for easier understanding, and so on. Therefore, I interviewed the Austrian physicist Hans Sallhofer, whose work on an electrodynamic model of the hydrogen atom exploits the structural similarity of the Maxwell postulates of electromagnetism and the Dirac equation. Sallhofer is also a keen observer of the state of modern physics, and you shall be provoked by the interview which concludes the first part of this book.

If your interest in detailed studies on the hydrogen atom is aroused, you may go on to *The Spectrum of Atomic Hydrogen* (Oxford University Press, 1957) written by G. W. Series. You may also find it profitable to borrow *Quantum Field Theory* (McGraw-Hill,

1980) from your library, the book's authors being C. Itzykson and J. Zuber.

I suppose you will read some essays completely, others partially. You may repair to a science library to get hold of advanced textbooks and monographs. But nothing beats the sheer joy of poring over a paper published by a great scientist many decades ago. Journals from even the 1950s may no longer be easily found in stacks because fearful librarians hide them in deep dark recesses. Librarians are rightly fearful that continued exposure to the readers shall decompose the volumes quickly, but scholars and enthusiasts are asphyxiated by the preservative steps librarians take. Nevertheless, you can have the thrill of reading seminal papers on the hydrogen atom because the facsimile reproductions of ten of them are included in the second part of this book. Enjoy!

In closing this note, let me express my appreciation of all nine essayists. They are an international group. Their styles are different but their commitment to science education is the same. You may prefer certain essays to others, as other readers shall appreciate a different set of essays.

I also thank Hans Sallhofer for flying down to Périgueux, France for the interview and Barbara Aman of World Scientific for looking after the book during the final production stage.

Contents

Preface
Patrick M. Lenahan v

To the Reader
Akhlesh Lakhtakia vii

PART I: NINE ESSAYS AND AN INTERVIEW

Chapter 1. Greek Metaphysics and Thales from a Modern Perspective
Dennis Rose 3

Chapter 2. J. J. Thomson's Atom: A Number of Negatively Electrified Corpuscles Enclosed in a Sphere of Uniform Positive Electrification
Máximo Barón 13

Chapter 3. The Hydrogen Atom: The Rutherford Model
Homer Benjamin Tilton 33

Chapter 4. Niels Bohr — Through Hydrogen Towards the Nature of Matter
Daniel Stewart Robertson 49

Chapter 5. Arnold Sommerfeld
William B. White 83

Chapter 6. Electron Spin and Samuel Goudsmit
Tarun Biswas 99

Chapter 7. Heisenberg and Hydrogen
James D. Edmonds, Jr. 117

Chapter 8. Erwin Schrödinger, Wave Mechanics, and the Hydrogen Atom
Owen L. de Lange 139

Chapter 9. Dirac and the Emergence of Quantum Mechanics
Vilas Y. Rajopadhye 169

Chapter 10. Hans Sallhofer Speaks
Akhlesh Lakhtakia 209

PART II: FACSIMILE REPRINTS OF TEN ORIGINAL PAPERS

Reprint 1. "On the structure of the atom", Joseph J. Thomson, *Philosophical Magazine and Journal of Science*, Series 6, Vol. 7, No. 39 (1904) 237–265 233

Reprint 2. "The scattering of α and β particles by matter and the structure of the atom", Ernest Rutherford, *Philosophical Magazine and Journal of Science*, Series 6, Vol. 21, No. 125 (1911) 669–688 263

Reprint 3. "On the constitution of atoms and molecules", Niels Bohr, *Philosophical Magazine and Journal of Science*, Series 6, Vol. 26, No. 151 (1913) 1–25 283

Reprint 4. "Zur Theorie des Zeeman-Effekts der Wasserstofflinien, mit einem Anhang über den Stark-Effekt", Arnold Sommerfeld, *Physikalische Zeitschrift*, Vol. 17, No. 20 (1916) 491–507 309

Reprint 5. "Spinning electrons and the structure of spectra", George E. Uhlenbeck and Samuel A. Goudsmit, *Nature*, Vol. 117, No. 2938 (1926) 264–265 327

Reprint 6. "Über das Wasserstoffspektrum vom Standpunkt der neuen Quantenmechanik", Wolfgang Pauli, Jr., *Zeitschrift für Physik*, Vol. 36, No. 5 (1926) 336–363 329

Reprint 7. "Quantisierung als Eigenwertproblem (Erste Mitteilung)", Erwin Schrödinger, *Annalen der Physik*, Series 4, Vol. 79, No. 4 (1926) 361–376 357

Reprint 8. "The quantum theory of the electron", Paul A. M. Dirac, *Proceedings of the Royal Society of London*, Series A, Vol. 17, No. 778 (1928) 610–624 373

Reprint 9. "The quantum theory of the electron. Part II", Paul A. M. Dirac, *Proceedings of the Royal Society of London*, Series A, Vol. 118, No. 779 (1928) 351–361 389

Reprint 10. "Hydrogen in electrodynamics. VI. The general solution", Hans H. Sallhofer, *Zeitschrift für Naturforschung*, Vol. 45a (1990) 1361–1366 401

Index 407

PART I: NINE ESSAYS AND AN INTERVIEW

Chapter 1. Greek Metaphysics and Thales from a Modern Perspective

Dennis Rose

Chapter 1

Greek Metaphysics and Thales from a Modern Perspective

Dennis Rose

Destouchesstrasse 67A,
80796 München, Germany

Abstract

Ruminations on waves and particles can be traced back to the Greeks. Thales of Miletus, a wave man, was apparently the first to think of water as the universal and the eternal substrate.

1. Pre-Socratic Natural Philosophy

The wave-particle dilemma of today's quantum physics and contemporary natural philosophy has its roots in the thinking of the pre-Socratic natural philosophers of Greece. Two antagonistic camps with divergent views may be identified among them.

The first camp is comprised of the Ionian-Eleatic natural philosophers who flourished *circa* 550 BC. It is most prominently represented by the Ionian Thales, Anaximander and Heraclitus, as well as the Eleatic Parmenides and Melissus. They thought of the world as the infinitely extended *being* according to the tautology: *the being is, the not-being is not.* Since empty space represented the embodiment of the *not-being*, it could not exist. The *being* was homogeneous, within itself free of any flow, immobile as a whole, infinitely extended, without beginning and end in time, invariable and indivisible, a dense and heavy mass.

Members of the first camp considered the large variety of perceptible phenomena as *illusions of the senses*, or even delusions.

Waves they knew only from perception: as surface phenomena on waters. They did not attach any importance to them, since waves may rock but cannot move ships. The *being*, however, possessed all the properties of a wave carrier that could be the medium for the fast moving light and for the matter waves of today's wave picture. All perceptible variety, all *illusions of the senses*, therefore could be comprehensible as progressing or standing perturbations of the *being*, especially since it had elasticity but not plasticity.

To the Ionian-Eleatic duality of *being* and *illusion* corresponded the duality of *medium* and *perturbation*. The pre-Socratic natural philosophers between Thales and Melissus thus offered a wave carrier with their *being*, the perturbations of which may make understandable the *illusion of the senses*, i.e., all perceptions. They were the true fathers of today's wave picture.

In the second camp congregated the founders of the philosophical school of Abdera in Thracia: Leucippus and Democritus chiefly, as well as their disciples. While the Ionian-Eleatic *being* was related to the medium and the illusions of the senses to the perturbations, the situation was completely reversed with the Abderan *atomists*. They regarded the material world (i.e., the perturbations) as the being and the medium as an illusion. Even more, they thought of the medium as not existent at all, and postulated instead the *void* or empty space. Only insofar as they considered empty space as something real, there was a vestigial medium: as a substrate. The Abderan *being* consisted of indivisible, eternally invariable and forever moving smallest particles, called the *atoms*, and between the atoms lay empty space. With this view the atomists were the true fathers of today's particle picture.

2. Particles or Waves?

A grave unsolved problem for the atomists was keeping distance: how can the atoms of a solid body remain at appropriate distance to each other? Why will they not drift apart in all directions or stick together completely? Democritus spoke of small hooklets on atoms. To those hooklets would correspond certain wavefields of today's wave picture, or the gluons of today's particle picture. Anyhow, the original concept of the atomists — *the world consists of empty space and the atoms, and nothing else* — cannot be sustained. On account of the hooklets, i.e., wavefields, the Ionian-Eleatic *being* reappeared on the scene in latter-day Europe.

Two thousand years after the passing away of the Greek natural philosophy at the end of the fifth century B.C., the natural philosophers of the (European) Age of Enlightenment became aware of wave phenomena and were in need of a *being* to carry waves. They called it *ether*. Because of the extremely high speed of light, ether had to be not only transparent but also, and above all, highly compact. Since the physicists of the Enlightenment followed the atomists with respect to perceptible objects, they considered planets as solid atomic aggregates. Planets orbiting the sun are inconceivable in a solid ether. Instead, that fact required an ether (Greek: *heavenly air*) that would not hinder the motion of matter: it would have to be extremely thin and mechanically imperceptible. The Enlightenment ether thus had both to be highly compact and imperceptible, and this conflict was seemingly unresolvable.

3. Particles and Waves

Earlier in this century, Louis de Broglie[1] and Paul Ehrenfest[2] resolved the wave-particle contradiction. De Broglie stated: Wherever there is a particle, there exists also a wave. Ehrenfest stated the converse: Wherever there is a wave, there exists also a particle (i.e., the center of energy of the wave).

There existed, however, still another circumstance that let the particle picture appear inadequate to many. The early atomic hydrogen models, culminating in Niels Bohr's 1913 model, were conceived purely within the particle picture. But even in 1915 we see with Arnold Sommerfeld[3] that the particle picture — despite relativistic treatment — was unable to produce one of the most important properties of particles, namely *spin*. But Paul Dirac, the wave protagonist, eventually managed to qualitatively express the spin of hydrogen by means of a hydrogen field (i.e., the matter wave).[4,5] Dirac's hydrogen model could not at that time show quantitatively the half-integer spin of hydrogen. The spinor components of this model did not then possess half-integer spin factors. However, fifty years later Hans Sallhofer applied the Maxwell–Dirac isomorphism to the hydrogen atom, and unambiguously showed it as having half-integer spin with the help of an electrodynamic wavefield.[6,7]

According to Sallhofer's electrodynamic representation of hydrogen, it is clear that any matter possesses — exactly in the same way as light — electrodynamic wave character, and that electromagnetic wave matter has to move through any ether, be it ever so solid, according to the law of inertia. The contradictory demands placed on the Enlightenment ether are thus eliminated. In retrospect, could this mean that the Ionians and the Eleatics were in possession of the more profound understanding as than to the atomists?

Had the Eleatics already known about quaternions, they would have been able to conceive a purely metaphysical cosmology as follows: The *being* is an infinitely extended substrate, without beginning and end in time, and stationary within itself. It is three-dimensional since one- or two-dimensional spaces are *not-beings.* The three-dimensionality is introduced necessarily and sufficiently by the requirement that the geometric structure of the *being* is determined by the quaternion group (i.e., the unique finite-dimensional algebra over the field of real numbers). From the inherence of the quaternions the validity of electrodynamics follows conclusively,[8] and from that the description of light and matter as perturbations of the being result.

4. Back to Thales and His Camp

Let us turn back once more to the Ionians. The first of them, Thales of Miletus, had to free himself, to begin with, of all influences of the Greek mythology of his time. Only then could he reflect on metaphysical and experience-determined cognition in irreligious openness. He arrived at the opinion: Water is the universal and eternal substrate, and all perceptible things originate from water. With that he had realized that water is contained in all matter. Of course, his understanding was only partially on a metaphysical nature, as it was derived — for the most part — from impressions of material experience. It makes Thales the first hydrogen researcher in the history of science. Nietzsche saw a threefold vision of truth in Thales' water thesis[9]: first, the question as to the material cause of all things; second, the demand that this question be logically answered; and third, the postulate that everything reduces eventually to one principle.

While Thales had assumed water as the substrate, his disciple Anaximander did not like the idea that perceptible water, an object of the illusions of the senses, should be identical to the purely metaphysically inferred substrate. He therefore postulated a substrate removed from sensory perceptions, which he called *apeiron*. For the first time we find here a medium removed from the senses, and this objectivity resounds to Anaximander's everlasting merit.

Another outstanding ancient was Heraclitus of Ephesus. He postulated, going beyond the substrate, an approach towards understanding those phenomena that his predecessors had viewed as sensory perceptions. He thus addressed the *perturbations* of the eternal *being*, i.e., objects characterized by birth and disappearance. This comprehensive program he approached with the three-component collective term of *logos*.

First, Heraclitus' logos contained a single god. This may have been probably as a reminiscence of the anthropomorphic polytheism he had abandoned. More likely though, his cosmological model was in need of a first causal trigger. Even within a modern cosmology one might possibly postulate neither a beginning nor an end in time for the *being*, but one cannot do so for the perturbations. Physicists today talk of a *big bang*: at a certain point in time a perturbation-to-be can free itself out of a point-like concentration and run off in all directions. And only by turning back this process and by a subsequent periodic repetition of the big bang can one free the perturbations from the tyranny of an originally initiating god. Thus an oscillating perturbation can be postulated, without beginning or end, provided a conservation law is assumed for it.

Second, logos contained a normative law. This common and general law was supposedly always and everywhere in effect. Thus, Heraclitus expressed his desire for a unified field theory.

Third, logos contained the unity of opposites. Thus, logos in the unity of the world was to be understood as the unity of opposite qualifications. Indeed, polar tendencies in tension against each other form a unit exactly because of their polarity. Heraclitus mentioned two examples:

(i) At any point of a bow string two oppositely directed forces (or *divectors*) act in both directions along the string. This divectorial force forms the potential that makes a bow suitable for shooting arrows.

(ii) In the same way, a divectorial force acts at any point of a lyre's string in both directions along the string. It forms the potential that enables the string to produce sound.

Translating into today's understanding, we may say, for example: in a dumb-bell made by by welding a rod to two disks, equal and opposing (divectorial) tensions act at any point in the rod. The unity of opposites is the foundation of the mechanical models of the hydrogen atom, and of matter in general. Sallhofer's wave picture is in complete agreement with Heraclitus' concept of the unity of opposites: in Sallhofer's electrodynamic hydrogen the electric and magnetic field strengths are divectors because of their half-integer spin factors. The electrodynamic hydrogen model therefore expresses explicitly the unity of opposites in any point of configuration space.

Finally, Heraclitus envisioned fire as the tangible substrate of logos. Nothing can be concealed from fire. It is rational and eternal. As the uniform principle of all things, fire was supposed to be the substrate out of which all phenomena originate and into which they return. Here we have Heraclitus' conservation law for perturbations.

It is appropriate in conclusion to recall Werner Heisenberg[9]: "... some statements of ancient philosophy are rather near to those of modern science. This simply shows how far one can get by combining the ordinary experience of nature that we have without doing experiments with the untiring effort to get some logical order into this experience, to understand it from general principles."

References

1. Louis de Broglie, "Sur le parallélisme entre la dynamique du point matériel et l'optique géométrique", *J. Phys. Radium* **1** (1926) 1–6.
2. Paul Ehrenfest, "Bemerkung über die angenäherte Gültigkeit der klassischen Mechanik innerhalb der Quantenmechanik", *Z. Phys.* **45** (1927) 455–457.
3. Arnold Sommerfeld, "Zur Theorie des Zeeman-Effkts der Wasserstoflinien, mit einem Anhang über den Stark-Effekt", *Z. Phys.* **17** (1916) 491–507.
4. Paul A. M. Dirac, "The quantum theory of the electron", *Proc. R. Soc. London* **A117** (1928) 610–624.

5. Paul A. M. Dirac, "The quantum theory of the electron. Part II", *Proc. R. Soc. London* **A118** (1928) 351–361.
6. Hans Sallhofer, "Hydrogen in electrodynamics. I. Preliminary theories", *Z. Naturforsch.* **A43** (1988) 1039–1043.
7. Hans Sallhofer, "Hydrogen in electrodynamics. VI. The general solution", *Z. Naturforsch.* **A45** (1990) 1361–1366.
8. V. Majerník and M. Nagy, "Quaternionic form of Maxwell's equations with sources", *Lett. Nuovo Cimento* **16** (1976) 265–268.
9. Werner Heisenberg, *Physics and Philosophy: The Revolution in Modern Science*, Harper & Brothers, New York, 1958, chap. 4.

Dennis Rose took his degree in philosophy and psychology at Cambridge University. After working in England for many years as a management consultant, he retired to his native Ireland to study Buddhist philosophy and its relationship to Greek as well as modern Western philosophies and science. For four years he lived in Sri Lanka as a Buddhist monk, studying part-time at the University of Peradeniya and conducting courses on Buddhism for Westerners. He now lives in Germany where he practises psychotherapy — Buddhist style — and management consultancy. He is currently engaged in writing a book on the relation of quantum theory and the Buddhist concept of "the void".

Chapter 2. J. J. Thomson's Atom: A Number of Negative Electrified Corpuscles Enclosed in a Sphere of Uniform Positive Electrification

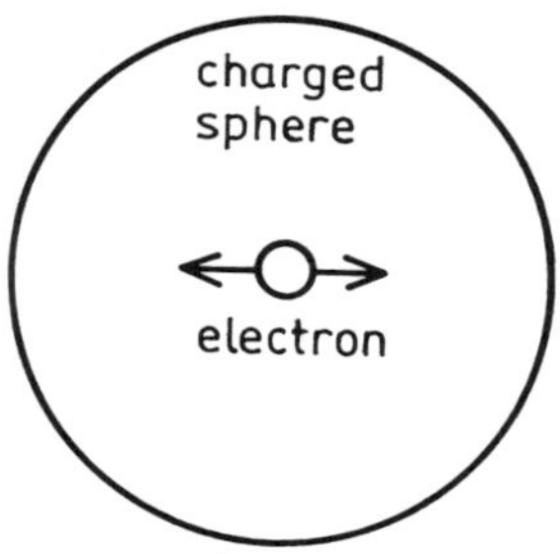

Máximo Barón

J. J. Thomson

Chapter 2

J. J. Thomson's Atom: A Number of Negatively Electrified Corpuscles Enclosed in a Sphere of Uniform Positive Electrification

Máximo Barón

Professor of Experimental Physics,
Departamento de Física,
Facultad de Ciencias Exactas y Naturales,
Universidad de Universidad de Buenos Aires,
Pabellón 1 - C. Universitaria,
1428 Buenos Aires, Argentina

Abstract

During the 1890s many facts regarding the nature of the atom emerged, the most important being that the atom was composed of even smaller particles. Theoretical and experimental discoveries of the time led to the identification of the first subatomic *corpuscle*, as J. J. Thomson originally called the electron. The measurement of it's e/m ratio, as well as of the properties of cathode rays, opened up a new perspective of the atom and helped to understand several then unexplained phenomena (e.g., atomic spectra). Reluctant to adopt a planetary view, Thomson deliberately chose an electromagnetic approach that has invoked the image of plum puddings and raisin muffins. Although Thomson's model was unsatisfactory for many reasons, it paved the way for Rutherford, one of his brilliant co-workers.

1. Prologue

Nearly one hundred years ago, just after the turn of the century, the Spanish writer Miguel de Unamuno (1864–1936) wrote a charming novel with the curious title of *Love and Pedagogy.* It is the story of a man who wanted to have a son with certain intellectual qualities, so he planned everything very carefully, including the choice of his bride. However reality fooled him constantly from the very start and nothing turned out the way he expected.

To make this fact absolutely clear and somewhat in jest, Unamuno divided the novel in three parts: a "prologue", a "logue" and an "epilogue". These are in line with the old recipe for a good presentation: say what you are going to talk about, say it, and then say what you have said. Therefore I will try to follow these directions.

The story of the structure of the atom in general, and that of hydrogen in particular, is a true relay race, with each contributor advancing the matter a little further and perhaps getting closer to a satisfactory solution. What is interesting about each step is that although initially the proposed answer seemed most promising very soon, reality — as in Unamuno's novel — started pounding in the facts that made it unacceptable. Even so valuable information slowly accumulated in the fruitful stepwise process.

For a long time the atom was considered to be the smallest fraction into which matter could be divided and still retain its properties. But after the middle of the 19th century, many phenomena were observed that were puzzling to no end. Explanations were attempted, but without much success because the existence of subatomic particles was not even speculated upon.

Then, during the last decade, discharges in gases and cathode rays led to the discovery of the electron and opened the road to the atomic interior. An attractive resulting atomic model was due to J. J. Thomson. The man, the time, the place and the model will be the subject of my essay.

So now that I have said what I will talk about, let me get to the details and tell you.

2. The Scene and the People

"*There are critical events in the history of the sciences which mark sharp breaks in men's thoughts*".

This is the opening sentence of the book entitled, *J. J. Thomson and the Cavendish Laboratory in His Day*, by his son George Paget Thomson.[1] It contains an age-old concept always applicable, but particularly pertinent in the case of what I will call hereafter *Thomson's atom.* For it to materialize, several ingredients were needed, among them: (i) the fact that viewing an atom as the smallest piece of matter was becoming inadequate, (ii) many known properties of matter (especially electromagnetic) required better explanations, (iii) there existed a wealth of unexplained experimental facts, and (iv) exceptional minds found themselves at the same time in the same place.

2.1. *Particles smaller than the atom and electromagnetic properties of matter*

The works of Volta,[2] Coulomb,[3] Ampère[4] and Oersted,[5] started to suggest that the phenomenon called electricity might be caused by a flow, or a current, of charged particles. Faraday's monumental *Experimental Researches in Electricity*[6] introduced the idea of lines of force, that Maxwell[7] used to propose the field concept. This brought forth the first field unification, that of electricity and magnetism. Maxwell also calculated the speed of electromagnetic radiation, that showed the connection between the electrostatic unit of charge — first determined by Weber and Kohlrausch[8] — and the velocity of light.

So, the idea of small charged bodies, initially proposed by Benjamin Franklin in the mid-18th century, started to emerge as a distinct possibility to explain both the flow of charged particles and the nature of atoms. (The latter became especially evident after Faraday's experiments on electrolysis, that showed quite clearly the existence of charged particles, (i.e. ions), that flow in the liquid where the electrodes are immersed).

The obvious question behind all these facts, although perhaps never stated in so many words, was: "What is the actual size of these charges?". In this respect it has been pointed out[9] that although Faraday did lay the basis for electrochemistry, he seemed to have missed a most important fact: that his laws implied the existence of a charge that each atom carried either as a unit or its multiple, so that electricity as well as matter could have a particulate nature.

The existence of a "unit of charge" was first proposed by Weber (1840) in an "electron theory",[8] that was later supported by

Helmholtz.[10] Stoney coined the word "electron" in 1894[11] for the "molecule of electricity".

2.2. *Unexplained experimental facts*

The scene was now set for the new observations that started to pour in during the latter part of the 19th century, as the result of work on electric discharges in gases. The pioneering work of Plücker[12] in 1859, completed by Hittorf[13] and Goldstein[14] on cathode rays, together with later studies by Crookes,[15] Hertz,[16] Lenard[17] and Perrin[18] ushered in a new perspective on electric charges.[19] They could be either portions of a fluid consisting of a large number of equal electrons or of certain stresses produced by charges in the "ether" surrounding them, stresses that bound to matter became observable. The latter being a rather involved way to define a wave propagating in matter. So, the crucial question of their being particles or waves was raised and the stage was set for the formal birth of contemporary physics as it is understood today.

There was, however, one additional discovery that was the crowning event of its time. On December 28 of 1895, Röntgen[20] communicated to the Physico-Medical Society of Würzburg the discovery he had made some eight weeks earlier on the properties of rays emitted from a Crookes tube. Röntgen found that the incidence of cathode rays on a metallic target inside an evacuated tube caused the emission of what he called X-rays. They are so well known that their name has now become a universal household word.

2.3. *The Cavendish Laboratory. Great minds and excellent hands in one place*

During most of the 19th century, physics — or Natural Philosophy as it was called — was taught using essentially mathematical methods. There were indeed a few teachers who used very good and appropriate instruments for lecture demonstrations, but the role of students was mostly a passive one: that of observers.

In the Americas there were at least two notable exceptions. At the College of William and Mary, Williamsburg, VA, USA, there still exist the remnants of excellent instruments used in the early years of the 19th century.[21] At the University of Buenos

Aires, Argentina, Ottavio F. Mossotti was provided with a complete laboratory,[22] that had been ordered from Paris for the Experimental Physics course; but nothing is left of it now, unfortunately.

However, the concept of an experimental physics laboratory, for both teaching and research, did not exist until the last decades of the 19th century. One of the first to be created, so to speak, was the Cavendish Laboratory[23] at Cambridge University, UK. The Duke of Devonshire, Chancellor of the University, offered to pay the cost. It was completed in 1874 and named after the Duke's scientist kinsman, Henry Cavendish (1736–1810), who among other things had established the difference between the quantity and the intensity of electricity.

The first professor of the Cavendish, as its Chairman or Head was called, was none other than James Clerk Maxwell (1831–1879).[24] After this death he was succeeded by Lord Rayleigh (1842–1919),[25] who strongly influenced the scientific culture of the Cavendish encouraging both theoretical and experimental work of the highest quality.

Rayleigh resigned the professorship at Cambridge in 1884 and the position was taken by J. J. Thomson. This came about at a time when profound changes in physics were at hand. Not many however, were aware of this fact, and some even considered that the time of the great discoveries was over, so much so that a belief prevailed that only more accurate measurements were left for physicists to accomplish. One of those that shared these views was, oddly enough, none other than Lord Kelvin.

The veritable revolution brought about by the introduction of the concept of discrete natural units in electricity, as had been earlier that of atoms in ordinary matter, started at this time when Thomson became in charge and had an entourage of exceptionally able scientists and technicians at the Cavendish. Among the former (and at various times) were: E. Rutherford, H. A. Wilson, C. T. R. Wilson, P. Langevin, J. S. E. Townsend, O. W. Richardson, W. Craig Henderson, A. J. A. McClelland, F. A. Aston, J. Zeleny and J. H. Poynting. As for the latter there were the demonstrators and technicians at the excellent workshop headed for seven years by W. G. Pye (who later started the well known commercial scientific instruments firm), G. F. C. Searle who was responsible for much of the practical teaching of physics, F. J. Lincoln that took

Pye's place, W. H. Hayles who often helped Thomson with routine administration, E. Everett who sometimes also operated the X-ray machine for local physicians, and J. M. Rolph who among many tasks looked after the liquid air plant.[25]

With staff of this caliber it was no wonder that much was accomplished during J. J. Thomson's tenure at the Cavendish.

2.4. *J. J. Thomson. The man*

Joseph John Thomson was born in a suburb of Manchester on December 18, of 1856.[26] He was called J. J. by those who knew him, both as friends and associates, and although I cannot count myself as a member of this group I will take the liberty, in this essay, to refer to him occasionally in the same way. Having read rather extensively about Thomson and seen numerous pictures in very human attitudes I somehow want to rescue the man, the warm human being that emerges from all the descriptions. Perhaps in this way, besides my own telling of the story I will be able to make J. J. talk to my readers, bridging the inevitable gap of time, space and generations.

It appears that he wanted to be an engineer but the waiting list for the apprenticeship was too long for his taste. So, at the age of fourteen he went to Owens College, that later became Manchester University. There, besides engineering, he studied mathematics, chemistry and physics. In two years he gained his certificate and turned to the latter subject because he had enjoyed it most. Under Professor Balfour Stewart, Thomson received his experimental training in physics in the new laboratory that had just opened on Quai Street. It was there that he started his lifelong friendship with J. H. Poynting.[26]

In 1876, J. J. went to Cambridge as a scholar at Trinity College where, after receiving the Adams Prize and publishing two essentially mathematical papers, he took his degree in 1880. At first he was interested in energy, the great theme of the 19th century, but soon he was attracted to electrical phenomena, like the fact that an electric charge on a drop of liquid diminishes evaporation and allows the drop to grow instead. This served C. T. R. Wilson ten years later to explain the appearance of a dense opaque cloud in his now famous chamber.

Reading Maxwell's theory, Thomson calculated the field due to a moving charge and published his first paper on the subject in 1881 in the *Philosophical Magazine*.[27] It contained the germ of his later ideas on a unit of charge.

Thomson started experimental work at the Cavendish under Lord Rayleigh. According to his son, J. J. was not too keen on being an experimentalist[28] but found exceptionally good assistants in W. G. Pye and G. F. C. Searle. When Lord Rayleigh resigned his professorship in 1884, Thomson succeeded him. This happened just at the time when the idea of discrete natural units for electricity began slowly seeping into physics, the same as when atoms had been used about 1803 by Dalton[29] in the realm of chemistry.

3. Work that Opened the Door

The concept of atomicity in electrical phenomena was by no means new. It had been somehow anticipated by Benjamin Franklin when he regarded matter as composed of negative charges grouped around a positive nucleus.[30] There was however no substantial evidence for it.

The view started to clear up, so to speak, with experiments on electric discharges in gases. Fluorescent lighting had already been observed by Hauksbee[31] as far back as 1709, when the air remaining in an evacuated glass vessel was "electrified" with a frictional machine.

Apparently during his Adams Prize-winning investigations, Thomson concluded that conduction of electricity through a gas was somehow connected with the dissociation of molecules. Evidently this placed him in the frame of mind for what was to come, because his readings must have made him aware of what had been done on the subject.

Much earlier, improved vacuum pumps had led Plücker to discover cathode rays in 1859,[12] finding also that they could be deflected by an ordinary bar magnet. But Plücker was unable to explain the phosphorescence they caused on the glass wall opposite to the cathode.

First Hittorf,[13] and especially Goldstein[14] later on, who gave the rays their present name (i.e. cathode rays), started regarding them as waves capable of casting shadows. But it was Crookes[15] who investigated most of the unusual properties of these rays from the late 1870s onwards and even arrived at the point of wondering whether they were waves or particles. Hertz in 1885[16] tried to solve this riddle by making the cathode rays pass through charged parallel plates, but considered his results to be inconclusive. Later on, his student Lenard[17] had the rays pass through very thin leaves of metal (first aluminum, then gold) to show that they did not behave like visible light: therefore, they were not waves but particles. In my opinion, most likely the gold leaf was not thin enough. It is a known fact that very thin foils of gold, usually made by beating gold sheets on leather anvils with wooden mazes, will allow to see through them the bulb of a flashlight. This is a well-known experiment used in the classroom to show the discontinuity of solid matter.

A little later, in 1895, Perrin[18] found that if the rays were caught in a metal bucket they gave it a negative charge. According to G. P. Thomson[32] this was considered to be an indication that cathode rays were particles.

Therefore by the end of 1895 the two views were held simultaneously, although under different national colors. In Britain they were particles while the Germans thought of them as waves. It is interesting to realize that this apparent duality, commonly believed to be a modern dilemma already existed in what can be considered the closing years of pure classical physics.

The year of 1895 is of particular importance because, as mentioned before, on December 28, Röntgen[20] announced the discovery of, what he called, X-rays. They were generated in the already mentioned Crookes tube. In essence what happened was that, due to the discharge in a highly evacuated tube, "waves" leaving the cathode hit the opposite wall causing the emergence from there of another set of rays with a much higher frequency and strong enough to go through most solid materials.

Just a few days later, on January 18 of 1896, in a letter to Oliver Lodge, Thomson expressed the opinion that X-rays "are waves so short, that the wavelength is comparable to a molecule".[33] Very few discoveries share with this one the remarkable honor of having been recognized in its true importance right from the very start. Röntgen's paper was widely distributed, read, discussed and the work repeated, especially, since he himself had proved the usefulness of his discovery in the field of medicine.

The first X-ray machines for diagnostic purposes were soon manufactured and became a standard tool well before the end of the 19th century. However X-rays were also of fundamental importance for J. J. Thomson's great discovery, because he himself found shortly after the publication of Röntgen's paper, and simultaneously with him, that the passage of X-rays made air conducting.[34] This was very important because it provided a method of "ionizing" a gas, as Thomson called it, that was much simpler and controllable than the other known procedures (discharges, etc.).

The stage was set and J. J. Thomson moved in firmly.

4. The Electron and Its Charge

X-rays came on the scene at a time when the existence of "an ether", a medium considered to be necessary for the propagation of any kind of waves, was a rather broadly accepted idea. It was hardly conceivable that electromagnetic waves could propagate without causing a disturbance in some kind of medium, the same as sound in air and ripples on a liquid surface. The proof that there is no ether still lay in the future but, in the meantime it was quite acceptable that X-rays could be considered either as longitudinal (compressible) as the former (i.e. sound) or transverse as the latter (i.e. ripples). A simple and interesting discussion on the ether was made by Einstein and Infeld in the 1930s.[35]

X-rays were, in fact, only a consequence of their originators: cathode rays. The nature of cathode rays also awaited a satisfactory explanation. Here again, as described earlier, was present the dilemma of waves or particles. One of their first properties to be looked at was their velocity, because it was expected that in the case of waves it would be that of the velocity of light if cathode rays are waves. In 1894, Thomson claimed that this was not what his experiments showed.[36] Although the numbers he got were too low, the central idea was essentially confirmed by others. Des Coudres[37] found the velocity of cathode rays to be greater than 2×10^6 m.sec^{-1}. Although faster than that of any other object known at the time, it is still less than 1% of the velocity of light.

In the three years that followed Thomson's 1884 paper, he continued experimental work and was able to report in 1897 his most important conclusions.[38] What he proved through electric discharges in gases was that electricity is transported by carriers of dimensions smaller that those of ordinary atoms and molecules, and that the carriers are the same regardless of the gas used in the discharge tube. These results he found consistently through both the electric and magnetic deflection observed on a beam of cathode rays passing through the space between parallel plates connected to either an electromagnetic or an electric source.

Thomson's experiment[39] had already been tried by Hertz more than a decade earlier with the conclusion that the rays remained unaffected. Evidently Thomson was not convinced feeling that perhaps, as in so many other cases, the original experimental conditions — the vacuum in particular — were not adequate.

With improved vacuum pumps a much lower pressure was reached and both deflections were unambiguously observed. Previous failures were then explained through the conductivity generated in the gases by the cathode rays that, of course, decreased substantially with decreasing pressure. In fact, cathode rays ionized the gas that then literally short circuited the plates. Thomson pasted a scale at the rounded end of the tube and found that "the deflection was proportional to the difference in potential between the plates". In this way he found that the rays, formed by a beam of particles, were attracted by the positive plate and repelled by the negative one, proving that they were negatively charged.

A further and very important result of Thomson's experiment was the value of the ratio (m/e) between the mass (m) of the electrified particle and its charge (e). This was a known concept, that had already been measured for a number of ions in solution, especially for the smallest of all: the well known positive hydrogen ions. Considering Faraday's natural unit, M/e (M = mass of the "charge atom" or ion) for hydrogen is roughly 10^{-4} in electrostatic units (or 1.67252×10^{-24} g/1.60210×10^{-20} $\text{cm}^{1/2}\text{g}^{1/2}$).[40]

To measure m/e, Thomson prepared the experiment in such a way that the electric and magnetic fields were opposed to each other, so that when the two were balanced no deflection occurred and he could calculate the velocity of the particles. When the current in the electromagnet was switched off, the deflection caused by the electric field alone could be measured and from it the value of m/e calculated. The original value was about 1000 times smaller than that for the hydrogen ion but 2000 times smaller compared to the actual value which is $(9.1091 \times 10^{-28}$ g/1.6021×10^{-20} cm$^{1/2}$g$^{1/2}$.[40] Upon refining the experiment further, Thomson came closer to the correct value and later on, in 1898, was able to calculate the electric charge on the "corpuscle" (as he called nature's unit of charge e)[40] first as 6.5×10^{-10} and later as 3.4×10^{-10} electrostatic units (e.s.u.). Although these numbers were not equal to the actual value (4.8×10^{-10} e.s.u.) they came close enough for those times.

Importantly, so Thomson found m/e not to depend either on the gas in the tube or on the metal of the cathode; and that, since each cathode ray particle balances electrically one hydrogen ion, the charges on both must be equal but opposite. This led Thomson to his sub-atom hypothesis in that "... the atoms of different chemical elements are aggregations of atoms ... of some unknown primordial substance X." For this primordial substance, or corpuscle, Stoney had already suggested the name of *electron*, a few years earlier.[11]

The purpose expressed by Thomson in the opening sentence of his 1897 paper were thus fulfilled: "The experiments described in this paper were undertaken in the hope of gaining some information as to the nature of the Cathode Rays."[41]

5. J. J. Thomson's Atom

In his father's biography, G. P. Thomson[42] states a number of questions before describing J. J. Thomson's atomic model. Some are: If electrons are not the only constituents of matter, why doesn't repulsion cause them to scatter? Why is all matter electrically neutral? How are positive and negative constituents neutralized? Is there a continuous spread of matter or is it concentrated at geometrically very small points?[43]

As a group, these questions remind us of basic principles in experimental science beautifully expressed by Justus von Liebig, the founder of experimental teaching of chemistry in the first half of the 19th century: "Nature speaks to us in her own language, it is composed of phenomena and we ask her questions in our experiments." Consequently, relevant questions will lead to informative answers. Those stated by G. P. Thomson describe the general situation of what was known and looked for regarding the problem of atomic structure, very specially around the turn of the century, just before Thomson's paper of 1904.[44]

There is a very curious fact related to Thomson's model. Two years earlier, Lord Kelvin had proposed an atomic model that consisted of a sphere of uniformly distributed positive electricity, within which discrete electrons were imbedded; equilibrium was reached when all charges were at rest.[45] This was a static model that anticipated future work. It must have been known in learned circles, probably having reached the point where it is generally considered to be an accepted fact, so that specific citations were not made in the scientific literature. But this is what can be concluded from the opening statement in Thomson's paper: "The view that the atoms of the elements consist of a number of negatively electrified corpuscles enclosed in a sphere of uniform positive electrification, suggests, among other interesting mathematical problems, the one discussed in this paper, that of the motion of a ring of n negatively electrified particles placed inside a uniformly electrified sphere."

Thomson abandoned the static view and, perhaps influenced by Maxwell's essay on Saturn's rings,[46] preferred to have the corpuscles (*electrons*) moving on a ring at right angles to the axis of rotation. The system so conceived he considered to be stable as the result of a balance between the attractive force of the positive sphere and the repulsion of the negatively charged corpuscles.

Of course Thomson's proposal *was not* limited to hydrogen and was intended for all elements. Therefore in discussing the details he arrived at the condition for steady motion as being a function of the number (n) of corpuscles present and his model was applicable to all values of n equal to or greater than 1. This also allowed him to calculate the frequencies corresponding to the displacements in any atom.

From his analysis he concluded that, for stability, when the number of negative corpuscles is larger than 5 ($n > 5$), the motion becomes unstable unless a number (p) of corpuscles is placed in the center of the sphere and that there is a relationship between n and p. In other words, he suggested that $p = 0$ for $n = 1, 2, 3, 4$

and 5. The interesting thing here is that when $p > 1^{47}$ the internal corpuscles cannot all be at the center of the sphere, so they will separate until their repulsions are balanced by the attraction of the positive electricity in the sphere.

As a result Thomson wrote: "We have thus in the first place a sphere of uniform positive electrification, and inside this sphere a number of corpuscles arranged in a series of parallel rings, the number of corpuscles in a ring varying from ring to ring." He then further added: "Each corpuscle is travelling at a high speed round the circumference of the ring in which it is situated, and the rings are so arranged that those which contain a large number of corpuscles are near the surface of the sphere, while those in which there are a smaller number of corpuscles are more in the inside."

To facilitate the described arrangement, Thomson indicated the need for concentric shells and gave detailed tables of how the rings and shells would be occupied. Immediately he considered the connections between this arrangement and the properties of chemical elements and it cannot be surprising that he found a very close relationship with Mendeleev's Periodic Table. He claimed that "We can divide the various groups of atoms into series such that each member of the series is derived from the preceding member by adding to it another ring of corpuscles."

An important consequence of this structure is that it would generate two sets of vibrations: those arising from the rotation of the corpuscles in their orbits and those arising from the distortion of a ring from its equilibrium circular shape. Both sets would lead to a group of vibrations with a sequence of frequencies, or lines in the spectrum, that are very similar for the various elements in a series.

At this point it is most important to remark that in J. J. Thomson's paper there is not a single figure attempting to represent *his* atom. This is curious in view of what is customary nowadays, possibly as a result of the maxim that a picture is worth a thousand words. But I wonder if in J. J.'s paper a figure is really necessary. In spite of the complicated algebra, the description is so clear and perfect that figures would be redundant. Possibly this is due to the fact that scientists in those days took their time to write, allowing thoughts and reasoning to flow slowly onto paper.

In any case, there it is: the atom as a positive sphere with electrons embedded in it. I have heard some teachers resemble it to a "plum-pudding" or "raisin-muffin", provided of course it were spherical. Be it as it may, the model was complete and ready to stand or fall on it's own merit.

6. Success, Failure and the Future

What was the importance of J. J. Thomson's atom? Was it successful?

To some extent the answer is yes, because J. J. viewed it as a system formed by electrical charges, leading away from conceiving the atom only as *the smallest lump of matter* that existed until his discovery of the electron. Furthermore, electricity and magnetism found their way into the microcosm for what was to come a few years later. Also, the connection established with the Periodic Table of Elements, provided a new means to understand the properties of matter. These facts alone are worthy of widespread recognition.

But apart from these auspicious results the model had its shortcomings. It failed to explain some phenomena, such as the regularities in spectral lines (the Balmer series for instance) and the deflection of electrons when passing through matter. Still, the model was indeed a step in the right direction.

Thomson's paper was published in March 1904, two months later a paper by H. Nagaoka, Imperial University of Tokyo was published in the same journal.[48] It contains a report read before the Physico-Mathematical Society of Tokyo on December 5, 1903, that essentially attempted to explain the vibrations described by the formulae of Balmer, Kayser, Runge and Rydberg. Nagaoka used equations derived from a system that somewhat approximated Maxwell's model of Saturn's rings. What appears to be rather strange when reading it is the fact that there is no mention whatsoever of Thomson's work. Certainly, information then did not move with the absolutely frightening speed possible today.

Finally, it was E. Rutherford, after he had already left the Cavendish for Manchester, who evaluated all previous work, Thomson's and Nagaoka's included. In 1911, Rutherford proposed for hydrogen the first model consisting of a small positively charged nucleus with an electron orbiting around.[49] It is now well known that this arrangement is also not satisfactory because the rotating electron as a continuous radiator would end up by falling into the nucleus, but this evidently does not happen. It was Bohr in 1913, after working with Rutherford in Manchester, who solved the puzzle showing the need for quantum theory.[50]

7. Epilogue

In science I believe that there are no such things as a beginning, an end, the beginning of an end, and the end of a beginning. There are only stages, much like in a relay race. If so, can anyone say who is the most important runner? All links are needed; with one absent, the chain breaks and everything is lost. J. J. Thomson's atom is an excellent example of this, because it acted as a bond between the past, where subatomic particles did not exist, and the future with its window towards the atomic interior.

Without Thomson's "corpuscle", the Rutherford and Bohr atoms would not have been possible. Neither would have quantum theory arisen, nor all the rest that followed suit. So, what I have attempted to give in this essay is the description of Thomson's stage, giving to it somehow the character of a watershed. This is only true in part because all important stages in the road of knowledge are exactly that: watersheds, sharing with the previous and the following the same property of being unique and irreplaceable.

Therefore what comes after J. J. Thomson's atom is just as fascinating in its own way but, as R. Kipling would say: that is another story.

References

1. G. P. Thomson, *J. J. Thomson and the Cavendish Laboratory in His Day*, G. P. Thomson, Thomas Nelson and Sons Ltd., London, 1965.
2. A. Volta, "Sur l'électricité dite animale", *Philos. Mag. IV* (1799) 59–68, 163–171, 306–312.
3. C. A. Coulomb, "Sur l'électricité", *Ann. Chim. Phys.* **II** (1789) 1–22; **III** (1790) 97–111, 112–133.
4. A. M. Ampère, "Sur l'action mutuelle de deux fils conjonctifs et sur celle qui a lieu entre un fil conjontif et un aimant ou le globe de la terre", *Gilbert Ann.* **67** (1821) 113–163, 225–228.

5. H. C. Oersted, "Experimenta circa effectum conflictus electrici in acum magneticam", *Ann. Chim. Phys.* **14** (1820) 417–425.
6. M. Faraday, *Phil. Trans. R. Soc. London*, 1832–1856. A total of Thirty Series under the general title of: *Experimental Researches in Electricity*.
7. J. C. Maxwell, *A Treatise on Electricity and Magnetism*, Dover Publications Inc., New York, 1954.
8. W. E. Weber and R. Kohlrausch, "Unipolare induction", *Poggend. Ann.* **52** (1841) 353–386 and "Über die Electricitätsmenge, welche bei galvanischen Strömen durch Querschnitt der Kette fliesst", *ibid.* **99** (1856) 10–25.
9. H. S. Lipson, *The Great Experiments in Physics*, Oliver & Boyd, Edinburgh, 1968, p. 129.
10. H. Helmholtz, "On the modern development of Faraday's conception of electricity", *Chem. Soc. J.* **39** (1881) 277–304.
11. G. Johnstone Stoney, "On the 'electron' or atom of electricity", Letter in *Philos. Mag.* **38** (1894) 418–420.
12. J. Plücker, Abstract of a series of papers and notes concerning the electric discharges through rarified gases and vapours, *Proc. R. Soc.* **X** (1859–1860) 256–269.
13. J. W. Hittorf, "Über die Electricitätsleitung der Gase", *Ann. Chim. Phys.* **21** (1886) 90–139.
14. E. Goldstein, "Über eine noch nicht unteruchte Strahlungsform an der Kathode inducierte Entladungen", *Berlin Ak. Sber.* (1886) 691–699.
15. W. Crookes, "On the attraction and repulsion accompanying radiation", *Philos. Mag.* **48** (1874) 81–95; **1** (1876) 245–249; **2** (1876) 145–151; **3** (1877) 231–235; **3** (1877) 467–477.
16. H. R. Hertz, "Über die Ausbreitungsgeschwindigkeit der Electrodinamischen wirkungen", *Ann. Chim. Phys.* **34** (1888) 551–564.
17. P. E. A. Lenard, "Über die Magnetische Ablenkungen der Kathodenstrahlen", *Ann. Chim. Phys.* **52** (1894) 23–33.
18. J. Perrin, "Nouvelles proprietés des rayons cathodique", *Paris Acad. Sci. C. R.* **121** (1895) 1130–1134.
19. Ref. 1, p. 37.
20. W. Röntgen, "Über eine neue Art von Strahlen", *Würzburg Phys. Med. Sber.* (1895) 132–141; (1896) 11–16, 17–19.
21. H. C. von Bayer, personal communication.
22. M. Barón, "La primera cátedra de física experimental en la Universidad de Buenos Aires", *QUIPU. Revista Latinoamericana de Historia de la Ciencia y la Tecnologia* **3** (1988) 340–400.
23. Ref. 1, p. 1.
24. Ref. 1, p. 2.
25. Ref. 1, pp. 25, 94 and 95.
26. Ref. 1, p. 22.

27. J. J. Thomson, "On the electric and magnetic effects produced by the motion of electrified bodies", *Philos. Mag.* **11** (1881) 229–499; "On some electromagnetic experiments with open circuits", *ibid.* **12** (1881) 49–60.
28. Ref. 1, p. 25.
29. J. Dalton, "On the absorption of gases by water and other liquids", *Manchester, Phil. Soc. Mem. I* (1805) 271–287.
30. Ref. 1, pp. 14 and 15.
31. Ref. 1, p. 28.
32. Ref. 1, pp. 37–38.
33. Ref. 1, p. 40.
34. Ref. 1, p. 39.
35. A. Einstein and L. Infeld, *The Evolution of Physics*, published simultaneously by Cambridge University Press, Cambridge, Sytholl's, Amsterdam and Simon and Schuster, New York, 1938. A historical review on ether can also be found in: *Subtle is the Lord ... The Science and Life of Albert Einstein*, Abraham Pais, Oxford University Press, 1982.
36. J. J. Thomson, "On the velocity of cathode rays", *Philos. Mag.* **38** (1894) 358–365.
37. Th. Des Coudres, "Electrodinamisches über Kathodenstrahlen", *Berlin Phys. Ges. Verh.* (1897) 157–162.
38. J. J. Thomson, "Cathode rays", *Philos. Mag. V* **44** (1897) 293–316.
39. Rom Harré, *Great Scientific Experiments. Twenty Experiments That Changed Our View of the World*, Oxford University Press, 1981, pp. 160 and ss.
40. Chemical Rubber Co., *Handbook of Chemistry and Physics*, CRC, 68th edn., 1987, pp. F-157.
41. Ref. 1, p. 52 (over). Contains the facsimile of the first page of J. J. Thomson's manuscript.
42. Ref. 1, Chap. 5.
43. Ref. 1, pp. 121 and ss.
44. J. J. Thomson, "On the structure of the atom: An investigation of the stability and periods of oscillation of a number of corpuscles arranged at equal intervals around the circumference of a circle; with application of the results to the theory of atomic structure", *Philos. Mag. VI* **7** (1904) 237–265.
45. Kelvin, Lord, "Aepinus Revisited", *Philos. Mag. VI* **3** (1902) 257–283; "On the weights of atoms", *ibid.* **4** (1902) 177–198.
46. J. C. Maxwell, "On theories of the constitution of Saturn's rings", *Edinburgh R. Soc. Proc. IV* (1862) 99–101.
47. Ref. 43, p. 254.
48. H. Nagaoka, "Kinetics of a system of particles illustrating the line and the band spectrum and the phenomena of radioactivity", *Philos. Mag. VII* (1904) 445–455.

49. E. Rutherford, "The scattering of α and β particles of matter and the structure of the atom", *Philos. Mag. VI* **21** (1911) 669–688.
50. N. Bohr, "On the constitution of atoms and molecules", *Philos. Mag. VI* **26** (1913) 1–25.

Máximo Barón, Professor of Experimental Physics (1968–1995) at the University of Buenos Aires, Argentina. Barón has been at the University of Toronto as an NRC Canada Scholar; at the University of Salford as a Visiting Lecturer; as well as at the Royal School of Pharmacy, Copenhagen, King's College, London, and the Universities of La Plata and San Luis in Argentina as a Visiting Scientist. A regular attendee of the Gordon Research Conferences since 1978, Barón's major interests lie in excess properties of liquid systems, chiral phenomena in cyclic molecules, optical properties of liquid crystals, dielectric resonance, thermodynamics, and history of science. The author of more than 40 papers and four books, he was awarded the 1981 History of Science Award of the National Academy of Science of Buenos Aires. He is a Titular Member of IUPAC Commission on Macromolecular Nomenclature, and the Director of the Library and Information Services of the Argentine Chemical Association. He also serves as a Member of the Editorial Board of the journals *Speculations in Science and Technology* and *Ciencia e Investigacion.*

Chapter 3. The Hydrogen Atom: The Rutherford Model

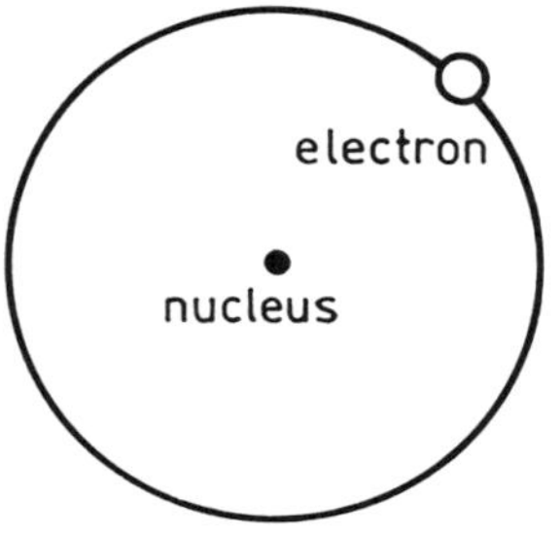

Homer Benjamin Tilton

E. Rutherford, c. 1919

Chapter 3

The Hydrogen Atom: The Rutherford Model

Homer Benjamin Tilton

Department of Mathematics, Astronomy, and Physics, Pima Community College East, Tucson, AZ 85709-4000, USA

Abstract

Early this century Ernest Rutherford established the nuclear model of the hydrogen atom, presently taught as representing the best visual model after modification by Niels Bohr and Arnold Sommerfeld. It replaced the so-called "plum pudding" model of J. J. Thomson which held sway previously. While the Rutherford model represented a large step forward in our understanding of the hydrogen atom, questions remained, and still do.

1. Things and Their Models

When talking about things that we can't inspect directly because they are too small or too big or for some other reason, we are faced with the need to construct models of them in our minds, then in our diagrams, and finally in our mathematics. These models relate the uninspectable thing either to familiar things or to abstract mathematics. Such is the case with atoms.

It is important to be aware at all times that *the model is not the thing*. Even though that may be most obvious, there is a human tendency to replace, in one's mind, "the thing" with "the model" once we have gained a close familiarity with the model. That old saying "familiarity breeds contempt" here takes this form:

Familiarity with the model breeds contempt for the thing. That is, once we have become intimately familiar with the model, there is a strong tendency to believe that we know the thing; but we almost never do. Albert Einstein is reported to have said concerning the photon model of light:

> "Every physicist thinks he knows what a photon is ... I spent my life trying to find out ... and I still don't know."

So you are cautioned that this essay discusses the Rutherford model of the hydrogen atom, not the hydrogen atom itself; an important distinction.

What good are models? What is their function? Models give us some measure of understanding of the things being modelled. In a practical sense that allows us to:

(1) Delve ever deeper into Nature's secrets;
(2) Harness, via technology, forces not previously available to us; and
(3) Develop ever better models by building on points 1 and 2.

Certainly those three points all apply to the Rutherford model of the hydrogen atom. In accordance with point 2, in 1919 Rutherford finally achieved that centuries-old dream of the world's alchemists: the transmutation of elements. That lead to the development of the atomic bomb and nuclear power only about two decades later.

Physics is about ideas and discoveries, but it is also about people; for without people there would be no one to appreciate the ideas and discoveries — nor would there be anyone to make them in the first place. Therefore it is appropriate that this essay includes a brief account of the central figures in the story that's about to unfold.

2. The Players

The year is 1897, the place Cavendish laboratory at Cambridge, England. The British physicist Joseph John (J. J.) Thomson is 41 and has just discovered the electron through his work with cathode rays. Thomson is Cavendish professor of physics; a post he is destined to hold until 1918.

J. J. Thomson's field is atomic modeling. He has suggested that the positive charge of the atom may be smoothly spread throughout the entire atomic volume with the electrons vibrating inside that bulk like plums in a pudding. Thomson has done much pioneering work in the field of atomic physics — work that others would build on. You have read of his hydrogen model in the previous essay.

One who would build on the work of J. J. Thomson is Ernest Rutherford, born at Nelson, New Zealand on August 13, 1871. He attended the University of New Zealand at Wellington, and then moved to England.

The year is now 1911. Thomson won the Nobel Prize in physics five years earlier. Rutherford is now 40; he's been to Canada and back to England, and is now at Victoria University in Manchester where he has been Langworthy professor of physics for the past four years. Three years ago he won the Nobel Prize in chemistry; and just this year, he seized onto an important clue supporting the idea of a nuclear structure for the atom, as opposed to one resembling a plum pudding. Rutherford will eventually move to Cambridge University where he will become Cavendish professor of physics in 1919.

Thomson is the older man, but he is destined to outlive Rutherford by nearly three years.

3. The Crucial Experiment

In 1910, many physicists believed in the "plum-pudding" model of the atom put forward by J. J. Thomson. Rutherford was well aware of Thomson's plum-pudding model of the atom, but he had a different idea, *viz.*, a nuclear atom. Not only that, he knew how to test his idea.

As is so often the case in the history of scientific development, others were also considering the idea of a nuclear atom at about the same time. One of those was Hantora Nagaoka[1] in 1904, and Rutherford acknowledged Nagaoka's contribution with this passage[2]:

> "It is of interest to note that Nagaoka has mathematically considered the properties of a "Saturnian" atom which he supposed to consist of a central attracting mass surrounded by rings of rotating electrons."

Another theorist who had considered the idea of a nuclear atom at about the same time was John William Nicholson, who wrote[3]:

> "The main conception involved in the structure of these simple atoms is that of the nature of positive electricity. This is supposed to exist in small spherical volume distributions of uniform density, whose radius is small in comparison even with that of an electron, a reversal of the more generally accepted view."

With a specified magnitude of electrical charge, one can determine the rough physical size of the charge carrier (i.e., the particle) by sensing its surface electrical field strength; a smaller particle producing a greater field strength. If charged particles of unknown size are bombarded by other charged particles having known energy, the resulting scattering angles can then be measured to reveal something about the field strength, and thus the size of the unknown particles. The order of magnitude of the particle size can then be inferred from the distribution of scattering angles.

An experimental team collaborating with Rutherford, composed of Hans Geiger and Ernest Marsden, performed such experiments.[4] Scattering angles well over 90° — with some near 180° — were obtained when gold was bombarded with alpha particles (nuclei of helium atoms). Rutherford saw such large scattering angles as a confirmation of his idea that the radius of the positively-charged region of an atom is some four orders of magnitude smaller than in the Thomson model.

4. The Rutherford Model of the Hydrogen Atom

In the plum-pudding model, the hydrogen atom was given a radius of about 10^{-10} m (100 000 fm). That overall approximate atomic radius is retained in the Rutherford model of 1911, but now the positively-charged portion of the atom no longer fills the entire atomic volume; rather it shrinks to occupy a central nucleus with a radius of about 10^{-14} m (10 fm). Thus the Rutherford model is a nuclear model.[2]

This view no longer requires that the electrons in an atom exist trapped in the molasses of a positively-charged material, but gives them free rein in practically free space — much as the planets in the space around the Sun. Indeed, popular accounts began to liken atoms to miniature solar systems. The science fiction literature for many years thereafter suggested that there may be tiny beings living on the electrons in an atom; and (at the other end of the scale) that perhaps our solar system is simply an atom in a much larger construct.

But of course those analogies are extremely loose if for no other reason than this: The tremendous difference in scale causes certain physical processes to predominate in one system and to be insignificant in another. As an example, the centripetal forces in the solar system are of gravitational origin whereas in the atom they are electromagnetic. Although the centrifugal forces appeared to be inertial in both cases, the origin of that inertial force in the atom might logically have been seen to be electromagnetic, not gravitational.

Thus the hydrogen atom as visualized in the Rutherford model is composed of a tiny nucleus about 10^{-14} m in radius which carries an atomic weight of 1 unit and an atomic number (electrical charge) of 1 unit; along with a single "orbital" electron scurrying around in a volume whose radius is about 10^{-10} m. We say the nucleus is a single proton whose charge is the same magnitude as that of an electron but of opposite polarity, and whose mass is the same as that of a neutron.

The important difference between the Thomson and Rutherford models does not lie in the obvious difference between plum puddings and planetary systems. Instead, it lies in the difference in electrical field strengths at the surface of the positively-charged region in the two cases. This is so because isotopes and trans-hydrogen atoms could probably be conceptualized using either model. But only the Rutherford nuclear model was able to lead to an eventual explanation of the makeup of atomic spectra.

The Rutherford 1911 atomic model seemed like a reasonable model; but there remained many unanswered questions.

5. The Problem with Classical Theory

There were many who objected to the planetary model of the hydrogen atom, saying things like this:

> "An orbiting electron is an accelerating electron, so it will radiate electromagnetic energy in accordance with Maxwell's equations. That will cause its orbit to gradually and continually decay while emitting radiation across a large region of the spectrum; but that doesn't happen. How can such a system possibly be steady state?"

Good question. An answer (of sorts) given and accepted by many is essentially this:

> "Orbital electrons in atoms simply don't lose energy by the mere act of orbiting; they must change orbits for there to be emission or absorption of electromagnetic energy."

The name of Niels Bohr is generally attached to that answer[5]; and it was an idea in harmony with Planck's ideas. You will find a discussion of Bohr's model and its refinement by Arnold Sommerfeld in the two essays following this one.

> Not everyone was satisfied with simply a decree; they wanted to know and understand the mechanism behind the action.
>
> Perhaps electrons "think" they are traveling in straight lines, along some sort of electromagnetic geodesic similarly to what Einstein said of light in relation to its path which is curved by the Sun's gravitational pull. But synchrotron radiation, predicted by Maxwell's equations, is a fact. In this, electrons traveling along a large circular path do emit radiation (in the direction of their motion). One might speculate that this is the mechanism behind some pulsar emissions. The obvious difference between synchrotron radiation sources and atoms is one of scale. Big orbits in one case, small in the other. Big and small relative to the size of electrons, as well as relative to each other.
>
> What about actual planetary systems? They exist, and they appear to be reasonably stable and in a steady state. Why don't planets emit gravitational radiation (not "gravity waves") — the gravitational analog of electromagnetic synchrotron radiation — and experience decaying orbits? It's hard to draw a strict parallel with electromagnetic radiation because gravitational radiation, proposed by Einstein, still has not been experimentally shown to exist and so the theory is incomplete. But perhaps orbiting planets don't emit gravitational radiation simply because planetary systems are too small relative to the sizes of planets!?
>
> Could (some may have secretly wondered) a sort of discrete, digital, gearmesh, or "quantum" effect be operating with both orbiting electrons inside atoms and orbiting planets inside planetary systems? In the case of planetary systems, terms like Bode's Law and de Broglie waves (matter waves, supposedly not the same thing as gravitational waves) suggest themselves. After all, it is easy to show that a standing wave of a particular spatial frequency will "fit" only onto circles having

particular radii, just as taut strings and drumheads can vibrate only in certain modes.

Those ideas suggested that a sort of celestial numerology might be acting in the case of planetary systems! Our experience with planetary systems is far too meager to really pin things down; but could there be a sort of Bode's Law operating with orbital electrons inside atoms? A "quantum" effect causing particular discrete electron orbits to be stable and steady state? A situation in which orbiting electrons lose energy only when an orbit is somehow caused to change? And could there be a gain in energy when the opposite change is forced?

At the suggestion of Rutherford, James Chadwick[6] performed experiments to show that the charge on the nucleus is equal to the atomic number and the law of force is the inverse square.

Little more was accomplished until Bohr, then Erwin Schrödinger, showed how a quantum mechanical model (wave mechanics) fits many of the previously unexplained facts about atomic spectra. The problem was that Schrödinger's model is almost entirely a purely abstract mathematical model. As a result, many could not accept it. You may be familiar with Einstein's oft-quoted objection to the probabilistic aspects of quantum mechanics:

"God does not play dice with the Universe."

You see, there is a probabilistic aspect to quantum mechanics in which both position and momentum of small particles cannot (it was said) both be known at the same time, which is the Heisenberg uncertainty principle. But along came Sallhofer[7] to suggest that the probabilistic aspect is an artifact due to a "lack of clarity" in the Schrödinger model which does not occur in Dirac's later electrodynamic formulation of wave mechanics!

When asked why electrons in certain atomic orbits do not radiate, one of the most recent answers comes via the Mach–Dirac–Sallhofer line of reasoning as discussed in the next section. What it says, essentially, is that a new model is overdue.

6. Whither the Rutherford Hydrogen Atom?

If the Rutherford–Bohr planetary model is to be swept away, is there to be no *intuitive*, *visual* model to replace it? Are we to be left with nothing more than an abstract mathematical construct?

Or, can that mathematical construct someday be given a visual meaning.

There is a growing discontent in the physics community with the *standard model* that has grown out of the "Copenhagen interpretation", it being likened to Ptolemy's epicycle theory of the universe in that — while it seems to mostly work — it is fast becoming intolerably complex, perhaps needlessly so. See, for example, Martin Macháček[8] and Silvan S. Schweber.[9] Also see "Hans Sallhofer Speaks" in this volume.

Let me report on two recent ideas now: one from Hans Sallhofer, one from Pieter van Heerden. Note how similar they are, even though developed independently.

6.1. *Sallhofer's idea*

Hans Sallhofer wrote, with an embedded quotation he credits to Ernst Mach[10]:

> "An alternative to particle matter is field matter... Two photon fields 'circle around each other, their centers of gravity forming a Kepler system... Here, the hydrogen atom is a standing disturbance of a special configuration, in essence a disturbance like stationary light'."

He further indicated that there is a simple, direct connection between Maxwell's source-free electrodynamics (E) and Dirac's formulation of wave mechanics (Q) via the Pauli vector (σ) according to this simple formula involving a scalar product[7]:

$$E \cdot \sigma \equiv Q$$

He then went on to opine:

> "Especially in an electromagnetic hydrogen model there is no room for a particle 'orbiting electron', because of div $E = 0$, i.e., electrodynamics cannot use the Rutherford model."

and

> "Here we face the fact that the formalism of [wave mechanics] which is developed from radiation theoretical and mechanistic basic concepts, now, after half a century, turns out to be a special case of the formalism of [Maxwell's] electrodynamics."

Finally, a "picture" began to emerge in Sallhofer's mind. He wrote in 1990[11]:

> "A first preliminary but basic picture of electrodynamic hydrogen is ... as a light wave impinging from the outside — from all directions — on a point, being reflected by itself at that point and running outward again ... forming a standing wave field with a high concentration of energy around the point of reflection, the center of the hydrogen field."

and

> "On closer inspection it becomes clear that the incoming and outgoing light wave breaks up into a field twin."

6.2. *Van Heerden's idea*

In an apparently independent discovery, Pieter van Heerden had this to say concerning matter[12]:

> "If material things are conceived as the superposition of ether waves traveling in opposite directions, [this gives some] of the results of the theory of relativity."

and this, concerning quantum mechanics:

> "Quantum mechanics has no picture."

He then proceeded to give a "picture for dynamics that may be extended to atomic dimensions," in terms of "a game of cards."

Elsewhere he had this to say concerning quantum mechanics[13]:

> "The fact remains that logically one has to admit that either the psi-function [of wave mechanics] is not real, or living conscious people outside of ourselves are not real. That is the very basis of quantum mechanics. When a positivist tells me calmly that reality has no meaning and that [my] argument therefore is of no value, I must confess that I have difficulty suppressing a white rage."

Most recently he wrote[14]:

> "... We can describe any relativistic particle with a de Broglie wavelength as given by two waves, traveling at the speed of light in opposite directions."

and

> "Photons are interference patterns between different energy states of particles."

Science does not progress linearly; it proceeds in fits and starts and sometimes even goes backwards. This is exemplified by van Heerden's outbursts of dissent[15]:

> "The [present] authorities in this field assure us that [pre-World War II knowledge of physics] is all a pitiful illusion. [But] where is the sound, simple, and certain basis that physics had for 300 years?"

He then invited readers to write him for more. I did, and he sent me a copy of his paper, "What is matter?" In a letter to me dated 14 July 1993 accompanying the paper, he wrote:

> "I send you herewith the talk I gave in Seattle [at a recent meeting of the American Physical Society]. ... The subject has had my lifetime interest, and I think I have thought well about it. But I am very much concerned about the almost total absence of opportunities to publish my work in *Physical Review*. The only time was in *Phys. Rev. D*, 1973, and then I was not far enough with my mathematics. Frankly, I really believe that my formulas are right on the button. ... I wish that some younger guys could become interested in the subject. I think that today they are getting brainwashed by the establishment... by the total one-sidedness of the literature. Nobody but me has been concerned about the wave nature of matter and the absolute demand that one should be able to understand how a large molecule, in a double slit experiment, will go 'through two separate holes simultaneously'. Einstein maintained all his life that quantum mechanics could not represent the real world. He was right. Q. M. is just a wave analysis of our experiments.
>
> Sincerely yours, Pieter J. van Heerden."

Van Heerden's theory says that a mass particle can be represented as a pair of waves of the type dé Broglie hypothesized. It is a highly quantitative mathematical theory, lending itself to exact testing by anyone who is so inclined.

Since van Heerden claims his equations automatically account for electromagnetic and gravitational forces, one should be able to find a subset within those equations which are Maxwell's electromagnetic equations. In addition, one should be able to find another subset there, perhaps resembling Maxwell's equations, which applies to gravitational radiation. When I queried him further, he answered in another communication dated 24 January 1994, reproduced here in its entirety except for some equations:

> Dear Homer,
>
> Electricity and gravity are obviously related, because they can be transformed into each other by machines. Also, light

> quanta are deflected in a gravity field according to $mc^2 = E = h\nu$, $m = h\nu/c^2$ .... There exists today no theory which links the two phenomena. That is what I am trying to accomplish, and I think my theory is logically inescapable and hopefully will be proved by real numbers.
>
> Sincerely yours, P. J. van Heerden

It is not completely clear what he meant by "they can be transformed into each other by machines". Does he mean, as an example, the fact that a weight can be lifted against gravity by an electric motor? If so, then that argument does not seem to me as strong as would, say, a machine that creates an artificial gravitational field by using some exotic gyrative motion of electric charges.

I have the impression that van Heerden's theory requires more work before it will actually do what he believes it potentially can. Judging from his letters to me, he has turned his attention now, at age 78, to attempting to convince the physics community that they are becoming a closed establishment which is unreceptive to non-mainstream ideas. From his letter of July 20, 1993:

> "...Right now, I want to pursue a different course... I want to confront the establishment, because I feel that it is becoming degenerate, and I want to revive it in the interest of future generations."

From his letter dated August 5, 1993:

> "I have traveled alone for 40-years with these ideas, and... I must rely on the thought that my writing has conveyed enough of the ideas to younger generations, so that they can carry the ball..."*

7. The Long Road Ahead

At various times throughout human history, there have been those who have declared that everything there is to know has been discovered. In every case they have been found to be wrong, leading to the reasonable speculation that it may not be possible to know "everything". Questions certainly remain today concerning the hydrogen atom. One concerns the possibility of low-temperature fusion.

*The above material appears here with van Heerden's permission.

Once a hope of low-energy alchemy, transmutation of elements by means of high-energy nuclear physics is now an accomplished fact. A recent "hot" topic is the expressed hope of some scientists of accomplishing cold nuclear fusion of deuterium and tritium. One theory has it that muons might catalyze such nuclear reactions at room temperature.[16]

Experiments towards that end by chemists B. Stanley Pons and Martin Fleischmann caused a stir in 1989 in the press and within the scientific community; but the modest energy production of their devices may represent nothing more than an exercise in electrochemistry. Indeed, the history of such attempts has been traced back to the abortive work by the two German scientists Fritz Paneth and Kurt Peters in 1926.[17] On the positive side, Arthur C. Clarke has commented as follows on that subject[18]:

> "I cannot help wondering if quantum fluctuations (also known as Zero Point Energy) (might) explain some of the baffling and bizarre results reported by... Pons and Fleischmann."

One reason such news items are interesting lies in the controversy they elicit within the scientific community. The high level of that controversy in this case may be an indication that even today there remains a profoundly incomplete common understanding of the hydrogen atom among scientists themselves.

References

1. Hantora Nagaoka, "Kinetics of a system of particles illustrating the line and the band spectrum and phenomena of radioactivity", *Philos. Mag.* **7** (1904) 445–455.
2. Ernest Rutherford, "The scattering of α and β particles by matter and the structure of the atom", *Philos. Mag.* **21** (1911) 669–688. (See the second part of the book for the facsimile reproduced.)
3. John William Nicholson, "The spectrum of nebulium", *Mon. Not. R. Astron. Soc.* **72** (1911) 49–64.
4. Hans Geiger and Ernest Marsden, "The laws of deflexion of α particles through large angles", *Philos. Mag.* **25** (1913) 604–623.
5. Niels Bohr, "On the constitution of atoms and molecules", *Philos. Mag.* **26** (1913) 1–25.
6. James Chadwick, "The charge on the atomic nucleus and the law of force", *Philos. Mag.* **40** (1920) 734–746.
7. Hans Sallhofer, "Maxwell–Dirac–Isomorphism. XI", *Z. Naturforsch.* **41a** (1986) 1087.
8. Martin Macháček, "Copernicus, Ptolemy and particle theory", *Phys. Today* **42**, 11 (1989) 13–14.

9. Silvan S. Schweber, "Physics, community and the crisis in physical theory", *Letter to Phys. Today* **46**, 11 (1993) 34–40.
10. Hans Sallhofer, "Elementary derivation of the Dirac equation. X", *Z. Naturforsch.* **41a** (1986) 468.
11. Hans Sallhofer, "Hydrogen in electrodynamics. V. Dirac-hydrogen by means of new operators", *Z. Naturforsch.* **45a** (1986) 1361.
12. Pieter J. van Heerden, "On the analogy between dynamics and a game of cards", *Am. J. Phys.* **27** (1959) 103–111.
13. Pieter J. van Heerden, *The Foundation of Physics: With a Proposal for a Fundamental Theory of Physics*, N. V. Uitgeveru Wistik; Wassenaas, The Netherlands, 1967.
14. Pieter J. van Heerden, "A new interpretation of quantum mechanics", Personal communication, 1994, to be published.
15. Pieter J. van Heerden, "Physics, past and present", *Optics Photon. News* (letter to the editor) **4**, 6 (1993) 3.
16. Johann Rafelski and Steven E. Jones, "Cold nuclear fusion", *Sci. Am.* (July 1987) 84–89.
17. Michael D. Lemonick, "Fusion illusion?", *Time* **133**, 19 (1989) 72ff.
18. Arthur C. Clarke, "Space drive: A fantasy that could become reality", *Ad. Astra.* **6**, 6 (1994) 38.

Homer Benjamin Tilton obtained a B.Sc. degree in engineering physics from the Montana State University. He worked as an aerospace engineer in the US defense industry till his retirement in 1989. He also pursued graduate work in engineering and physics at the University of California at Los Angeles and the University of Arizona; and was also a faculty member at the University of Arizona Engineering Research Laboratory in 1960–61. Currently, he is an adjunct faculty member in the Department of Mathematics, Astronomy and Physics of the Pima Community College, Tucson. He has published numerous scientific and technical papers in journals and state-of-the-art review books. He is the author of *Waveforms: A Modern Guide to Nonsinusoidal Waves and Nonlinear Processes* and *The 3-D Oscilloscope: A Practical Manual and Guide*, published by Prentice-Hall in 1986 and 1987, respectively.

Chapter 4. Niels Bohr — Through Hydrogen Towards the Nature of Matter

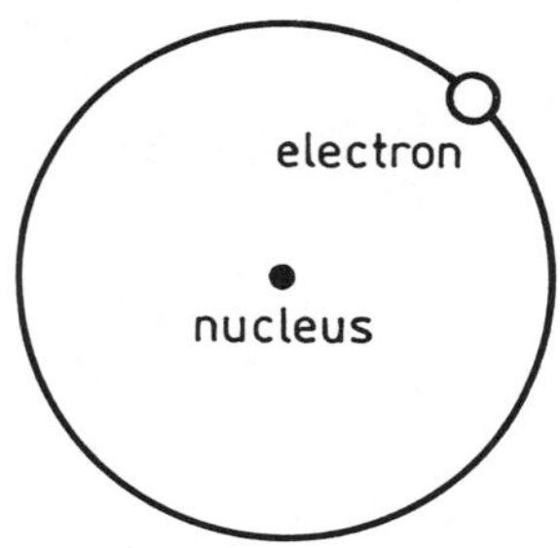

Daniel Stewart Robertson

N. H. D. Bohr

Chapter 4

Niels Bohr – Through Hydrogen Towards the Nature of Matter

Daniel Stewart Robertson

205, Pickersleigh Road,
Malvern, WR14 2QS, Worcs, Great Britain

Abstract

This essay on the work of Niels Bohr as "modeller of hydrogen" outlines the state of knowledge of the atom in the late 19th and early 20th centuries and details the particular background and discoveries which influenced Niels Bohr in developing his concept of the structure of the atom of hydrogen. It shows that although he was not the first to propose a structure of the atom, he was the first to successfully use the experimental and theoretical developments of the day to produce an atomic structure from which results could be obtained in agreement with experiment. Based on the original paper, the development of the structure of the hydrogen atom is discussed and the difficulties encountered detailed. The original paper by Bohr can be seen as one of the works at the beginning of the application of the quantum concept of Max Planck to natural phenomena other than the emission of light. The limitations of the Bohr model which resulted in its gradual replacement with other models of the atomic structure are discussed.

1. Introduction

By the year 1885, when Niels Hendrik David Bohr was born in Denmark, the humans had amassed a huge amount of knowledge concerning their environment. However, continuing discoveries regarding the properties of the natural surroundings were beginning to upset relationships within this knowledge. For example, prior to the observations made by Georges Cuvier, who proved the extinction of species and Charles Darwin, who proved the evolution of species, the human race had believed that it was not only unique, but immortal. These discoveries indicated that the human race was by no means one of the original life forms on Earth and, even more disturbing, the race could not consider itself to be immortal. Much the same situation existed with respect to the knowledge of other parts of the surroundings. Observations of the effects of manipulating varieties of matter, particularly by chemists, had led to the conclusion that there existed the smallest part of matter which was not capable of further subdivision. The discovery of radioactivity seemed to indicate that even the smallest part of matter could be further divided. The examination and application of electrical and magnetic effects, which had been known since the time of Thales of Miletus twenty five centuries ago, had lead to the conclusion that these phenomena had properties which resembled light. However they were also observed to behave like the force of gravity and could affect matter at a distance. When Niels Bohr was ten years old a new form of radiation was discovered. This was named X-radiation and it could pass through any matter, a property which light clearly did not possess. Thus the natural world had been shown to contain effects that appeared to upset the relationships between parts of knowledge, some of which had been accepted as certainties for at least two centuries. This was disturbing to a species whose evolution had brought it to rely on knowledge of its surroundings as the principal means of survival, particularly in numbers that were not possible without knowledge. Bohr was one of the research workers in the early part of this century who advanced proposals as to how the confusion concerning the nature of matter in particular might be resolved. The properties of the elemental gas, hydrogen, played an important part in his efforts.

2. Niels Bohr — Influences on His Future

Niels Bohr was the first son of Christian and Ellen Bohr and he had an older sister, Jenny and a younger brother Harald. His father was a professor of physiology at the University of Copenhagen, thus ensuring his familiarity with the recent discoveries. His education began at Gammelholm school in Copenhagen. Information about his school days is scanty and it is not known if his ability at school gave any indication of his future abilities as a physicist.[1] From school he proceeded to the University of Copenhagen where he obtained his first degree. He was granted his second degree, a Master's degree, for research on the surface tension of liquids. For this research he was awarded the Gold medal of the Royal Danish Academy. He gained his doctorate in 1911 with a treatise on the electron theory of metals. All these steps directed Bohr towards an interest in the nature of matter. After obtaining his doctorate, he first joined Joseph John Thomson at Cambridge University, England and shortly afterwards he went to work with Ernest Rutherford at Manchester University, also in England. Thus when Bohr arrived in England in the year 1912 it was clear that he wished to be involved in research directed towards the understanding of the nature of matter. The research groups he joined were two of the leading European scientific groups of the period whose programmes of research were directed towards this end.

3. The Nature of Matter

The nature of matter is the oldest continuous study in human history. It effectively began with the manufacture of the first tools by humans. To make such tools it was necessary to learn how to manipulate materials such as wood, bone and skin. Stone, which produced improved tools, required the acquisition and transfer of knowledge from generation to generation. For example it was necessary to establish which stones were the most effective, to learn to recognise these stones amongst others and to remember where they occurred. From this point onwards other forms of matter, clay, glass, and metals, were studied and manipulated.

Twenty-six centuries ago human knowledge concerning matter lead to speculation about the existence of the smallest piece of matter. One member of the Greek civilisation of the time, called Leucippus, advanced the proposal that "sensible matter is produced or constituted by the concourse of substantial or concrete monads moving more or less freely about one another".[2] Monads mean fundamental parts. This proposal, in much the same form, had been also generated by Hindu philosophers at a much earlier date. One of the students of Leucippus, called Democritus, applied the name *atomi*, meaning not divided, to these monads.

This had been converted, over time, to the word atom for the smallest part of any matter. The important basis of this doctrine was that these atoms could not undergo further subdivision. This was accepted throughout history until the discovery of radioactivity by Antoine-Henri Becquerel in the year 1896.

Radioactivity is the emission of penetrating particles and radiations by certain elements. The nature of these particles ultimately became known. First the negatively charged electrons, discovered by Thomson in 1897. These are emitted as beta particles from radioactive elements. A year later the positively charged proton was discovered by Wilhelm Wien. When Bohr joined the team lead by Rutherford at Manchester University, one of the principal interests of the researchers was the manner in which these two particles were arranged inside an atom. In pursuing this objective Bohr became involved in another subject of great interest in the early part of the 20th century. This was the nature of light.

4. The Nature of Light

The nature of light has been debated for centuries and was one of the chief interests of Isaac Newton. He demonstrated that white light, familiar every day on Earth, was composed of a set of coloured lights, called a spectrum. He also advanced the proposal that the nature of light was the movement of small rigid bodies. This proposal was opposed by Christian Huygens who, in 1678, proposed that light moved as waves and that each colour in a spectrum was related to a particular length of wave. The latter proposal was further reinforced by the work of Thomas Young and Augustine Fresnel in the early 19th century and given theoretical support by James Clerk Maxwell in the middle of the same century.

The particular study of light that was of importance to Bohr was spectroscopy. This had originated with Gustav Kirchhoff and Robert Bunsen in the middle of the nineteenth century when the former discovered that each pure substance produced its own characteristic spectrum of light. In the year that Bohr was born, Johann Balmer had shown that the wavelength of lines in the characteristic spectrum of hydrogen gas could be fitted to a simple formula. This formula contained a constant whose value was to be of the greatest interest to Bohr. The constant was known as the

Rydberg constant, after a discovery by Johannes Rydberg in 1890. He found that the formula could be used to describe the spectra of other elements and that the constant retained the same value. The value of the constant can be obtained experimentally to a high degree of accuracy. Rydberg was interested in how his formula was related to the position of elements in the Periodic Table, and it was agreed that it was important in the production of light from matter. However the manner in which it was involved was then unknown.

Spectroscopy was not the only study of light that influenced the work of Bohr. The wave theory of light had received a setback in the late 19th century. This was the result of difficulties encountered in using it to explain the energy distribution in the spectrum of light emitted by heated matter. This difficulty was overcome by Max Planck in 1900, who succeeded in producing a formula describing these emission spectra. This formula contained a constant, which is invariably denoted by the letter **h** and came to be known as Planck's constant. The value of **h** could also be experimentally determined. The unit of this constant is energy multiplied by time and it also came to be known as the action constant. Planck was not at all sure of the meaning of this constant, but finally proposed that light was emitted and absorbed in finite units which he named quanta and that the energy carried by these quanta was related to the constant **h**.

Albert Einstein applied this proposal to produce an explanation of the emission of electrons from metallic surfaces when irradiated with light. This effect is known as the photoelectric effect. The relationship he produced was that the energy of a quanta of light is given by $\mathbf{h}\boldsymbol{\nu}$ where $\boldsymbol{\nu}$ is the frequency or colour of the light being considered. The emission of light as quanta seemed to give light a nature closer to the corpuscles of Newton and was thus at odds with the wave theory. However it was generally accepted that Planck's constant was involved in the emission of light from atoms. Once again the manner in which it was involved was unknown.

All of the developments described above influenced Bohr when he came to consider the nature of the atom, however he was not the first to advance a description of the structure of the atom.

5. Structures of the Atom in the Early 20th Century

Researchers specifically involved in studies of the nature of atoms in the early part of this century had three interests,

(i) the origin of the particles and radiations emanating from the atom.
(ii) the nature of ionised atoms.
(iii) the relationship of atomic structure to the Periodic Classification of the elements. In the year 1912 the Periodic Table was of recent origin. It had been proposed independently by Dmitri Mendeleeff in 1869 and Lothar Meyer in 1870, so in 1912 it would be as if the classification had been developed in the 1950s in relation to the present day, and it was of great interest to both chemists and physicists.

The evidence available to research workers of the time specifically concerning the atom can be reasonably described as scanty and confusing. Atoms were thought of as spherical and having diameters of the order of 10^{-10} metres. Electrons were considered to originate from atoms. Their size was not known although they could, of course, be taken as smaller than an atom if they came from inside an atom. Electrons displayed a negative electric charge and were deflected by a magnetic field. In mass at least, protons were larger than electrons, their mass being one thousand eight hundred and fifty times greater. They were also considered as part of an atom. However their linear dimensions were also not established. They displayed a positive electric charge and were deflected in the opposite sense to electrons by a magnetic field. This convention of negative and positive charge indicates that electrons would repel electrons as would protons repel protons. It also indicated that electrons were attracted to protons. Other particles, called alpha particles, were emitted by atoms. These were eventually identified as the nuclei of the helium atoms. The atom was also a source of penetrating radiations called X-radiation and gamma radiation.

In these early years of the 20th century, both Thomson and Rutherford were investigating the way in which particles emitted by atoms travelled through matter. In working for both of these men Bohr was in contact with the very latest developments in this field.

The former used electrons in his studies and the latter used alpha particles. Both were to make suggestions as to the arrangement of electrons and protons in an atom, as has been described in the respective chapters on these two scientists. Thomson[3] proposed that an atom consisted of a positively charged sphere of space about 10^{-10} metres in diameter. In this sphere the electrons moved in circular orbits all in the same place. The number of electrons in each atom was also unknown at the beginning of the 20th century. Thomson had concluded the major mass of an atom was the result of the presence of electrons only. This meant that atoms of high atomic weight would contain many thousands of electrons. In turn this meant that the dimensions of an electron had to be very much less than the dimensions of an atom.

On the other hand Rutherford was led to the conclusion, by the results he obtained, that inside the atom there existed a great deal of empty space.[4] This meant that the electrons and protons inside the atom had to occupy the minimum space possible. He advocated a structure that had been proposed by Hantaro Nagaoka a few months after Thomson advanced his model early in the year 1904.[5] The model of Nagaoka was described by Rutherford as a "Saturnian model". It consisted of a central attracting positive mass surrounded by negative electrons rotating in rings without changing their separation from one another. The rings were all in the same plane in analogy to the rings around the planet Saturn. Nagaoka showed that such a system would be stable if the attracting force was large. Rutherford proposed that all atoms consisted of a group of protons in very close proximity at the centre of a sphere of space of atom dimensions. This was surrounded by an atmosphere of electrons containing the exact number of electrons required to "neutralise" the charge of the central group of protons, such that all atoms were electrically neutral. The much greater mass of the proton compared to that of the electron meant that the major part of the mass of an atom was therefore concentrated in the proton core.

The concept that particles which were known to repel each other under experimental observation could form stable groups when placed close together inside an atom was by no means easy for the researchers of the period to accept. This was especially so considering the fact that the attraction and repulsion of these particles was governed by an inverse square law identical to that governing the force of gravity. There was however no alternative apparent at the time. The acceptance of Rutherford's concept was one of the factors which eventually lead to the present widely held belief that a pictorial description of the inside of the atom is not possible.

Rutherford established the important fact that the number of electrons in any atom was equal to the chemical atomic weight divided by two and that this corresponded to the number indicating the position of the element in the Periodic Table. The latter number is now called the Z-number and is considered to be the number of positive charges in the atomic nucleus. This was the first link ever made between the nature of the atom and the elements.

There remained, however, the part played by the Rydberg and Planck's constants in the structure of atoms and it was the introduction of these constants into the mathematical description of the atomic structure of Rutherford which Bohr finally achieved.

6. Hydrogen and the Rydberg Constant

The element which featured most prominently in the work of Bohr was the gaseous element, hydrogen. Hydrogen gas was first known as inflammable air and was probably initially observed by Van Helmont prior to 1644. However it was isolated and described by Robert Boyle in 1642, who produced it from the reaction between iron nails and dilute sulphuric acid. The reaction of burning hydrogen in air to produce water was observed by Pierre Macquer in 1776. This work was refined by Henry Cavendish, Antoine Lavoisier and others. Beginning in the 1780s the chemical properties of this gas, particularly the relationship to water, were to be studied in detail for twenty years. Out of these and many other chemical studies chemists defined various laws of chemical combination, the culmination of which was the atomic theory proposed by John Dalton in 1803. In this theory Dalton retained the principle, originating with Leucippus and Democritus that atoms were indivisible, and stated that when they combined they did so as whole units. The theory of Dalton was not, of course, a theory of the structure of atoms. Since study of the physical properties of hydrogen had shown it to be the lightest gas known, Dalton assigned to it the atomic weight of one unit and against this all other atomic weights were measured relatively.

The study of spectra had continued since the days of Kirchhoff and Bunsen and by 1912 the spectra of many elements were known. Among the other physical properties of hydrogen which had been studied was the spectrum of the light emitted by the gas when subjected to electric discharge or extreme heating. The spectrum of hydrogen is a series of sharp lines divided into groups. These are called *series*, and are named after the researchers who discovered them. Thus there is the Lyman series, the Balmer series, the Paschen series, the Brackett series and the Pfund series. Each series occurs in a different part of the full spectrum, for example the Lyman series occurs in the ultraviolet part of the spectrum and the Pfund series occurs in the infrared part of the spectrum. In the visible part of the spectrum of hydrogen there appear the H_α, H_β, H_χ, and H_δ spectral lines. These lines appear in the spectrum of the Sun and their wavelengths have been measured with great accuracy. Balmer obtained a mathematical relationship between the wavelengths of these spectral lines. This allowed the calculation of the wavelength of lines present. The Balmer formula is $\lambda = k\tau^2/\tau^2 - 4$, where λ designates the wavelength of the light involved, k is a constant and τ is a whole number which takes the values 3, 4, 5, 6, and so on respectively for the lines in the above series. The formula is usually written in terms of the reciprocal of the wavelength, $1/\lambda$. The reciprocal is known as the wave number, η, and is the number of wavelengths per centimetre of the light involved. The formula for hydrogen then becomes $\eta = R(1/2^2 - 1/\tau^2)$. In the formula τ, is a whole number and the constant R is the Rydberg constant.

Using this formula to calculate the wavelengths of lines in the Balmer series the first number in the brackets of the Rydberg equation is 2 and the second is 3, 4, 5 and so on. To perform the same calculation for the Lyman series the first number is 1 and the second is 2, 3, 4 and so on. The formula dictates that as the second number becomes larger the calculated wavelength becomes smaller, the frequency becomes larger and the spectral lines become closer together or converge. The limit of a series is called the *convergence limit* and the value of this limit is obtained by giving the value of infinity to the τ-value of the second term in brackets, which makes it equal to zero.

7. Planck's Constant and the Structure of the Atom

The measured value of Planck's constant is 6.624×10^{-27} erg seconds. This is the same order of dimensions as the electron mass with a value of 9.1×10^{-28} grams and this observation encouraged the conclusion that the constant was in some way related to the structure of the atom. Research workers at a conference in 1912, called the Solvay Conference, discussed the relationship of Planck's constant to the line spectra. They were trying, in particular, to decide whether the full value of h, a fraction of the value of h or h divided by 2π was the relevant factor to be used in the description of line spectra. The first attempt to involve the constant in the structure of the atom was that of Arthur Haas[6] in 1910. Using the atomic structure of Thomson he obtained a relationship for h by considering the case of a single electron orbiting inside a positive Thomson sphere. He arrived at the condition that $h = 2\pi e(ma)^{1/2}$, in which e designates the value of the electric charge carried by the electron, m designates the mass of the electron and a the radius of the positive Thomson sphere.

However it was the work of John Nicholson[7] which first related Planck's constant to a Saturnian atomic structure. He was examining the possible stability of ring models of theoretical atoms and showed they were not stable if the electrons, rotating in their fixed ring positions, vibrated in a direction along a radius of the ring. They were however stable for vibrations in a direction perpendicular to the plane of the orbit. He then postulated that the latter vibrations give rise to a wave that travelled around the ring. This wave was a light wave. As a result he computed the frequencies or colour of light that would be generated by such oscillations and was able to relate some of these to the spectral lines in some stars. Nicholson calculated the total energy (J) of his theoretical atoms and found that the division of this value by the frequency of the radiation (ν), namely J/ν; was equal to the value of h multiplied by a whole number.

8. Niels Bohr and the Structure of the Atom

Bohr was twenty-seven years old when he started work with Rutherford in 1912 and by this date he had published five papers on topics involved with the nature of matter. He was only to stay in Manchester for a short period before returning to the University of Copenhagen, first, as an assistant and then a lecturer. In 1912 he met and married Margrethe Nørlund whose support was to prove invaluable. She organised and dealt with much of the correspondence that her husband attracted as a result of his work on the structure of the atom and other topics.

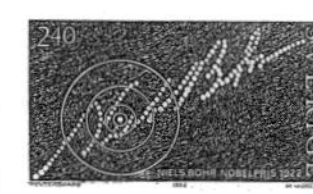

When Bohr came to consider the structure of the atom in the years 1912 and 1913, he, like all other scientists at this time, was aware of the following considerations.

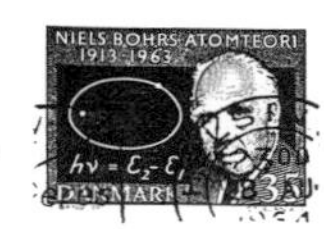

(i) It was accepted that the hydrogen atom consisted of one proton and one electron.

(ii) The atomic structure of Rutherford placed the protons at the centre of the space occupied by the atom and these were surrounded by electrons. Although it seems to have been accepted that the core was composed of individual protons, no explanation had been advanced as to the manner in which a group of closely associated, mutually repelling charges was stable against disintegration.

(iii) Electrons in the atom would repel one another.

(iv) The number of electrons in any atom was equal to half the numerical value of the chemical atomic weight.

(v) The atom was electrically neutral, hence the number of protons was the same as the number of electrons.

(vi) There was no explanation of how the weights of the particles, known to be in the atom at that time, gave rise to the chemical atomic weights assigned to the elements.

(vii) Consideration of the mechanics of the Saturnian atom, by Nicholson and others, had already indicated that the electrons orbiting in rings would not be stable if subjected to small disturbances (vibrations) in a direction along the radius of the plane.

(viii) Planck's concept of light emission in quanta was becoming accepted and his constant, h, was in some manner involved in the structure of atoms.

(ix) The Rydberg constant was considered to be involved in the emission of light from atoms. Curiously Bohr was not, at first, familiar with this constant.

(x) The emission of light was, on the basis of classical physics, the result of a charged particle being subjected to accelerations and a moving electron was proposed as the origin of the line spectra.

Bohr was at first more interested in an atomic structure which could explain the classification of elements into the periods of the Periodic Table, influenced no doubt by Thomson who was interested in this topic. He was not successful in this project at this time but he was to return to the topic later in his career. Careful historical research indicates that Bohr did not arrive at his concept of the structure very easily. It would seem that at one point he considered that other researchers were much more advanced in developing a possible structure for atoms. He concluded that he would not be the first to arrive at an acceptable structure for an atom of hydrogen.[6] This shows the intense interest that there was at the time in developing an atomic structure which gave an explanation of the recent discoveries.

He presented his concept of the structure of atoms in a three part paper entitled "On the constitution of atoms and molecules" published in the *Philosophical Magazine* and *Journal of Science* in 1913.[8] This paper is the best known of all the papers published by Bohr and, of the three parts, only Part 1 is ever dealt with to any extent. It can be seen from the copy of the original paper included in this volume that the paper is not exclusively concerned with the structure of the hydrogen atom.

He began his paper with the calculation of the mechanical frequency, ω, of an electron orbiting a proton in an elliptical orbit on the basis of the planetary orbit laws of Kepler. These laws require that the orbit be elliptical. He then showed that values of the mechanical frequency (ω) and the radius of the orbit a cannot be computed unless the work (*energy or effort* W) required to separate the electron and proton is known. It is to be noted that E is used in the paper for the charge on the atomic nucleus and not energy.

This is followed by a description of the result of the electron orbiting a proton if classical electrodynamic theory is applied to the system. Since the charged electron orbiting around the proton is under constant acceleration it should, according to classical electrodynamics, emit light continuously, giving a spectrum known as a continuous spectrum. As stated previously the spectrum of hydrogen consists of groups of sharply defined lines and in no way resembles a continuous spectrum. As the result of the loss of energy by radiating the kinetic energy in orbit would decrease and the orbiting electron would move into lower and lower orbits towards the proton. This was therefore the dilemma that faced physicists of the time in developing a model of the atom.

On the basis of a Saturnian atomic structure the hydrogen atom would consist of an electron in orbit around a proton. The electron maintained a fixed distance from the proton as a result of the balance between the force of attraction of the proton at this distance and the centrifugal force of the orbiting electron trying to fling it clear of the proton. The force of attraction (ϕ_a) is derived from Coulomb's law of attraction between oppositely charged bodies. This is given by the expression $\phi_a = Ze/a^2$, where Z designates the number of charges involved, e designates the charge on an electron and that of the proton is taken to be the same. The radius of the orbit is designated by a. The centrifugal force is derived from Newton's law that mechanical force is the product of the mass of a body and the acceleration of a body. For a body in orbit around another the force is given by $\phi_r = mv^2/a$ where m indicates the mass of the body and v the velocity in orbit. The expression v^2/a is known as the angular acceleration. This then represents the hydrogen atom and the forces involved as examined by Bohr.

However a charged particle moving around a closed loop is the same as an electric current moving in a closed loop of wire and could be expected to generate a magnetic field. It was therefore necessary to consider any effect this might have on the electron orbit. Bohr assumed that the velocity in orbit was very much less than the velocity of light and therefore a significant magnetic field was not produced. This assumption allowed him to present a simple version of his model.

He started his development of atomic structure from the position of an electron being drawn from infinity to an orbit around the proton and in this case the orbit is taken to be circular, not elliptical. This change does not at this stage affect the structure of the atom he proposed, however in later papers a return to the elliptical orbits was found to be essential. In being drawn to the proton the electron loses energy during the process of formation of the hydrogen atom. This arises from the fact that work has to be done *on* a proton-electron system in order to separate these attracting particles and therefore when they are in the process of being brought together work will be done *by* the system, equivalent to energy lost. He proposed that this energy is lost as light of a single frequency, or colour, designated by ν. Introducing the Planck concept, he assumed that the loss involves a fixed number of quanta that is $W = \tau h\nu$, where τ is a whole number. The argument for making this assumption is simple, if all light is emitted as quanta then there is no reason to suppose that any light emitted during the process of binding of the electron to the proton should not also be emitted as quanta. At this point in the paper he introduced for the first time an expression that would become familiar to later researchers, namely the *binding energy.*

A very interesting implication can be drawn from this description of atom formation and it is that atoms can be or are formed from their component parts and that they may not have existed forever. The component parts, the proton and electron, were implied to have infinite existence in the early part of the twentieth century. Even today where and how atoms form is a matter of controversy.

There is an important point in this model of the way in which the electron and proton are associated with one another. A body orbiting another passes a particular point in the orbit frequently and is therefore said to have a frequency.

This frequency is the number of times that the orbiting body passes this same point per second. Since the distance in a circular orbit is $2\pi a$, a being the radius of the orbit, and giving the velocity in orbit the symbol v, then the time for one complete orbit is $2\pi a/v$ seconds and the number of times the orbiting body passes the same point per second is then the frequency. This is symbolised as ω and is given by the formula, $\omega = v/2\pi a$. This frequency is called the mechanical frequency and it was expected that the frequency or colour of the light emitted from an atom would be related to this mechanical frequency.

Once the electron was in orbit Bohr had to make an assumption concerning the mechanical frequency. By taking the average of the mechanical frequency at infinity (ω_0) and that in the final orbit (ω) he stated that the frequency of the radiation by which energy was lost during binding was $\nu = \omega/2$. This was and is a critical assumption since it meant that the frequency of light radiated by an orbiting electron was no longer that of the mechanical frequency of the electron in orbit, the two had been separated or decoupled. Using the relationship he has derived he presented values for the radius of the hydrogen atom, the wavelength of light emitted during the binding process and the ionisation potential of the hydrogen atom, which are all in reasonable agreement with the experimental values. The last value, the ionisation potential, is the voltage required to remove the electron from a hydrogen atom and is considered as being equivalent to removing an electron to infinity from a positive charge.

However it was and still is very difficult to envisage the manner in which an electron orbiting a proton can oscillate or vibrate and thereby experience accelerations, in addition to the mechanical acceleration of orbiting, and thus give rise to radiation. The vibrations could be to and fro in a direction at right angles to the plane of the orbit, as was proposed by Thomson and Nicholson, but this was not accepted by Bohr. This being so, he then had to introduce a mechanism which allowed the separation of the mechanical and radiation frequencies.

Having discussed the problems surrounding the structure of the atom Bohr then introduced the concepts for which he was to become widely known. These can be summarised as:

> When in a stationary state the actions of an electron can be described by the principles of mechanics but in changing from one state to another such principles cannot be used and further it is only during the change from one state to another that radiation is emitted. The emission of the radiation during the change of state is governed by the quantum principles of the theory of Planck.

At this point in the paper he also introduces another term which would become familiar and this is *stationary state.* These are orbits which the single electron of the hydrogen atom can occupy and in which there is no radiation of energy in spite of the fact that it will be experiencing acceleration.

Although a mechanism allowing the separation of mechanical and radiation frequencies had been advanced the implication was that the "origin of light was the physical movement, that is, a change of coordinates, of electrons in space" not the acceleration of a charged particle. Since the electron was not vibrating in any way this change should not produce a wave of light. If this was correct, then it had to be accepted that classical physics was no longer capable of describing the events that occurred within the atom. The proposals that an electron did not radiate while experiencing mechanical acceleration in an orbit, along with a change of orbit brought about by an unknown means, was to arouse great controversy among research workers in this field of atomic physics at the time. An attempt at justification of the statements was given by Bohr later in the paper. It can be taken that Bohr had realised, while writing the paper, that the calculation of values of constants of the hydrogen atom, which were within the experimental error of the measurement of these constants, would not be accepted as being adequate justification of his requirement that the concepts of classical physics be abandoned. He therefore searched for further support for his concepts.

9. The Linking of Planck and Rydberg Constants to the Structure of the Atom

It is recorded that Bohr only took an interest in the spectrum of hydrogen as a result of contact and conversation in 1912 with Hans Hansen, the spectroscopist at the University of Copenhagen.[6] It does seem strange that someone who became intimately involved in the question of emission of light from atoms should not have been aware of the work of Balmer. It is possible that this arose from his original interest in relating the structure of the atom to the element arrangement in the Periodic Table and which had occupied his early efforts in dealing with atomic structure. This work required knowledge of the properties of elements rather than light emission. It can also be supposed that the work of Balmer did not form part of the degree course which Bohr studied. Certainly he himself is recorded as saying that on seeing the Balmer formula for the first time "everything was made clear".[6] This statement indicates that there was something in the formula which "struck a chord". This was the fact that the formula contains a term in which there is one unit subtracted from another. The connection was therefore that this was exactly the same as saying radiated energy arose as a change from one stationary state to another and he states as much in the paper.

He took the value of the charge on the proton as equal to that on the electron and the orbit as circular and derived a formula for the energy radiated when an electron enters a stationary state orbit round a proton, this being the energy of this particular state. Applying his concept for the change in energy when an electron moves from one stationary state to another he derived a relationship which has the same mathematical form as the Balmer relationship. This is presented in the paper as a frequency relationship. Then, by equating the change to the energy of a quanta of emitted light, $h\nu$, he was able to obtain a value for the Rydberg constant. It will be noticed that he gave the value of the latter as a frequency, not as a wave number. To convert the result to the wave number value, the value in the paper is divided by the velocity of light since $\nu\lambda = c$. This gives rise to the wave number, $\eta = 1/\lambda$, with a value of 103,402.27. This he compared with the value of 109,739.83 as measured by Friedrich Paschen in 1908. The agreement between the two values was convincing, especially since the values of the constants he used in the calculation were not as accurate as later values.

The mechanism whereby the lines are formed in the Balmer series is therefore that electrons have been moved into orbits whose radii are related by the τ-values 3, 4, 5, 6, and so on and they return to an orbit whose radius is related to the τ-value 2. In doing so they emit radiation of a frequency calculated from the formula. This process is exactly the same as that of an electron losing energy on becoming bound to the proton. The actual radius values can be obtained from the expression Bohr presented earlier in the paper. Bohr then explained that certain lines in some spectra, which should exist on the basis of his derivations, may never be observed either as the result of too great a radius of the orbit necessary to produce them or too low an intensity under conditions where they might be observed. The numbers indicated by the symbol τ are known today as the principle quantum numbers and a diagrammatic representation of the process is shown in Fig. 1.

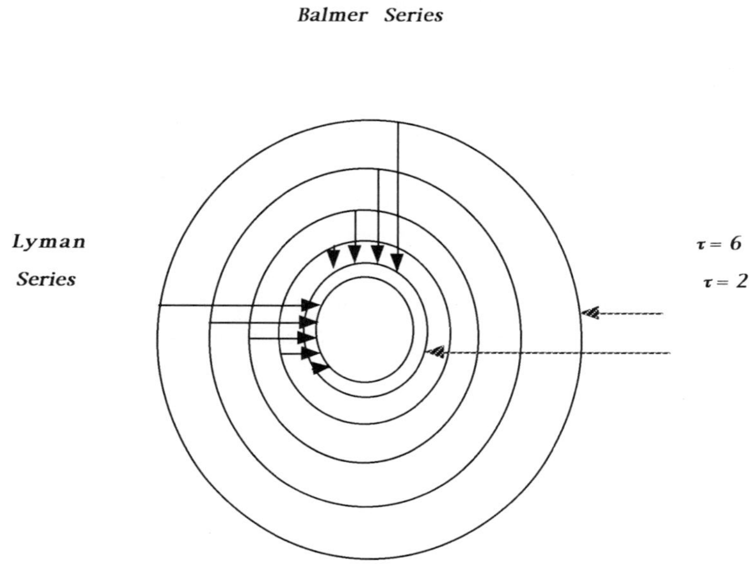

Fig. 1. The emission of light by electron transition between orbits.

In spite of the above results Bohr searched for further support of his proposals. This support he found in the spectrum of light from a star. Observation of the spectra of stars was and still is important research work. The research was aimed at identifying the elements present in stars. Nicholson, who was effectively the rival of Bohr at the time, developed his ideas of atomic structure to this end. The star Zeta-Puppis possessed a spectra which was assigned to hydrogen but in which some lines were not accounted for by the spectrum of hydrogen. A similar spectra was also observed in laboratory experiments from a mixture of hydrogen and helium gases. The structure of the atom proposed by Nicholson had allowed him to calculate the wavelengths of some, but not all, of the spectral lines observed but he was not able to identify his theoretical atomic structure with any particular element.

The Rydberg equations of the observed lines in the spectra of Zeta-Puppis were, $\eta = R[1/1.5^2 - 1/\tau_1]$ and $\eta = R[1/2^2 - 1/(\tau_1 + 1/2)^2]$. The proposals of Bohr with regard to the emission of radiation were dependent on the binding energy being a whole number of quanta. As far as he was concerned, the above relationships were obviously in error. He therefore rearranged them to give $\eta = 4R[1/3^2 - 1/(2\tau)^2]$ and $\eta = 4R[1/4^2 - 1/(4\tau+2)^2]$. The τ-values are now whole numbers. Singly charged helium was considered to consist of a positive charge of two protons and a negative charge of one electron. The single electron therefore made it like the hydrogen atom. As a result E becomes $2E$ in Bohr's equations, and the factor outside the brackets becomes $2\pi^2 me^2 R(2E)^2 = 4R(2\pi^2 me^2 E^2)$. Since the above rearrangement had produced the value 4, which is the square of the double charge on the helium proton core, Bohr concluded that the unknown lines must belong to the helium spectrum and not to the hydrogen spectrum. In this way he showed that the spectrum of Zeta-Puppis contained lines originating from helium. In doing so he had found a solution that Nicholson had failed to obtain. It has been reported that on hearing this result Einstein declared that the ideas of Bohr must be correct.[6]

The Bohr proposal concerning the manner in which light is generated by atoms is taken to be the same for elements whose atoms are like the hydrogen atom. That is, they consist of a positive charge and one electron, the so-called hydrogenic elements. For example helium with one electron removed (He^+), lithium with two electrons removed (Li^{++}) or beryllium with three electrons removed (Be^{+++}). The positive charge numbers (E) of course change from 1 for hydrogen to 2, 3 and 4 because these are the charge values of the proton core of these elements. The situation with respect to spectra for elements which consist of a positive charge and two or more electrons is different. For example lithium with one electron removed (Li^+) which then consists of a positive core of three protons and two electrons in orbit.

In 1908, Walter Ritz extended the ideas of Rydberg to the spectra of all the elements. He concluded that the frequency of lines in all spectra can be obtained from a formula which involved the difference of two numbers or terms. The first is constant and he named this the *series term*, the second is variable, which he named the *current term*. In practice the experimental relationship found for most elements is $\eta = R/(\tau + \alpha_1)^2 - R/(\tau + \alpha_2)^2$ where R is again the Rydberg constant and the α-values are fractions with values less than one.

Bohr did not attempt to give a complete explanation of the emission of line spectra from elements with more than one electron. He was not in fact in a position to do so because at this time he had not dealt with the manner in which electrons were arranged in multi-electron atoms. In dealing with the Ritz equation, Bohr showed that he did not accept the fraction terms and rearranged the equation to remove the fractions. He contended that the difference of two terms in the Ritz equation showed that the mechanism of light emission in these cases was the same as for hydrogen, that is, it involved changes between stationary states. He considered the case of an atom where there were already bound electrons, but where there was one extra proton capable of binding an electron, the final atom being neutral. This situation is the same as that for the hydrogen atom and hence the electron binding energy will also be the same as that for the hydrogen atom. He then showed that his rearranged Ritz equation would behave mathematically in exactly the same manner as the Rydberg equation when large τ-values were involved, namely it would approach closer and closer to the value $2\pi^2 me^4/h^3$. This he presented as proof that mechanism of light emission in all atoms was the same as that he had proposed for the hydrogen atom.

10. Justification of the Model

The atomic structure advanced by Bohr was effective in allowing the calculation of measurable parameters connected with the atom. However there was no justification for the radical assumptions he made. He was also aware that Nicholson had shown that it is possible to calculate the frequency of light emitted from a Saturnian atom using classical electrodynamics and without making such assumptions. To deal with this situation he presented what he considered a more rigorous derivation of his earlier relationships.

The assumed relationship $W = \tau h\nu$ implied that the energy is controlled by the quantum number and for different stationary states different numbers of quanta are emitted. In the Planck relationship this is not the case, the energy is a function of frequency

Energy $= h\nu$, the greater the amount of energy the higher frequency of the emitted light. Therefore he had not strictly applied the concepts of Planck in obtaining his results. He adopted a new relationship $W = f(\tau)h\omega$ and gave the function $f(\tau)$ the value $c\tau$ in equations for mechanical frequency of the orbiting electrons and the frequency of light emitted, as obtained from the Balmer formula derived from his model. He then demonstrated that the change in mechanical frequency is negligible when a large τ-value changes by one unit, that is, the ratio of the mechanical frequencies before and after such a change is close to unity. Classical electrodynamics leads to the conclusion that the mechanical frequency and the frequency of the emitted light would be the same, that is their ratio (ν and ω_{N-1} in the original paper) would also be unity. This will be the case for high τ-values when the coefficient c has the value 1/2. When this value is substituted in the above relationship it becomes the same as the one he used originally. Further, since ν, as derived from the Balmer formula is related to W, which is in turn related to ω, then ν is related to ω, that is $\nu = n\omega$, n being a whole number. From this, the frequency of the radiation emitted when an electron changes orbit is a multiple of $\omega/2$. He has again achieved the separation of mechanical and radiation frequencies. There is however the question of the value to be attached to large τ-values, after all he had made it clear earlier in the paper that for large τ-values radiation might not be seen. This point was not discussed.

Having revised his derivation in this manner it is fair to ask what had been achieved. He felt that he had shown that,

(i) The energy emitted on the change of orbit $W = hn\omega = h\nu$ and had therefore met the conditions of the Planck relationship.

(ii) Shown that the mechanical and emission frequencies need not always be the same.

(iii) Shown that at large τ-values his model and the classical model give the same results.

Hence it was quite possible for Nicholson to have obtained the correct results using classical electrodynamics. Bohr evidently felt that all the above gave the proposed atomic structure firmer support and justified his model. So much so in fact that he stated that if the model he has developed is correct then there should be exact agreement between the calculated and experimental values of the Rydberg constant. Since this is not so, he concludes that the values of the measured parameters he has used in his calculation of the value of the constant must be in error.

11. The Momentum of the Orbiting Electrons

In the final paragraphs of the third section he discussed angular momentum of the electron in orbit. However before he did so he made the following statement, "While there obviously can be no question of a mechanical foundation of the calculations given in this paper, it is, however, possible to give a very simple interpretation of the result of the calculation on page 5 (of his paper) by the help of symbols taken from ordinary mechanics". This statement is only understandable when it is realised that the concept that most interested, in fact fascinated, research workers of this period was that of quanta, introduced by Planck. Light "moves" and light is emitted as quanta, that is, it is quantised. Physicists of the day began to consider whether or not movement of matter inside the atom, governed to this date by the laws of Newton, was also quantised.

With regard to this possibility, Nicholson made the following statement in one of his published papers,[7] "If therefore the constant h of Planck has, as Arnold Sommerfeld has suggested, an atomic significance, it may mean that the momentum of an atom can only rise or fall by discrete amounts when electrons leave or return." Following the Sommerfeld suggestion Nicholson had demonstrated that the angular momentum of the electrons in his models had a whole number relationship to the value of h.

The significance of this relationship was that the angular momentum of an orbiting electron seemed to be quantised, indicating that the laws governing interactions *inside* the atom were very different from those governing interactions *between* atoms. Physicists were faced with a situation that not even the classical Newtonian laws of mechanics might apply inside the atom. Therefore a new type of mechanics, "quantum mechanics", was required to describe events inside an atom. These had not yet been developed but from this time on much effort was expanded in developing these new mechanics.

The atomic structure of the hydrogen atom proposed by Bohr involved one body orbiting another. Under these conditions the orbiting body possesses angular momentum (the moment of momentum). The angular momentum of an electron in an orbit of radius a is given by mva. From Newtonian laws, a force acting on a body for a given time results in a change in the angular velocity and causes a change in momentum. The nature of the force, acting for whatever time, and causing the change in the electron orbit was never clearly identified by Bohr. As the paper shows he derived a relationship for angular momentum of the electron in a stationary state orbit and demonstrated that it was a whole number multiple of $h/2\pi$. One consequence of his model was that the angular momentum of an electron in any stationary state orbit must always be a whole number multiple of $h/2\pi$.

He ends this section of the paper by observing that most stationary states can only be observed from spectra and for all other physical conditions "atoms of matter" are all in the stationary state which represent the greatest amount of energy that was liberated during the formation of atoms. This he names the *permanent state of the atom* or the state at Normal Temperature and Pressure and for hydrogen atoms this is the *stationary state with radius a.*

12. The Application of the Model

Bohr applied his model to various atomic phenomena with varying degrees of success. The discussion of phenomena is largely performed by means of analogies. He was, at this stage, unable to proceed in any other way since, as stated previously, he had not yet considered the arrangement of electrons in complex atoms. He was to encounter difficulties, particularly with the absorption of light and the results obtained in this area by Robert Wood. Since Bohr was intent on explaining the manner in which light was emitted from atoms he could not ignore the process of absorption of light. He had explained the former as the case of an electron moving from a stationary state orbit with a particular radius to another with a smaller radius. Therefore absorption must be the reverse of this mechanism, namely that the absorption of light energy, by the electron, increases the kinetic energy. This, in turn, leads to increase in the centrifugal force which results in the electron moving to an orbit of larger radius. Having acquired this energy, the nature of the mechanism which then causes the electron to move to an orbit of smaller radius and emit radiation remained unknown.

He concluded that his model allowed explanations of phenomena not possible using the theories available at the time and that his model did not involve assumptions which were in conflict with classical theories of dynamics and the wave theory of light. This paragraph in his paper clearly indicates that Bohr did not want to abandon classical physics and he was to spend considerable effort in his later work in attempts to reconcile his proposals of atomic structure with classical physics. These comments are an excellent demonstration of the crisis that had appeared in physics in the early years of this century caused by discoveries concerning the atom. The evidence suggested that classical physics was not able to give an explanation of these new discoveries, yet many of the physicists of the time were very reluctant to accept this conclusion.

13. Other Atoms and Molecules

At this point it is necessary to examine how the atom was seen at this time. It was regarded as a positive core orbited by electrons but there was no decision concerning the actual arrangement of the orbits. Bohr mentions both electrons in individual orbits and the "ring as an entity". There are two possibilities for the structure of an atom with more than one electron which could meet the requirements of orbit change. These are shown in Fig. 2. In the Saturnian atom it was considered that there could be more than one electron in an orbit and that the orbits were all in the same plane. The electrons were supposed as being situated at equiangular intervals around the ring and the ring rotated without the electrons changing their distance from one another. The number of electrons that could be in the ring was not known, although Thomson had concluded that the limiting number was seven or eight. Any arrangement of electron orbits with more than one electron in each orbit does not seem to be compatible with the principle of radiation resulting from changes in orbit. This is because the question arises as to whether or not all the electrons in the common orbit underwent the orbit change together. In addition any orbit change would have to be small compared with the separation of the orbits in the common equatorial plane or there would arise the possibility of electron–electron interaction.

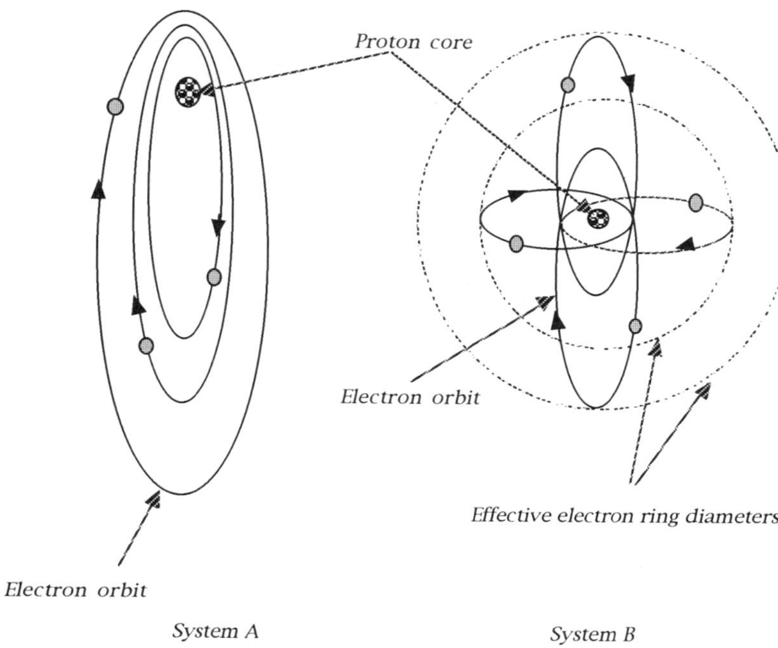

Fig. 2. Possible arrangements of electron orbits in many electron atoms.

However in the system A with only one electron in each orbit, again all in the same plane, then the orbit changes required by the concept above are possible, provided these do not approach the inter-orbit separation distance. In order to accommodate all the 119 electrons of a uranium atom requires orbits in different planes around the proton core. However there are only 360° of angle available for all the planes around the proton core and for a uranium atom the angle between the orbit planes would be about three degrees. Only if the distance between electrons in orbit is fixed could there be no electron–electron interaction. This is not possible since the orbit velocities would all be different as a result of the balance of forces. Under these conditions the distance apart of two electrons in adjacent orbits at their point of closest approach would be the length of the arc of a circle with diameters drawn every three degrees around the circumference. Such a length could be less than the closest approach of electrons to the positive atomic nucleus. Therefore the force of repulsion between two such electrons would be greater than the attraction exerted on them by the nucleus. In addition no two orbits could have the same dimensions even if they were in different planes, because this would again lead to possible electron–electron interaction. It is of course possible to have two interlinking systems of this type in the form of a chain link but this does not really release the arrangement from electron–electron interaction and these interactions would make the system unstable.

The system B with only one electron in each Keplerian elliptical orbit and all the orbits being in the same plane allows the orbit changes as envisaged by Bohr. However there is the possibility of electron–electron interaction in this case because the orbits are crossing one another and again the orbit velocities cannot be the same. This of course would be avoided, but not eliminated, if the orbit planes were all different. There is, in addition, the manner in which electron motion continued under normal conditions of non-emission of light. Nevertheless the structure B seems to have been the one that Niels Bohr preferred from the records of his notes. It can be seen from these structures that for Li^+ with two electrons the question arose as to whether or not the two electrons changed orbits independently during the emission of light.

In the final section of Part I of his paper, Bohr discusses the above topic along with the total energy of an atom and the stability of the structure of the atom of hydrogen that he has proposed.

The total energy of an atom, which is the sum of the potential and kinetic energies involved can be used to demonstrate the loss of energy on binding an electron to an atom. The potential energy (P) in an electron in orbit of the radius is a and is given by the product of the total charges of the protons and those of the electrons divided by the distance between the charge centres. For the proton the charge is $+e$ and for the electron the charge is $-e$ and for the hydrogen atom there is only one of each and $P = e^2/a$. When a has the value infinity, P has the value zero. This is the same as saying that at infinity the attraction or repulsion between charges is zero.

Thus as the electron approaches the proton the potential energy it acquires increases from zero to the value $-e^2/a$, the negative sign implying that energy must be supplied to the electron to move it to an infinite distance from the proton core. The kinetic energy in orbit is $T = mv^2/2$. From the equations given above for the Coulombic and for the Newtonic forces, $(-e)^2/a^2 = mv/a$, from which $mv^2/2 = e^2/2a$. The total energy of the orbiting electron is the sum of the potential and kinetic energies, that is $J = -e^2/2a$. When an electron enters and orbits around a proton it has lost half the energy it would have had in moving from infinity to a distance a from the proton. This is the point that Bohr started from, namely that the electron loses energy when becoming bound into an orbit around a proton.

Bohr was well aware that the Saturnian atom had been shown to be unstable for the very changes in electron orbit that he was proposing as the basis of light emission. In this section of his paper he restates that the electrons are in the permanent state when the "electron ring" has been formed with the loss of the greatest amount of energy by each electron during the binding process and the angular momentum of electrons in this permanent state is $h/2\pi$. These conditions he contends makes an otherwise unstable system into a stable system. His reasoning seems to have been that these conditions have allowed him to theoretically describe the manner in which light is emitted from hydrogen atoms and accurately reproduce the experimentally obtained Balmer relationship. This being so, electrons orbiting in the permanent state orbits under these conditions must be stable or else he could not have achieved these ends.

14. Molecular Hydrogen

All of the above considerations are, of course, the ones that are related to the hydrogen atom. However hydrogen gas occurs as a diatomic molecule, and as a "modeller of hydrogen" Bohr had to attempt a description of this molecule. This he did in Part 2 of his paper which is not reproduced in this volume but a brief description of his solution will be given. The hydrogen molecule was considered by Bohr to consist of two protons situated in line with one another. The two electrons required to neutralise the charge on the protons were stationed in an orbit between these protons and the plane of the orbit was orientated at right angles to the line joining the centres of the protons. To give support to this configuration he calculated the heat of formation of a gram molecule of hydrogen gas from hydrogen atoms. This value can be obtained experimentally by chemical and physical methods. The theoretical value was 60 kilocalories per gram molecule and the measured value was 130 kilocalories per gram molecule. Bohr never really reconciled the difference between the two values.

15. Conclusion

It cannot be stated that the model received universal support when first published. Otto Stern is credited with the comment "If this nonsense is correct, I will give up being a physicist".[9] The principle that radiation is emitted only when the electron orbiting the proton in a hydrogen atom undergoes a change of orbit was to be debated for many years. No satisfactory explanation was ever to be advanced as to the cause of the change of orbit or the manner in which it took place. Bohr has been reported as insisting that it is not possible to give a picture of the way in which an electron undergoes this transition.[10] In the year 1926 it was reported that Bohr and Erwin Schrödinger were involved in an extended, and reportedly heated discussion, concerning the nature of the electron transitions from one stationary state to another. Schrödinger is stated to have maintained that the whole idea of such transitions leads to nonsense.[10] There was also the question concerning the proposal that electrons do not radiate when being subject to acceleration in stationary orbits. These points were, in time, to result in research workers looking elsewhere for an explanation of the phenomena associated with the atom. In his efforts to define the structure of the hydrogen atom Bohr had derived an expression which gave the values for various parameters of the hydrogen atom which were in agreement with experimental values, particularly the radius of an atom and the Rydberg constant. However it has been pointed out that the efforts of Haas also gave rise to the same expression as that obtained by Bohr for the calculation of the value of the radius of the atom.[6] Haas was using an entirely distinct atomic structure. In addition Nicholson had obtained correct results for the frequency of emitted radiation of electrons in orbit using classical electrodynamics. It is of interest to consider why this should be the case and it almost certainly arises from an assumption concerning the atom, which has not changed since Bohr wrote his paper. This is that electrons and protons are separated within the atom and do not interact. Planck's constant, which it is agreed is involved in the nature of the atom, is a constant of angular momentum. This could legitimately mean that one or more of the bodies within the atom are rotating around their axes rather than rotating in an orbit. It would seem that the only orbital motion that can satisfy the requirements of classical theory of light emission and the theory of Planck is one in which the electron is orbiting in an intermittent and not a continuous manner. All of this means that there may yet be other arrangements of the particles, known to be inside the atom, which have not yet been advanced and studied.

Bearing in mind the objections to the theory of the structure of the hydrogen atom proposed by Bohr the question can be posed as to how he advanced the understanding of the nature of matter. Without doubt Bohr contributed greatly to the present situation where existence of the atom is no longer a subject for debate. It is only how matter is arranged and behaves inside an atom that is open for discussion. He also advanced one way which could account for some of the known facts concerning the atom. Through his work and that of others "quanta" and "quantisation" were to become the basis of a new type of mechanics.

However these developments were to lead to the conclusion that the atom can have no pictorial description and can only be described through the medium of mathematics. This asserts that the nature of the atom is beyond human reason. Human reason and human knowledge identified the possible presence of both atoms and molecules, "the unseen parts of the surroundings", long before there was any indirect or direct evidence to support the existence of these entities. It is unlikely that human reason will never be able to deal with the manner in which affairs are arranged inside the smallest organised part of matter. Niels Bohr played a prominent part in laying stepping stones on the path to this end. However he and his contemporaries seemed to have forgotten the words of Isaac Newton, written nearly three hundred years ago,[11]

> "We are to admit no more causes of natural things than such as are both true and sufficient to explain their appearances."

To which he added,

> "To this purpose the philosophers say that Nature does nothing in vain, and more is vain when less will serve; for Nature is pleased with simplicity, and affects not the pomp of superfluous causes."

References

1. Stefan Rozental, *Niels Bohr, His Life and His Work as seen by His Friends*, North Holland Publishing Co. Ltd., London, 1967.
2. Joseph W. Mellor, *A Comprehensive Treatise on Theoretical and Inorganic Chemistry*, Longman, London, 1932.
3. Joseph J. Thomson, "On the structure of the atom", *Philos. Mag.* **7** (1904) 238–265.
4. Ernest Rutherford, "The scattering of α and β particles and the structure of the atom", *Philos. Mag.* **21** (1911) 669–688.
5. Hantora Nagaoka, "Kinetics of a system of particles illustrating the line and band spectrum and the phenomena of radioactivity", *Philos. Mag.* **7** (1904) 445–455.
6. Antony P. French and Peter J. Kennedy, *Niels Bohr, Centenary Volume*, Harvard University Press, London, 1985, pp. 33–49, 365.
7. John W. Nicholson, "The spectrum of nebulium, the constitution of the solar corona", *Mon. Not. R. Astron. Soc.* **72** (1911) 49–64, 139–150, 677–693, 729–739.
8. Niels Bohr, "On the constitution of atoms and molecules", *Philos. Mag.* **26** (1913) 1–26.
9. Gerald Holton, *Thematic Origins of Scientific Thought — Kepler to Einstein*, Harvard University Press, London, 1988, chap. 12, p. 461.
10. Walter Moore, *Schrödinger, Life and Thought*, Cambridge University Press, Cambridge, 1989, chap. 6, pp. 226–229.
11. Isaac Newton, "Regulae Philosophandi", *Mathematical Principles of Natural Philosophy*, 1686, translated by Andrew Motte, 1729, p. 398, revised edition by F. Cajori, University of California Press, London, 1934.

Daniel Stewart Robertson received his early education at schools in England. He joined the Atomic Energy Research Establishment at Harwell, England in 1952 to assist in a programme on the chemistry of the element polonium. After two years of compulsory military service with the Royal Air Force, he returned to Harwell in 1955 and continued as part of the same research team. Between 1956 and 1959 he attended the University of Durham, England and gained a Double Honours Degree in Chemistry and Physics. He spent two years in industry applying the technique of radioactivation analysis to industrial problems.

In 1961, he married Maxine Hart who accompanied him to Paris when he was granted first, the de Broglie Research Fellowship and second, the Joliot–Curie Research Fellowship at the Institute of Radium in the University of Paris. His research in France concerned the preparation of the element protactinium in the metallic state. On returning to England, he joined the Royal Radar Establishment (now the Defence Research Agency), Malvern. At RRE he performed research on the identification and preparation of advanced electronic crystals for potential and actual use in semiconductor electronics, optoelectronics and superconductive applications. He studied crystal growth processes and examined the properties of the crystals of some fifty different chemical compounds. He and his assistants prepared the crystals using the majority of the known methods for the preparation of crystals along with methods he developed himself. From the results obtained during the above studies he proposed the first unified theory of crystal growth. For the development of crystals for computerised axial tomography, used in medical diagnosis, he, his team and industrial collaborators were granted the Queen's Award for Technology. His work with the recently discovered high temperature superconductors allowed him to advance a new theory of the origin of superconductivity. He has published a total of sixty-two papers in the open literature. In 1992 he left RRE and is presently occupied in writing on matters of scientific interest.

Chapter 5. Arnold Sommerfeld

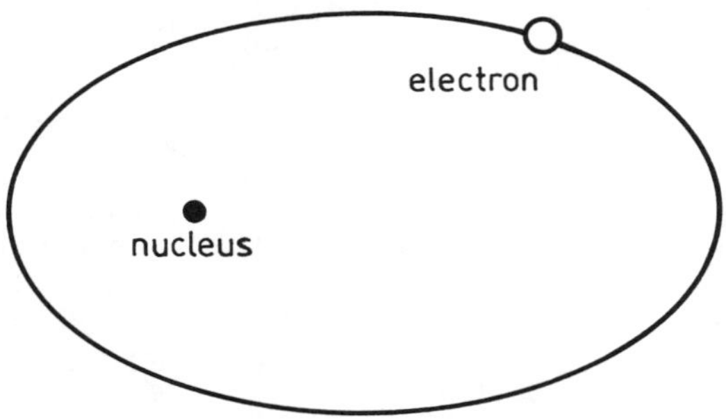

William B. White

A. J. W. Sommerfeld, c. 1949

Chapter 5

Arnold Sommerfeld

William B. White
Materials Research Laboratory,
The Pennsylvania State University,
University Park, PA 16802, USA

Abstract

Arnold Sommerfeld's part in the hydrogen atom story is in large part that of teacher and synthesizer. His book, *Atomic Structure and Spectral Lines* and it's supplementary volume, *Wave Mechanics* provided the foundation for most other researchers in the field. His principal original contribution was rewriting the old quantum theory of Niels Bohr into relativistic form and in so doing explained the origin of the fine structure of the hydrogen lines and discovered the fine structure constant.

1. Introduction

The hydrogen atom story touches upon some of the greatest names of late nineteenth and early twentieth century physics. These towering intellects may be broadly (and perhaps inaccurately) divided into two categories: the great conceptualizers and the great synthesizers. The great conceptualizers, for example Albert Einstein, Niels Bohr, Erwin Schrödinger, and Werner Heisenberg, were those who broke the paradigms. They presented world views that were radically different from the visions that had gone before.

Arnold Sommerfeld was one of the great synthesizers, and also one of the great teachers. He contributed relatively little in the way of really innovative new ideas of his own but he quickly seized on the new ideas of others and forced them into a consistent theoretical framework. His lasting monument in the evolving understanding of the hydrogen atom was his attempt, in the formalism of the old quantum mechanics, to describe Bohr's electron orbits in properly relativistic terms and in so doing discovered the fine structure constant. Overall, however, his most lasting monuments in early 20th century physics were his students. An amazing number of those who created modern physics — Debye, Pauli, Bethe, Heisenberg, to name only the Nobel prize winners — were students of Sommerfeld's.

2. Sommerfeld: A Short Biography[1,2]

Arnold Johannes Wilhelm Sommerfeld was born on December 5, 1868 in what was then Königsberg, Prussia, later Kaliningrad, USSR. His father, Franz Sommerfeld, was a physician with a great interest in natural history. His mother, Cäcile Matthias, was a vigorous and energetic woman who had a strong influence on young Sommerfeld.

From a basic schooling at the Altstädtisches Gymnasium from 1875 to 1886 Sommerfeld entered the University of Königsberg in the September of 1886. College life was a great deal of fun. Fraternity life, drinking parties, and saber duels competed with a wide-ranging study of languages, history, literature, and mathematics. In time, mathematics won out thanks to excellent mathematics instruction by Adolf Hurwitz, Ferdinand Lindemann, and David Hilbert. Hilbert, who was to become one of the greatest early 20th Century mathematicians was at the beginning of his own illustrious career as Privatdozent at the University of Königsberg.

In the style of the German universities, Sommerfeld proceeded directly to his doctorate degree. This he achieved in 1891 with a dissertation entitled, "The arbitrary functions of mathematical physics". The work was really applied mathematics. It dealt with the representation of arbitrary functions by trigonometric functions such as cylinder and spherical harmonics. The work was not profound either as mathematics or as physics but it laid for Sommerfeld the foundations for the powerful mathematical methods that he was later to bring to bear on a diverse collection of applied and theoretical problems.

Following a year spent in preparation for the examination to qualify as a Gymnasium teacher, and another year fulfilling his obligatory military service, Sommerfeld came to the University of Göttingen in October, in 1893. He was drawn to Göttingen, he said, as "the seat of mathematical high culture" although he spent the first year as an assistant in the Mineralogical Institute. During the following two years he was the assistant to the mathematician Felix Klein who was at the time systematizing the entire subject of mathematical physics. Sommerfeld also prepared his Habilitationsschrift, the second dissertation required in the German system for a teaching post in the universities. His second thesis was a deeper and more significant work than the first. "The mathematical theory of diffraction", made use of complex integrals in closed form suitable for numerical analysis, a focus toward the practical that would characterize much of his later work. He became Privatdozent at Göttingen in March, 1895 and lectured on advanced topics in physics and mathematics for the next five terms.

In October, 1897, at age 29, Sommerfeld accepted a professorship in mathematics at the Bergakadèmie Clausthal. The teaching was more elementary than at Göttingen but the salary and a regular position allowed him to marry Johanna Höpfner. She was the daughter of Ernst Höpfner, the curator of the University of Göttingen. Clausthal was sufficiently close to Göttingen to allow Sommerfeld to maintain not only family ties but also a collaboration with Klein. At Clausthal Sommerfeld applied his mastery of boundary value problems to such practical problems as the propagation of electromagnetic waves along wires and the diffraction of X-rays by a wedge-shaped slit.

April, 1900 saw Sommerfeld as a full professor of technical mechanics at the Technische Hochschule of Aachen. His position at Aachen required much interaction with the engineers as their resident theoretician. He attacked practical problems with the same enthusiasm and mathematical skill as he had applied to mathematical physics. As a result, his work on resonance phenomena in the vibration of bridges and ships, on problems of locomotive construction, on oscillations in coupled dynamos, on brake retardation in railroads and on the theory of lubricant friction was first rate and established his reputation in what we would now call engineering science or perhaps theoretical mechanics. He did not neglect physics and produced a series of papers on the dynamics of electrons that established him as a name to be reckoned with among theoretical physicists.

In the summer of 1906, when he was 38 years old, Sommerfeld's growing reputation earned him a call to the Chair of Theoretical Physics at the University of Munich where he was to spend the remainder of his career. The chair had been vacant for some years and was reinstated only after vigorous politicking by Wilhelm Röntgen, who occupied the Chair of Experimental Physics at Munich. It was apparently not an easy battle. Other faculty at the university saw no value in theoretical physics, which perhaps says something about the status of the subject at the turn of the century. The chair was first offered to Hendrik A. Lorentz, the leading Dutch theoretician, and when Lorentz turned it down, the offer went to Sommerfeld.

Once installed in Munich, Sommerfeld launched into an ambitious program to teach all of modern physics. He also had a small experimental program including fundamental work on the diffraction of x-rays by crystals. Sommerfeld was perhaps a bit slow to take up the quantum hypothesis following Max Planck's discovery of the radiation law in 1900, but by about 1910 he was convinced of its importance. When Bohr's quantum model for the hydrogen atom appeared in 1913, Sommerfeld embraced it eagerly with the results discussed in some detail below. Always close to experiment and always the teacher, he recognized that the best experimental check on the various models of the atom were the spectra that these atoms produced when excited by arcs, sparks, or electrical discharges. After making his own mark on the Bohr quantum model, Sommerfeld decided to organize all of the available information. The result was one of the most influential books in the history of physics. *Atomic Structure and Spectral Lines* first appeared in 1919, a second edition in 1920, and a third in 1922.[3] By the time of the fourth edition, the old quantum theory had been completely superseded by the new quantum mechanics of Schrödinger and of Heisenberg, Born, and Jordan. Although Sommerfeld was not one of the primary inventors of the new theory, his response was characteristic. He prepared a complete volume, *Wave Mechanics*,[4] published in 1928 as a "supplement" to the fourth edition of *Atomic Structure and Spectral Lines*.

In the 1930s Sommerfeld turned to the application of Fermi statistics to the electron theory of metals.[5] And he continued teaching. The final years of his career coincided with the rising power of the Nazis and the period of German ultranationalism. The campaign to cleanse physics of any Jewish taint and promote a pure "German science" engulfed many of the prominent physicists but Sommerfeld resisted. His continued friendship with Einstein, his support for other Jewish scientists caused him to be attacked as a "white Jew of science" and there was an effort to uproot him from his Chair as he passed retirement age in 1935. Rather than a prominent theoretical physicist — Werner Heisenberg was the leading candidate — his appointed successor in 1940 was Wilhelm Muller whose primary credential was being one of the stalwarts of the Nazi movement.

Sommerfeld occupied the war years in preparation for publication of his course notes on theoretical physics. After the war, in the late 1940s, he resumed for several years, the directorship of the Institute of Theoretical Physics. At the age of 83, while strolling with his grandchildren, he was struck by an automobile and died on April 26, 1951.

3. The Theoretician

Although steeped in the mathematical methods of classical physics, Sommerfeld quickly accepted the new paradigms of the conceptualizers and immediately set out to weave the new concepts into a consistent and mathematically rigorous system. Sometimes, his insights were ahead of their time as in the case of electron theory and later, as will be seen, the case of the fine structure constant.

Electrons and their properties particularly as beams in vacuum tubes were the center of attention for experimental physicists in the late 19th century. It was logical that Sommerfeld, always looking for theoretical problems that were close to experimental science, undertook, in the years 1903 and 1904 a series of calculations on the dynamics of electrons and in particular on the dynamics of electrons that were moving faster than light. The timing was exceptionally bad. The next year, 1905, saw the publication of "Special Relativity" in which Einstein showed quite clearly that neither electrons nor anything else could have a velocity greater than that of light. Sommerfeld's papers should have vanished into obscurity as the old physics was buried under the new. Then, decades later, came the discovery of Čerenkov radiation. The eerie blue glow seen in swimming pool type nuclear reactors indeed arises from particles moving faster than the velocity of light, possible because the velocity of light in water is less than the velocity of light in a vacuum. Sommerfeld's pre-relativity calculations could be applied to the new phenomenon.

While some physicists bickered over relativity for years, Sommerfeld was among its first converts. Never mind that the new theory had consigned his hard work on electron dynamics to the trash heap.

In 1913, came another major new concept with the publication by Niels Bohr of a quantum model for the hydrogen atom. Bohr had introduced two new postulates: that electrons moving in stationary orbits were stable against energy loss by radiation and that energy gain or loss by radiation was in the form of quanta. The great success of the Bohr postulates was that they led immediately to a derivation of the Balmer formula

$$\nu = R\left(\frac{1}{n^2} - \frac{1}{m^2}\right)$$

For $n = 2$, $m = 3, 4, 5\ldots$ gave the well-known Balmer series of sharp lines in the visible spectrum of hydrogen. $n = 1$ gave the ultraviolet Lyman series, and $n = 3, 4, 5$ gave the infrared Paschen, Brackett, and Pfund series, respectively. Further the Bohr theory produced a theoretical interpretation and quantitative estimate of the Rydberg constant

$$R_\infty = \frac{2\pi^2 me^4}{ch^3}$$

The agreement with experimental value was within the experimental error of the day.

Bohr's interpretation of the hydrogen atom spectrum was a conceptual triumph. However, he had arbitrarily used only circular orbits and had not set the rules for quantization. The model of particle-like electrons moving in well-defined orbits around the nucleus looks a lot like classical mechanics except for the quantum condition on the frequency of the emitted radiation. Further, the velocities of the electrons in their orbits would approach the velocity of light indicating that relativistic effects must also be taken into account.

Just as he had earlier recognized relativity as an important new concept, Sommerfeld had an immediate admiration for Bohr's 1913 trilogy. In characteristic style, he undertook the task of improving the theoretical substructure and of recasting the theory in relativistic form.[6–8]

Sommerfeld's quantization conditions are a generalization of Bohr's second postulate

$$\oint p_\phi dq_\phi = n_\phi h$$

$$\oint p_r dr = n_r h$$

where ϕ is the angular position of the electron in its orbit and r is the radial distance from the nucleus to the electron.

For circular orbits, there would be only the angular momentum equation. Admitting r as an additional variable brings in the possibility of elliptical orbits of which the circular orbits are only a special case. Elliptical orbits also bring in a second quantum number. This generalization has become known as the Bohr–Sommerfeld quantization rules. Curiously, the energies of the elliptical orbits remain the same as for circular orbits

$$E = -\frac{2\pi^2 m e^4 z^2}{n^2 h^2}$$

Thus the elliptical orbits, the more general quantization rules, and the second quantum number do not introduce any new spectral lines and thus do not disturb the good agreement between the theory and the Balmer formula.

The hydrogen atom, however, still had some surprises. Measurements of high resolution spectra of hydrogen dating back at least to Albert A. Michelson in 1891 showed that each of the lines was made up of a number of closely spaced components. This additional fine structure is not described by the Balmer formula.

When Sommerfeld generalized the quantization of the hydrogen energy levels by allowing elliptical orbits, he also solved the fine structure problem. Electrons moving in circular orbits move at constant velocity but electrons moving in elliptical orbits move with varying velocities much as the planets move with varying velocities as they follow elliptical orbits around the sun. Because the orbital velocities are near the velocity of light, there must be relativistic corrections to the energy.

Sommerfeld calculated the relativistic corrections for elliptical orbits in the winter of 1914–1915. His relativistic generalization of the Balmer formula is

$$E_{n,k} = -R'\left[\frac{Z^2}{n^2} + \frac{\alpha^2 Z^4}{n^3}\left(\frac{1}{k} - \frac{3}{4n}\right)\right] + O(\alpha^4)$$

$$n = 1, 2, 3, \ldots; \quad k = 1, 2, 3, \ldots, n$$

Here R' is the Rydberg constant $(\mathrm{sec}^{-1}) = \mathrm{cR}(\mathrm{cm}^{-1})$, Ze is the nuclear charge and, k is related to the orbital angular momentum quantum number by $k = l + 1$. The hydrogen emission lines have fine structure because relativistic effects lift the degeneracy in l.

Appearing quite naturally in the relativistic analysis of elliptical orbits is the coefficient, known ever since as the fine structure constant

$$\alpha = \frac{2\pi e^2}{hc}$$

The fine structure constant connects the most important constants of physics. The fundamental electron charge, e, is electrodynamics; Planck's constant, h, is quantum mechanics, and the velocity of light, c, is relativity. Further, the ratio is dimensionless; α is a pure number with a value of 1/137.0359895 according to the most recent measurements.[9]

There is more.[10] Although the relativistic equation gave the energies of the fine structure lines quite precisely, there was a great deal of uncertainty about the selection rules which determine which of the predicted energy levels actually appear as spectral lines. Bohr and independently Adalbert Rubinowicz, a member of Sommerfeld's group, settled on $\Delta l = \pm 1$. It was not until 1925 that George Uhlenbeck and Samuel Goudsmit in Leiden found a way out of the difficulty by introducing a new definition of k: $k = j + 1/2$ with a selection rule of $\Delta j = 0, \pm 1$. The j quantum number was shortly after interpreted by Uhlenbeck and Goudsmit as a total momentum quantum number equal to the sum of the orbital angular momentum and the angular momentum due to electron spin.

The idea that the electron had spin was completely new. The final and exact derivation of the hydrogen atom energy levels had to await the Dirac equation in 1928 but it is remarkable that taking account of relativistic effects, even within the framework of the old quantum theory had the existence of electron spin as one of its consequences.

This is by no means the end of the story with regard to the fine structure constant. It was to play a central role in the development of quantum electrodynamics.[11] A very tiny, about 3%, discrepancy in the position of the hydrogen fine structure lines, the Lamb shift (the measurement of which won Willis Lamb the Nobel Prize in Physics in 1955) was a central point of discussion in the first Shelter Island Conference on the foundations of quantum mechanics on 2–4 June, 1947. First with respect to classical physics, then with respect to the Bohr–Sommerfeld formulation of the old quantum mechanics, then with respect to Schrödinger's non-relativistic quantum mechanics, and finally with respect to the Dirac equation, better and better measurements of the hydrogen spectrum showed that the theory was still inadequate. The final resolution of all details required the invention of quantum electrodynamics and more Nobel Prizes awarded to Richard Feynman, Julian Schwinger, and Sin-Itiro Tomonaga.

The fine structure constant is still very much with us. It appears in the Grand Unified Theories of particle interactions and in spite of intense effort, its numerical value is no better understood today than it was in 1915. Among other side effects, the fine structure constant has spawned on unending stream of papers purporting to show either (i) that the constant is really the reciprocal of an integer, 1/137 and that the either h, e, or c must be wrong or (ii) to derive the exact numerical value by methods that for the most part can only be described as pure numerology. Arnold Sommerfeld has cast a very long shadow.

4. The Teacher

For 32 years at the University of Munich Sommerfeld lectured on theoretical physics. There were six of these courses each of which met four hours a week supplemented by a two-hour problem period. It is worth quoting from Martin O. Stern, who prepared the English translation of Mechanics:[12]

> "From the student's point of view the great attraction of Sommerfeld's lectures lay in their clarity: the approach from the physical side, the formulation of the mathematical problem, the simple and yet general explanation of the mathematical methods used, and the thorough discussion of the result again in terms of physical experiment. His firm, well-distributed writing on the blackboard, and the evidence of his diagrams, helped the student to survey at the end of each period all the subjects that had been covered. Besides, the standard of the course was high enough to tax the powers of the better students and to demand vigilant cooperation. This was all the more important in a university system where there was no check on attendance and only a voluntary one on performance. In giving an original discussion of a problem in the exercises even the beginner would attract the attention of Sommerfeld or of the assistant in charge and he would be stimulated by the appreciative understanding his effort received."

By far, Sommerfeld's greatest impact was through his graduate students and post-doctoral visitors. Until Max Born took the Chair of Theoretical Physics at Göttingen in the 1920s, Sommerfeld had the most students and produced the most Ph.Ds in theoretical physics of any institution in Europe. Many of the revolutionaries who changed the face of physics in the 1920s and 1930s received their graduate education from Sommerfeld. A partial list[13] of those who spent a year or more at Sommerfeld's institute includes Gregor Wentzel, Walter Heitler, William Houston, Ernest C. G. Stueckelberg, Peter P. Ewald, Philip M. Morse, Linus C. Pauling, Otto Laporte, Karl Herzfeld, Edward Teller, Leon Brillouin, Carl Eckart, Wolfgang Pauli, Edward U. Condon, Alfred Landé, Fritz London, Hans A. Bethe, Peter J. W. Debye, Werner Heisenberg, Max von Laue, and William L. Bragg. Ph.D. students who went on to win the Nobel Prize include Peter J. W. Debye (chemistry, 1936), Werner Heisenberg (physics, 1932), Wolfgang Pauli (physics, 1945), and Hans A. Bethe (physics, 1967). Sommerfeld himself never won the Nobel prize although some have argued that his contribution to the old quantum mechanics was of Nobel quality.

Sommerfeld's success in inspiring his admittedly brilliant students came in part from his being always at the cutting edge of physics and in part to a personal interaction uncommon in the formal and stuffy world of German academics. A quote from Sommerfeld himself[14]:

> "Personal instruction in the highest sense of the word is best based on intimate personal acquaintanceship. Ski trips with my students offered the best opportunity for that. Munich is situated so close to the mountains that one can reach the ideal ski terrain of the Sudelfeld by rail in two hours. There I had a ski hut... In the evening when we were gathered around the stove, it was inevitable that our conversation should turn from snow and weather to the subject of mathematical physics."

After formal retirement, Sommerfeld undertook the task of bringing his lecture notes into a form suitable for publication. The days when one person could know it all — could hope to teach the entire content of theoretical physics as a single integrated sequence of courses — was drawing to a close. The age of specialists was dawning. So here was Sommerfeld, in his early seventies, composing a set of books. The books were published in the same order as the courses:

1. *Mechanics*
2. *Mechanics of Deformable Bodies*
3. *Electrodynamics*
4. *Optics*
5. *Thermodynamics and Statistical Mechanics*
6. *Partial Differential Equations of Physics*

The first German edition of Mechanics appeared in 1943 published by Akademische Verlagsgesellschaft Geest und Portig in Leipzig. A second edition appeared in 1944. These copyrights were vested in the Attorney General of the United States after World War II. The books were translated into English and published by Academic Press with *Mechanics* appearing in 1952. They enjoyed use as popular textbooks in the United States in the 1950s and thus Sommerfeld's influence as a teacher extended beyond his death.

References

1. Paul Forman and Armin Hermann, in *Dictionary of Scientific Biography*, C. C. Gillispie, ed., Charles Scribner Sons, New York, vol. 12, 1975, pp. 525–532.

2. Fritz A. Bopp, Opening address, in *Physics of the One- and Two-Electron Atoms*, Fritz A. Bopp and Hans Kleinpoppen, eds., American Elsevier, New York, 1968, pp. 1–7.
3. Arnold Sommerfeld, *Atomic Structure and Spectral Lines*, 3rd edn., English translation by H. L. Brose and E. P. Dutton, New York, 1923.
4. Arnold Sommerfeld, *Wave Mechanics*, Supplement volume to the 4th edn. of *Atomic Structure and Spectral Lines*, English translation by H. L. Brose and E. P. Dutton, New York, 1929.
5. H. Welker, "Impact of Sommerfeld's work on solid state research and technology", in *Physics of the One- and Two-Electron Atoms*, Fritz A. Bopp and Hans Kleinpoppen, eds., American Elsevier, New York, 1968, pp. 32–43.
6. A. Hermann, "Sommerfeld's role in the development of early quantum theory", in *Physics of the One- and Two-Electron Atoms*, Fritz A. Bopp and Hans Kleinpoppen, eds., American Elsevier, New York, 1968, pp. 17–20.
7. B. L. van der Waerden, "The history of quantum theory in the light of the successive editions of Sommerfeld's Atombau und Spektrallinien", in *Physics of the One- and Two-Electron Atoms*, Fritz A. Bopp and Hans Kleinpoppen, eds., American Elsevier, New York, 1968, pp. 21–31.
8. Max Jammer, *The Conceptual Development of Quantum Mechanics*, McGraw-Hill, New York, 1966.
9. Richard E. Cohen and Barry N. Taylor, "The fundamental physical constants", *Phys. Today* **44** (1991) 9–13.
10. Abraham Pais, *Inward Bound*, Oxford University Press, New York, 1986.
11. Robert P. Crease and Charles C. Mann, *The Second Creation*, Macmillan, New York, 1986.
12. Arnold Sommerfeld, *Mechanics*, English translation by Martin O. Stern, Academic Press, New York, 1952.
13. Paul Kirkpatrick, *Am. J. Phys.* **17** (1949) 312–314.
14. Arnold Sommerfeld, "Some reminiscences of my teaching career", *Am. J. Phys.* **17** (1949) 315–316.

William B. White is Professor of Geochemistry with a joint appointment between the Department of Geosciences and the Intercollege Materials Research Laboratory at the Pennsylvania State University. He received a B.S. degree in chemistry from Juniata College in 1954. Between 1954 and 1958, he was a research assistant and research associate in the Department of Chemical Physics at the Mellon Institute in Pittsburgh, PA during which time he also pursued part-time graduate study in physics at the University of Pittsburgh. He came to Penn State as a graduate student in 1958, received his Ph.D. in geochemistry in 1962 and joined the faculty in 1963. He was made full professor in 1972. White's research has ranged over many subjects including crystal chemistry, aqueous chemistry, phase equilibria, and infrared, Raman, optical, and luminescence spectroscopy. He has an ongoing interest in the history and sociology of science.

Chapter 6. Electron Spin and Samuel Goudsmit

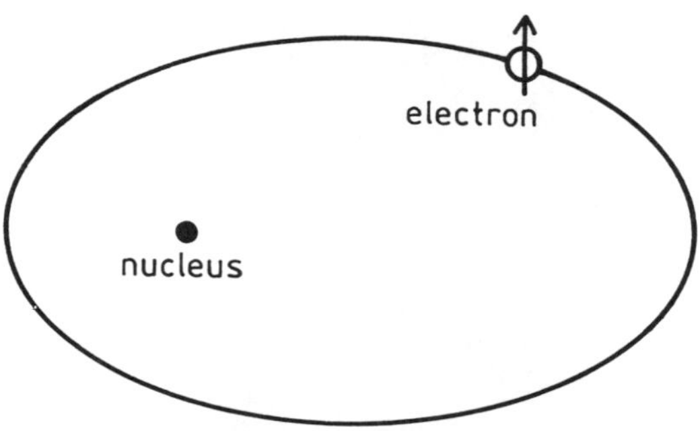

Tarun Biswas

S. A. Goudsmit

Chapter 6

Electron Spin and Samuel Goudsmit

Tarun Biswas
Physics Department,
State University of New York at New Paltz,
New Paltz, NY 12561, USA

Abstract

Samuel Goudsmit reminds us of an exciting period in the history of physics — a period of sweet confusion in the experimental data available and in the theories that attempted to explain them. Hence, physicists were busy — so busy that they did not notice an oddball like Goudsmit was working amongst them. Goudsmit had unusual interests and unusual methods of approaching problems. This is what helped him and Uhlenbeck to come up with the concept of electron spin. Although present day physicists might find electron spin a perfectly natural idea, in 1925 it was strange and unusual. An attempt is made here to follow the historical development of the concept of electron spin as seen by Goudsmit. A current mathematical model is also presented. Finally, Goudsmit's persistent concern about the current directions of young physicists is discussed.

1. The Man for the Job

There have been many attempts at painting the picture of the archetypical scientist. The motion picture industry has been particularly prolific in such attempts. The standard opinion is that a scientist is absent-minded, childlike, carelessly dressed and above all anti-social. In reality, scientists are found to have as varied characteristics as people involved in any other direction of human endeavor. Samuel Goudsmit (1902–1978) is an example of such variance. His background and approach to physics was so non-standard, that historians and psychologists would have a difficult time if ever they tried to "explain" his aptitude in physics. Goudsmit himself knew of this and seemed to relish the idea of confounding attempts at classifying his character. In his own words[1]: "They want to know why you've become a physicist, why you've done everything the way you have. They want to know all about your ancestors, and they hope that your grandfather was a great chemist or a great mathematician. They are always very disappointed when they get to me."

In 1925, Goudsmit and Uhlenbeck introduced the idea of electron spin which has since been seen to be a key piece in the puzzle of atomic spectra and other quantum phenomena. Uhlenbeck is known to have said of Goudsmit[2]: "Sem (Goudsmit) was never a conspicuously reflective man, but he had an amazing talent for taking random data and giving them direction." This unique talent that Goudsmit possessed was exactly what was needed to decipher the effects of electron spin in volumes of spectroscopic data. His mother knew of this talent early on. She owned a fashionable hat shop in The Hague and consulted her son on the esthetics of new hat designs! Hats and atomic spectra may have very little in common. But the esthetics of both arise from small details that affect the big picture. So one may conclude that the secret of Goudsmit's talent was in the understanding of small details while still keeping the big picture in mind.

Samuel Goudsmit was born in 1902 in The Hague. His father was a wholesale dealer in bathroom fixtures. His interest in atomic spectra was piqued as early as the age of 11, when he learnt that such spectra show distant stars to be made up of the same elements as the earth. However, his interests were not restricted to physics alone. For example, his fascination for deciphering codes and unravelling mysteries led him to learn about forensic methods and the decoding of hieroglyphics.

A wide range of interests made Goudsmit look more "human" to non-physicists. Mrs. Regina Brown, his administrative secretary once said,[3] "They (physicists) go around talking their special language, like monks in India, ... (But) Sam (Goudsmit) talks the language the rest of us talk." Other physicists have been known to say things like,[3] "That fellow Goudsmit! He talks physics, but he talks a lot of other things, too. Gives you the feeling that we've been infiltrated by a layman." Goudsmit saw physics as a part of a bigger and more complete picture where the interactions of different human interests bring a higher level of fulfillment. Once, while visiting Uhlenbeck at the Institute for Advanced Studies in Princeton, Goudsmit was amazed to find that T. S. Eliot was also there but his friend, Uhlenbeck, was not particularly interested in talking to the man. He says,[3] "...What a wretched situation it was, I thought, that two men of such enormous curiosity as George (Uhlenbeck) and Eliot could spend a couple of months in the same small community and never even get around to passing the time of day."

A breadth of interests also allowed Goudsmit to communicate better with non-physicists. It was probably for this ability that he was chosen by the US Defense Department in 1944 to serve as the uniformed civilian head of a highly secret mission to Europe (named Alsos) that was to collect evidence of the German atomic bomb. There seems to be some controversy on Goudsmit's role in this mission and in the post-war treatment of German scientists.[4,5] I will not discuss this matter any further here as most public accounts on the subject are emotionally charged (and understandably so), making it difficult to find an objective viewpoint. Goudsmit's own parents had been gassed at Auschwitz.

Prior to the historic papers of 1925 and 1926 that reported the discovery of spin,[6,7] Goudsmit had done a tremendous amount of ground work that led to this discovery. He had developed a good understanding of the vector model (*"das Rumpf-Modell"*) which, at that time, provided the best known explanation of atomic spectra. He taught Uhlenbeck all he knew about the model and then, together, they wrestled with the inconsistencies that still needed explanation.

2. The Paradox — Atomic Spectra

Atomic spectra have provided one of the earliest verifications of the quantum hypothesis. The discrete nature of a spectrum suggested that only certain frequencies of electromagnetic radiation are allowed for a given atom. Planck[8] had shown that a blackbody produces electromagnetic energies in "lumps" or "quanta" that are proportional to the frequencies i.e. $E = h\nu$ where E is the energy, ν the frequency and h a universal constant now known as Planck's constant. Einstein[9] suggested that these quanta are physically observable as particles — namely photons. Later, this was experimentally verified by Millikan.[10] Hence, it was clear that the discrete nature of atomic spectra is related to the discreteness of possible energies of the atom. However, the then-existing classical theory of electromagnetic radiation required that an orbiting electron of an atom must radiate continuously and hence, lose energy continuously. So classical mechanics had to be abandoned in some way. The mechanism suggested by Bohr and Sommerfeld was the quantization of angular momentum. It was assumed that electrons could have angular momenta that were only integer multiples of $\hbar$ ($\hbar = h/2\pi$). This produced a corresponding quantization of electron energies. For the hydrogen atom, these quantized energy levels were easily computed. For more complex atoms, the potential energy of an electron due to the other electrons is rather complex and, hence, finding the corresponding energy levels was difficult. However, even when actual energies could not be found, the energy level structures could still be predicted. Observed atomic spectra were expected to correspond to the differences in these energy levels as the photons were understood to be taking away the balance of the energy when an atom jumped from a higher to a lower energy level. For the hydrogen atom, the agreement of computed and observed spectra was dramatic. For alkali elements some experimentally fitted parameters were needed for acceptable predictions as the potential energy was not known. It was assumed that the outermost electron is bound loosely to the rest of the atom (including the other electrons) which was called the core. The core was assigned some phenomenological properties. For non-alkali elements, one could still assume a core, but there would be more than one outer electrons (i.e. the outer shell of electrons). Such theoretical models generally agreed with experiment and hence, the Bohr–Sommerfeld hypothesis was accepted.

The sticking point of the Bohr–Sommerfeld theory appeared when atomic spectral lines were seen to have a fine structure. The lines predicted by theory were, very often, experimentally seen to be a group of lines with very small separations. A plausible explanation of this fine structure came from the dipole–dipole interaction energy due to the magnetic dipoles created by orbiting electrons. But the single electron of the hydrogen atom and the effectively single electron of an alkali atom also showed a fine

structure! Besides, even for atoms with multielectron outer shells, such a dipole–dipole interaction could not produce the correct number of lines in the fine structure. However, the diligent works of Landé, Heisenberg, Pauli, Goudsmit and others showed that such an idea could still be made to work if the atomic core itself was assigned an angular momentum and hence, a magnetic dipole moment. This gave rise to the vector model. For example, for alkali atoms, the vector model assumed the atomic core to have an angular momentum of $\hbar/2$ ($\hbar = h/2\pi$) and the corresponding magnetic moment to be $e\hbar/2m$ (in SI units where e is the electron charge and m the electron mass).[a] However, these assumptions led to several difficulties. First, the original Bohr–Sommerfeld hypothesis allowed angular momenta only in integer multiples of $\hbar$. Second, there was no explanation for the assumed magnetic moment of the core (in particular, its dependence on the *electron* mass rather than the whole mass of the core). Third, some lines of hydrogen seemed to be exceptions to the rule as they still followed the original Bohr–Sommerfeld theory. Finally, the assigned angular momentum of the core seemed too arbitrary.

3. The Solution — Electron Spin

Although the idea of a spinning electron was not new,[1] the first hint with an experimental basis came from Pauli's famous paper on the exclusion principle.[11] He assigned four quantum numbers to an electron. In Goudsmit's simplified notation these numbers are n, l, m_l, and m_s. The first three were related to known physical quantities — energy, magnitude of angular momentum and one component of angular momentum, respectively. But m_s seemed to have no physical meaning. Nonetheless, Pauli showed that the periodic table could be explained by an exclusion principle of the four quantum numbers, i.e. no two electrons could have the same four quantum numbers.

Although Pauli's scheme of quantum numbers worked, Uhlenbeck was unhappy as the quantum numbers seemed quite unrelated to a physical picture. He said,[12] "...it seemed like abracadabra to

[a]This also correctly explained the splitting of spectral lines due to an external magnetic field. This is the so-called Zeeman effect.

me." But soon he realized that three of the quantum numbers (n, l and m_l) are related to the three translational degrees of freedom of the electron. Hence, the fourth quantum number m_s must be related to some fourth degree of freedom. This could be a rotational degree of freedom or spin! It was already known to Goudsmit and Uhlenbeck that m_s could take values of only $\pm 1/2$. So it occurred to them that the $\hbar/2$ angular momentum attributed to the atomic core really belonged to the electron. The corresponding magnetic moment should also belong to the electron.[6]

The assigning of spin to electrons explained a lot. But there were still many unanswered questions. First, for the electron spin to show its effect in the spectrum, it must produce some energy. This it could do only if placed in a magnetic field. Eventually, Einstein came up with the source of this magnetic field within the atom.[12] The relative motion (velocity $\mathbf{v}$) of the electron with respect to the electric field $\mathbf{E}$ of the core produces a magnetic field of $\mathbf{E} \times \mathbf{v}/c^2$ (c is the speed of light in vacuum). Second, the level scheme of the hydrogen atom spectrum was worked out using electron spin.[7] This explained some transitions that were not allowed by previous models. But there was still an unexplained factor of 2. This was later explained by Thomas[13] as a relativistic precessional effect.

4. Another Paradox — Is Spin Rotational Motion?

Now let us consider some of the remaining problems that still needed ironing out. Although it looked like rotational motion, electron spin had some peculiar properties that distinguished it from the classical concept of rotational motion.

First, electron spin had inherited the magnetic moment problem from the earlier idea of the core angular momentum in the vector model. This problem arises from the classical relationship of angular momentum and magnetic moment of a charged object. If the electron were a rotating charged object (rigid or otherwise), then it could be considered to be a collection of infinitesimal point objects each having a charge to mass ratio of e/m, the same as that of the whole electron. The magnetic moment of such an infinitesimal point object is[14]

$$d\boldsymbol{\mu} = \frac{1}{2}\mathbf{r} \times \mathbf{J}dv, \qquad (1)$$

where $\mathbf{r}$ is the position, $\mathbf{J}$ the current density and dv the volume of the infinitesimal object. If $\mathbf{J}$ is produced by charges of the same sign, $\mathbf{J} = \rho\mathbf{u}$ where ρ is the charge density and $\mathbf{u}$ the velocity. So $\mathbf{u} \equiv d\mathbf{r}/dt$ where t is time. Hence, Eq. (1) can be written as

$$d\boldsymbol{\mu} = \frac{1}{2}\mathbf{r} \times \mathbf{u}\rho dv. \tag{2}$$

As e/m is the charge to mass ratio, the charge density may be written as $\rho = \rho_m e/m$ where ρ_m is the mass density. Then Eq. (2) gives

$$d\boldsymbol{\mu} = \frac{e}{2m}(\mathbf{r} \times \mathbf{u}\rho_m dv). \tag{3}$$

In the above equation, the quantity in parentheses can be seen to be the infinitesimal angular momentum $d\mathbf{L}$. Then integration over volume will give

$$\boldsymbol{\mu} = \frac{e}{2m}\mathbf{L}. \tag{4}$$

This classical relationship between the magnetic moment and angular momentum does not work in the case of electron spin (as seen in its effect on atomic spectra), and one needs to replace it by

$$\boldsymbol{\mu} = g\frac{e}{2m}\mathbf{L}, \tag{5}$$

where g, the so-called gyromagnetic ratio, is equal to 2. One may try to explain a factor of g greater than 1 by choosing the mass and charge distributions to be different. For example, if the mass were a uniform volume distribution and the charge a uniform surface distribution for a spherical electron, then it can be shown that[15] $g = 5/3$. A value of $g = 2$ can be obtained by a uniform surface charge distribution and a volume mass distribution of $\rho_m = A/r$ for a spherical electron where A is a constant and r the distance from the center. But this is too contrived an explanation. Besides, the inclusion of relativistic effects would further complicate this explanation. For example, for an angular momentum of $\hbar/2$, there would be a minimum possible size of electrons required to keep the surface speed less than that of light.

Second, when Uhlenbeck decided that the quantum number m_s was due to a rotational degree of freedom, he ignored the fact that there are 2 more rotational degrees of freedom that have no corresponding quantum numbers. If such quantum numbers were to exist, they might be called n_s for the total spin energy and s for the total spin angular momentum. n_s would give an infinite spectrum of possible internal energies of the electron. This is not observed in experiment. s would give an infinite spectrum of possible total spin angular momenta and a corresponding infinite spectrum of angular momentum components that would be related to m_s. The only experimentally observed value of s is $1/2$ and those of m_s are $\pm 1/2$. Hence, we see another problem with spin being rotational motion.

Finally, the electron angular momentum of $\hbar/2$ still violated Bohr's hypothesis that all angular momenta must be in integer multiples of $\hbar$. One may argue that Bohr's hypothesis itself was quite *ad hoc* and hence, one could just as well hypothesize angular momenta to be in half-integer multiples of $\hbar$. But one would still have to explain why there should be any quantization of angular momenta at all.

5. What's in a Name? — Spin Is Not Rotational Motion!

The solutions to these problems of electron spin were to be found in group theory. They are seen primarily in the works of Pauli,[16] Dirac,[17,18] and Wigner.[19] The argument is roughly as follows.

From the explanations of atomic spectra, we know spin produces angular momentum. Classically, this would at once mean the existence of rotational motion as angular momentum is $\mathbf{r} \times \mathbf{p}$ where $\mathbf{p}$ is linear momentum. However, there is a more general definition of angular momentum. For this we need to consider symmetry operations of a rotationally symmetric system. A rotation operation is achieved for any system by rotating the coordinate system by some angle θ about some axis given by a unit vector $\hat{\mathbf{n}}$. A general form of the operator that will produce the effect of such a rotation on any mathematical object (maybe a vector, a function of vectors or something else) is

$$R(\theta, \hat{\mathbf{n}}) = \exp(i\theta\hat{\mathbf{n}} \cdot \mathbf{L}/\hbar) \tag{6}$$

where $i = +\sqrt{-1}$ and each component of the vector $\mathbf{L}$ is an operator suitable for operation on the mathematical object in question. This vector $\mathbf{L}$ is the generalized definition of angular momentum. For

spatial degrees of freedom it has the classical form $\mathbf{r} \times \mathbf{p}$. But the spin degree of freedom is seen not to be spatial. Hence, for spin, $\mathbf{L}$ takes a different form and has no relation to rotational motion. So, the name spin is somewhat unfortunate as it seems to imply rotational motion. Spin is a degree of freedom that produces angular momentum without rotational motion! It is represented by the so-called Pauli spinors which are two-component objects that can take only two physically observable values. The corresponding representations for the components of $\mathbf{L}$ are 2×2 matrices (Pauli spin matrices) that operate on the Pauli spinors. Dirac generalized the Pauli spinors to combine both particle and antiparticle states. This resulted in the 4-component Dirac spinors.

Using the 4-component spinor representation, Dirac came up with an equation for relativistic quantum mechanics. This equation produced $g = 2$ as a natural consequence. Both the Pauli and the Dirac approach showed electron spin to be a single discrete degree of freedom that could take only two possible values. Now it could also be proved that, unlike Bohr's hypothesis, angular momenta must be half-integer multiples of $\hbar$. I shall discuss these ideas of modern quantum mechanics in more detail in the next section. Thus finally, all aspects of electron spin were explained.

6. A Modern Mathematical Model

Let us first briefly go through some of the principles of quantum mechanics. Here, we shall use the generalized notation introduced by Dirac.[18]

In classical physics a complete description of a system is its trajectory in coordinate space as a function of time. In the microscopic scale of quantum mechanics, such a trajectory is experimentally meaningless. The experimentally meaningful and complete quantum description of a system is its *state*. The state of a system is represented, equivalently, by both its *ket* and its *bra* vectors. The ket is denoted as $|s\rangle$ and the bra as $\langle s|$ where s is a set of labels (could be position, momentum, energy etc.) that identify the *state*. For the present purposes, a ket may be considered to be a column vector with, in general, an infinite number of complex components. The bra corresponding to a ket is its complex conjugate transpose (*hermitian adjoint*). So a bra is a row vector with complex components. Every observable quantity (position, momentum, etc.) has a linear operator attributed to it. In the present picture, such an operator is an infinite-dimensional square matrix and its operation on a ket is like a matrix multiplication from the left and that on a bra is like matrix multiplication from the right. So, for an operator Q, $Q|s\rangle$ is a ket and $\langle s|Q$ is a bra. The *dot product* or the *inner product* of a bra $\langle s|$ and a ket $|r\rangle$ has the usual matrix product meaning. It is denoted as

$$\langle s||r\rangle \equiv \langle s|r\rangle, \tag{7}$$

which is a complex number. This is the *bracket* notation from which the terms *bra* and *ket* have been derived.

Every Q, corresponding to an observable, has a set of *eigenstates* (represented by $|q\rangle$ or $\langle q|$) and a set of corresponding *eigenvalues* q such that

$$Q|q\rangle = q|q\rangle, \quad \langle q|Q = q\langle q|. \tag{8}$$

The set of possible eigenvalues q is the set of possible results of experiments measuring the observable corresponding to Q. Hence, all q must be real. If a system is in some state $|s\rangle$, the probability of finding a value q as a result of measuring Q is proportional to $|\langle q|s\rangle|^2$.

In particular, for the position observable of a single particle, the three components are represented by three operators given by the vector operator $\mathbf{R}$. The corresponding eigenvalues and eigenstates are $\mathbf{r}$ and $|\mathbf{r}\rangle$. Here $\mathbf{r}$ is expected to take values in all space as every position in space is measurable. And, the probability of finding the particle at a position $\mathbf{r}$ is $|\langle \mathbf{r}|s\rangle|^2$ if the system is known to be in some state $|s\rangle$. It was originally assumed that the function of position $\psi_s(\mathbf{r}) \equiv \langle \mathbf{r}|s\rangle$ (called the wavefunction) is a single-valued function.[22] This was to go along with the classical idea of the trajectory. But the concept of spin shows that an electron can have two different states for the same position.[b] Hence, one needs another label for the position eigenstate $|\mathbf{r}\rangle$ to distinguish the two spin

[b]This is the spin-half case. Other particles may have different spin values (in half-integer multiples), and in general, a particle of spin p has $2p+1$ different states for the same position.

states. A convenient notation is $|\mathbf{r}, m_s\rangle$ where $m_s = \pm 1/2$. So, the wavefunction can be pictured as a two-component function:

$$\psi_s(\mathbf{r}) \equiv \langle \mathbf{r}, m_s|s\rangle \equiv \begin{pmatrix} \langle \mathbf{r}, +1/2|s\rangle \\ \langle \mathbf{r}, -1/2|s\rangle \end{pmatrix} \equiv \begin{pmatrix} \psi_{s+}(\mathbf{r}) \\ \psi_{s-}(\mathbf{r}) \end{pmatrix}. \tag{9}$$

This is the Pauli spinor.

At the present time it is clear to physicists that quantum mechanics describes systems in microscopic detail and classical mechanics is its approximation. However, our understanding of classical mechanics is significantly better than quantum mechanics. So, very often, it is fruitful to know some standard method of transition from classical to quantum. Clearly, this is a problem because it is a transition from the approximate to the exact! But fortunately, there exists such a method. In classical mechanics, every observable can be written as a function of position coordinates q_n and the corresponding momenta p_n (n is a label that identifies different coordinates). Then, a structure called the *Poisson Bracket* (PB) is defined to depend on a pair of observables J and H as follows:

$$\{J, H\} \equiv \sum_n \left(\frac{\partial J}{\partial q_n}\frac{\partial H}{\partial p_n} - \frac{\partial J}{\partial p_n}\frac{\partial H}{\partial q_n} \right). \tag{10}$$

In quantum mechanics, observables being represented by operators, the sequence of observations is important. If H were observed first and then J, it corresponds to the (matrix) product JH. So, reversing the order of observation makes a difference as, in general, $JH \neq HJ$. Hence, a non-trivial meaning is given to the following *commutator bracket* (CB) of two observables:

$$[J, H] \equiv [JH - HJ]. \tag{11}$$

Now the classical to quantum transition can be stated as

$$\{J, H\} \longrightarrow \frac{[J, H]}{i\hbar}. \tag{12}$$

For the angular momentum components L_x, L_y, and L_z the classical PB's are:

$$\{L_x, L_y\} = L_z, \quad \{L_y, L_z\} = L_x, \quad \{L_z, L_x\} = L_y. \tag{13}$$

Reversing the order of the observables in the PB introduces a negative sign and hence, the PB of an observable with itself is always zero. The set of Eq. (13) is quite straightforward to derive from the definition in Eq. (10) if it is noted that the position coordinates q_n are the cartesian components x, y, and z of the position vector $\mathbf{r}$ and the corresponding components of the linear momentum $\mathbf{p}$ are p_x, p_y, and p_z. The angular momentum components are given by the components of the cross-product $\mathbf{r}\times\mathbf{p}$. Now, Eqs. (12) and (13) would give the corresponding quantum CB's to be

$$[L_x, L_y] = i\hbar L_z, \quad [L_y, L_z] = i\hbar L_x, \quad [L_z, L_x] = i\hbar L_y. \tag{14}$$

However, it should be noted that in quantum mechanics the angular momentum is not given by the orbital part ($\mathbf{r} \times \mathbf{p}$) alone. From group theory we know that the set of Eqs. (14) makes sure that the rotation operator $R(\theta, \hat{\mathbf{n}})$ (see Eq. (6)) produces physical rotations. Hence, to produce the effect of rotation on 2-component Pauli spinors, we need 2×2 matrix forms of the angular momentum components that will still satisfy the set of Eqs. (14). Such a set of matrices is given as the components of a vector $\mathbf{S}$:

$$S_x = (\hbar/2)\sigma_x, \quad S_y = (\hbar/2)\sigma_y, \quad S_z = (\hbar/2)\sigma_z, \tag{15}$$

where the following definition of the so-called Pauli spin matrices is used:

$$\sigma_x = \begin{pmatrix} 0 & 1 \\ 1 & 0 \end{pmatrix} \quad \sigma_y = \begin{pmatrix} 0 & -i \\ i & 0 \end{pmatrix} \quad \sigma_z = \begin{pmatrix} 1 & 0 \\ 0 & -1 \end{pmatrix}. \tag{16}$$

Now, the eigenvalues of the matrix operators in the set of Eqs. (15) are $\pm\hbar/2$. Hence, the measured values of electron spin angular momenta must also be $\pm\hbar/2$.

More generally, the set of Eqs. (14) can be consistent only if the eigenvalues of the operators L_x, L_y, and L_z are integer multiples of $\hbar/2$.[20] So, observed values of angular momentum components must also be integer multiples of $\hbar/2$. However, orbital angular momentum ($\mathbf{r} \times \mathbf{p}$) components can be only integer multiples of $\hbar$ (as was originally assumed by Bohr).

7. Some Pointers for Young Physicists

Having discussed the concept of spin in some detail, I feel obliged to write about another equally important lesson that Goudsmit tried to teach us. For quite some time, he showed concern about physicists losing direction. Ironically, he blamed the easy availability of research money for physics after the Second World War. Many present day physicists will surely disagree. As early as 1953 he had said,[3] "Money and engineering have diverted us to the point where I don't know if we're gadgeteers or scientists. There are easily three times as many physicists now as there were before the war, but only ten per cent of them are in the fields where important discoveries are made." One may wonder about his opinion on present day mammoth proposals like the Superconducting Super-Collider. In 1953, he had complained,[3] "Lord, the expensive equipment they (contemporary physicists) expect! ... Right now, it seems, everybody there (Brookhaven National Lab.) wants a new type of oscilloscope that sells for thirty five hundred dollars."

To learn from Goudsmit's advice, today's young physicists must understand the world of physics in the 1920s — the "springtime of modern atomic physics".[21] Physicists then were a small group of people motivated, primarily, by their need to discover the secrets of the physical world. Communication of ideas within the group was quick and straightforward. This is seen in the rapidity with which the ground-breaking ideas on spin and the hydrogen atom appeared in the literature — Pauli exclusion principle in 1925,[11] electron spin in 1925,[6] spin and the hydrogen spectrum in 1926,[7] Thomas precession in 1926,[13] the Schrödinger equation in 1926,[22] Pauli spinors in 1927,[16] and the Dirac equation in 1928.[17] This also shows that, in those days, there was a lot to be done in physics. The benefits of such work were in the satisfaction of seeing how nature works. Money and fame for doing physics was virtually non-existent. It was after the Second World War that physicists acquired the respect of the general public. Then governments (particularly the United States) started handing out large sums of money for physics research and youngsters in droves wanted to do physics. Hence, a new generation of physicists was created. Most of these new physicists spent less and less time on physics and more and more time on gathering research funding. They considered it the duty of the public to provide funding for all physics research at any cost and without question of purpose.[23] This, of course, has dramatically changed the public opinion of physicists. In 1976, Goudsmit wrote,[21] "The world has lost confidence in scientific and rational reasoning; physics is now hard to sell." Let us hope this change in attitude of the public (and the consequent withdrawal of public research funding) will change physicists so that, once again, they will look for very little other than the satisfaction of understanding nature.

By way of advice for young physicists, Goudsmit repeatedly reminded us to focus on the physics and not the fame or the money. In 1953, in reference to selection of graduate schools by students, he said,[3] "Back in the twenties, we looked for the man (teacher) rather than the machine (expensive equipment)." On considering the importance of one's own work he wrote that,[1] "... the discovery of the electron spin was an important contribution. I know it now, though we didn't then. ... Even if you make only a small contribution, however unimportant, it will give you enormous pleasure. I don't believe you always have to tackle the big stuff. It's enough to have the continuing pleasure of working in physics and getting whatever results you get."

Many of us (including I) have often mourned the human failings of present day physicists. But Goudsmit pointed out that such failings are not unique to our times; although their effects may be exaggerated by present day higher stakes — government funding, the Nobel Prize etc. On the subject of plagiarism he wrote,[21] "This had happened to me a couple of times with work on spectral lines, but there was so very much left to do that it was merely a disappointment, not a catastrophe."

There is also the unfortunate controversy of who discovered spin first. It was said by Thomas in a letter to Goudsmit,[1] and later by Uhlenbeck,[12] that Ralph Krönig had formulated the idea of electron spin before Goudsmit and Uhlenbeck. Krönig had shown his work to Pauli who discouraged him from publishing it. But, what is unique about those times is that authorities like Pauli could own up to their mistakes without the fear of losing credibility in other matters. In a conversation with Uhlenbeck in the 1950s, Pauli took the blame for his mistake and said,[12] *"Ich war so dumm wenn ich jung war!"* ("I was so stupid when I was young!") Present day physicists of any calibre would not dare own up to the smallest of mistakes for fear of losing credibility among peers.

So the final word to aspiring physicists could be summarized as — Focus on the physics and do not be distracted by fame, money or the paranoia of being beaten by competitors.

References

1. Samuel A. Goudsmit, "Guess work: The discovery of the electron spin", *Delta* **15** (1972) 77–90.
2. Abraham Pais, "George Uhlenbeck and the discovery of spin", *Phys. Today*, Dec. 1989, pp. 34–40.
3. Daniel Lang, "A farewell to string and sealing wax", *The New Yorker*, Nov. 1953, pp. 47–72; Nov. 1953, pp. 46–65.
4. Mark Walker, "Heisenberg, Goudsmit and the German atomic bomb", *Phys. Today*, Jan. 1990, pp. 52–60.
5. Jonathan L. Logan, Helmut Rechenberg, Max Dresden, and A. van der Ziel, "Heisenberg, Goudsmit and the German 'A-bomb' ", *Phys. Today*, May 1991, p. 13.
6. George E. Uhlenbeck and Samuel Goudsmit, "Zuschriften und vorläufige Mitteilungen", *Naturwissenschaften* **13** (1925) 953–954.
7. Samuel Goudsmit and George E. Uhlenbeck, "Spinning electrons and the structure of spectra", *Nature* **117** (1926) 264–265.
8. Max Planck,"Ueber das Gesetz der Energieverteilung im Normalspectrum", *Ann. Phys.* **4** (1901) 553–563.
9. Albert Einstein, "Über einen die Erzeugung und Verwandlung des Lichtes betreffenden heuristischen Gesichtspunkt", *Ann. Phys.* **17** (1905) 132–148.
10. Robert A. Millikan, "A direct photoelectric determination of Planck's 'h' ", *Phys. Rev.* **7** (1916) 355–388.
11. Wolfgang Pauli, "Über den Zusammenhang des Abschlusses der Elektronengruppen im Atom mit der Komplexstruktur der Spektren", *Z. Phys.* **31** (1925) 765–783.
12. George E. Uhlenbeck, "Personal reminiscences", *Phys. Today*, June 1976, pp. 43–48.
13. L. H. Thomas, "The motion of the spinning electron", *Nature* **117** (1926) 514.
14. John R. Reitz, Frederick J. Milford, and Robert W. Christy, *Foundations of Electromagnetic Theory*, Addison-Wesley, 1979.
15. Max Abraham, "Prinzipien der Dynamik des Elektrons", *Ann. Phys.* **10** (1903) 105–179.
16. Wolfgang Pauli, "Zur Quantenmechanik des magnetischen Elektrons", *Z. Phys.* **43** (1927) 601–623.
17. Paul A. M. Dirac, "The quantum theory of the electron", *Proc. R. Soc. London* **A117** (1928) 610–624; [See second part of the book.]
18. Paul A. M. Dirac, *The Principles of Quantum Mechanics*, Oxford, New York, 1958.
19. Eugene P. Wigner, *Group Theory*, Academic Press, New York, 1959.

20. Leonard I. Schiff, *Quantum Mechanics*, McGraw-Hill, New York, 1968.
21. Samuel A. Goudsmit, "It might as well be spin", *Phys. Today*, June 1976, pp. 40–43.
22. Erwin Schrödinger, "Quantisierung als Eigenwertproblem", *Ann. Phys.* **79** (1926) 361–376; [See second part of the book.]
23. Tarun Biswas, "Finding novel niches in a poor job market", *Phys. Today*, Jan. 1994, p. 15.

Tarun Biswas was born in India in 1956. He graduated from high school in New Delhi and did his bachelor's and master's degrees in physics at the Indian Institute of Technology at Delhi. In 1978, he came to the United States to do his Ph.D. in physics at Syracuse University. His dissertation was on relativistic many particle theories (classical and quantum) using constraint dynamics and direct interactions. In 1984, he received his Ph.D. in physics and also an M.S. in computer science. The same year, he accepted an assistant professorship in physics at the State University of New York, The College at New Paltz, where he became an associate professor in 1989. Some of his interests are in quantum field theories of composite objects (hadrons), special relativistic Newtonian gravity that can explain experimental results that are generally accepted as tests of general relativity, computer simulation of physical systems (e.g. a stretched string with large amplitude oscillations that bring out nonlinear effects) and the use of computers in physics education.

Chapter 7. Heisenberg and Hydrogen

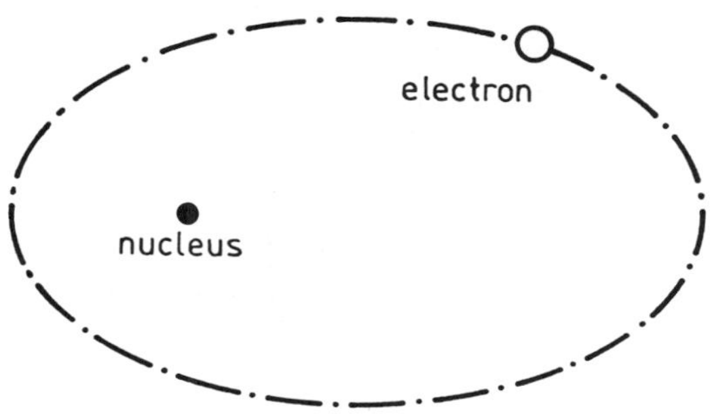

James D. Edmonds, Jr.

W. K. Heisenberg

Chapter 7

Heisenberg and Hydrogen

James D. Edmonds, Jr.
Department of Physics,
McNeese State University,
Lake Charles, LA 70609, USA

Abstract

The hydrogen atom is approached from classical physics, starting with basics and evolving into Poisson brackets. Heisenberg entered the quantum world from there. This approach is perhaps better for scattering theory than bound state energy levels, so Heisenberg did not contribute much of direct value to the understanding of quantum hydrogen structure. Schrödinger showed that his and Heisenberg's formalisms are equivalent, so in principle hydrogen could be analyzed using Heisenberg's approach, but not easily.

1. Introduction

To really understand Heisenberg's connections to the quantum hydrogen atom, we need to start with the basic classical description and then modify it into more complicated but equivalent notation. Finally, we reach the classical description in the Poisson bracket notation. This was then guessed by Heisenberg to be only an approximation of a more accurate, quantum theory where $\hbar$ is not zero. The classical approximation follows here "easily" from the limit $\hbar \to 0$. We shall move as directly as possible to Poisson brackets from $\mathbf{F} = m\mathbf{a} = e\mathbf{E} + e\mathbf{v} \times \mathbf{B}$, which is familiar to students at the introductory level.

We shall follow Corben and Stehle,[1] if you wish to see a more general development. We must introduce the vector potential

$\{\mathbf{A}, \phi\}$ and the conjugate momentum $\mathbf{p}$. Then we rewrite $\mathbf{F} = m\mathbf{a}$ into a form which shows the Lagrangian and the Hamiltonian. This Hamiltonian is central to the Poisson bracket concept, when applied to writing an equation totally equivalent to $\mathbf{F} = m\mathbf{a}$. In our development, the classical variables $\mathbf{r} = r\hat{\mathbf{r}} + 0\hat{\theta} + 0\hat{\phi}$, in spherical coordinates, and conjugate momentum, $\mathbf{p} = p_r\hat{\mathbf{r}} + p_\theta\hat{\theta} + p_\phi\hat{\phi}$, play equal roles in everything. But the quantum world has the particle m, with charge e, spread out all over the hydrogen atom at the same time. Thus $\mathbf{r}$ loses most of its meaning there, and $\mathbf{p}$ takes over in importance. Quantum is a strange mixture of classical thinking about a particle taking a classical path and also imagining several paths (infinitely many, really) all occurring *simultaneously*. That is why it is so difficult. It makes no intuitive sense to us huge creatures, who only experience huge objects behaving simply and moving along only one path or another. Now we begin the long evolution from $\mathbf{F} = m\mathbf{a}$ into a form that is unrecognizable but equivalent. You would not believe it unless you saw all the details, so we skip almost nothing here.

2. Energy and Momentum

We start with energy $dH \equiv \mathbf{F} \cdot \mathbf{dr}$, where $\mathbf{F} = e\mathbf{v} \times \mathbf{B} + e\mathbf{E}$. We play with this as follows

$$\mathbf{v} \times \mathbf{B} \cdot d\mathbf{r} = \left(\frac{d\mathbf{r}}{dt} \times \mathbf{B}\right) \cdot \frac{d\mathbf{r}}{dt}\, dt = 0dt = 0 \tag{1}$$

This is because $(\mathbf{A} \times \mathbf{B})$ is perpendicular to the plane of $\mathbf{A}$ and $\mathbf{B}$, so $(\mathbf{A} \times \mathbf{B}) \cdot \mathbf{A} = 0$. The magnetic field does no work since it pushes only *sideways* on any particle. However, it does contribute in a complicated way to the momentum change of the system as we shall see. The other force term gives

$$dH = e\mathbf{E} \cdot \mathbf{dr} \equiv e(-\nabla\phi) \cdot \mathbf{dr} = e(-d\phi) \tag{2}$$

where voltage, ϕ, has been *invented* because it is a useful concept, easier to deal with than its vector equivalent $\mathbf{E}(\mathbf{r}, t)$. Now replace $\mathbf{F}$ by $m\mathbf{a}$ and obtain:

$$\begin{aligned} dH &= m\mathbf{a} \cdot \mathbf{dr} = m\,\frac{d\mathbf{v}}{dt} \cdot \mathbf{dr} = m\mathbf{dv} \cdot \frac{\mathbf{dr}}{dt} \\ &= m\mathbf{dv} \cdot \mathbf{v} = \frac{1}{2}\,md(\mathbf{v} \cdot \mathbf{v}) = d\left(\frac{1}{2}\,mv^2\right) \end{aligned} \tag{3}$$

If we invent $H \equiv \frac{1}{2}mv^2 + e\phi$, then $dH = d(\frac{1}{2}mv^2) + d(e\phi) = -d(e\phi)+d(e\phi) = 0$. Thus $H = H(v,\, \phi) = H(v_0,\, \phi_0)$ at $t = 0$, when the particle had initial velocity $\mathbf{v}_0$ and position $\mathbf{r}_0$, and where the voltage was $\phi_0(\mathbf{r}_0,\, t = 0)$. H allows us to determine the latter speed v at an *unknown* time $t_1 > 0$, if the object manages to eventually reach a position $\mathbf{r}_1$. We cannot know when it gets there, nor its direction, only its speed once there. H seems rather limited but it is the Hamiltonian and has deeper insights than just predicting future speeds or whether a particle can get over a force barrier or not.

The hydrogen atom has an electron $(m,\, e)$, attracted to an "infinitely" heavy (simplification) nucleus with charge Ze. The voltage here is Coulomb potential and we can write:

$$H = \frac{m\mathbf{v} \cdot m\mathbf{v}}{2m} + -|e|\,\frac{|e|Z}{r} = H(|\mathbf{v}|,\, |\mathbf{r}|) \tag{4}$$

We take e to always be positive so the electron in orbit has charge $-|e| = -e$ here. The particle (kinetic) momentum is *defined* as $\mathbf{P} = m\mathbf{v}$.

You have been taught that Newton's second law says that action and reaction are equal and opposite so $\mathbf{F}_A = m_A\mathbf{a}_A = -\mathbf{F}_B = -m_B\mathbf{a}_B$, thus $d\mathbf{P}_A = -d\mathbf{P}_B$. Therefore, $\mathbf{P}_A + \mathbf{P}_B = \mathbf{P}_{\text{total}}$ has $d\mathbf{P}_{\text{total}} = 0$, but this is not true! One exception is $e\mathbf{v}\times\mathbf{B}$, and there may be others. Consider e_A located above e_B, with A moving right and B moving up, toward A. Use the Biot–Savart law to find the $\mathbf{B}$ field at A due to B and the $\mathbf{B}$ field at B due to A. Next use $\mathbf{v} \times \mathbf{B}$ to find $\mathbf{F}_A$ and $\mathbf{F}_B$. See that $\mathbf{F}_A$ is zero but $\mathbf{F}_B$ is not zero and to the left. Thus $\mathbf{F}_A \neq -\mathbf{F}_B$. Momentum conservation *fails* unless we include more momentum sources! To find these sources, we need to rewrite the magnetic field contribution. We *invent*:

$$\mathbf{B} \equiv \nabla \times \mathbf{A}|_t \qquad \mathbf{E} \equiv -\nabla\phi\bigg|_t - \frac{1}{c}\frac{\partial\mathbf{A}}{\partial t}\bigg|_{\mathbf{r}} \tag{5}$$

Here $\mathbf{A} = \mathbf{A}(\mathbf{r},\, t)$ and $\phi = \phi(\mathbf{r},\, t)$. These are well defined over a wide range of values of $\mathbf{r}$ at any time, t, over a long span of time. The partial derivatives on $\mathbf{A}$ and ϕ simply mean that we take ordinary derivatives with respect to one variable, x, y, z, t, while considering all other variables as ordinary *constants*. This $\mathbf{A}$ concept seems awful and messy at first, but of course it is done with lots of hindsight. It makes things simpler later on. We start to see this by using the general vector identity

$$\mathbf{v} \times (\nabla \times \mathbf{A}) = \nabla(\mathbf{v} \cdot \mathbf{A}) - (\mathbf{v} \cdot \nabla)\mathbf{A} \tag{6}$$

where $\mathbf{v} \equiv d\mathbf{r}/dt$ and $\mathbf{A} = A_x\hat{\mathbf{i}} + A_y\hat{\mathbf{j}} + A_z\hat{\mathbf{k}}$. It is much easier to verify such things than to find them. Expand both sides and compare terms, using

$$\nabla = \hat{\mathbf{i}}\,\frac{\partial}{\partial x} + \hat{\mathbf{j}}\,\frac{\partial}{\partial y} + \hat{\mathbf{k}}\,\frac{\partial}{\partial z} \tag{7}$$

and using the dot and cross product rules. Our $\mathbf{F} = d(m\mathbf{v})/dt$ equation then becomes

$$\frac{d}{dt}\,(m\mathbf{v}) = e\left[-\nabla\phi - \frac{1}{c}\left(\frac{\partial \mathbf{A}}{\partial t} + (\mathbf{v} \cdot \nabla)\mathbf{A}\right) + \frac{1}{c}\,\nabla(\mathbf{v} \cdot \mathbf{A})\right] \tag{8}$$

(Talk about making a mountain out of a molehill!) We next introduce the total differential of *any* vector field which exists at points along the path of a particle:

$$\begin{aligned}
\frac{d\mathbf{A}}{dt} &\equiv \frac{\partial \mathbf{A}}{\partial x}\,\frac{dx}{dt} + \frac{\partial A}{\partial y}\,\frac{dy}{dt} + \frac{\partial \mathbf{A}}{\partial z}\,\frac{dz}{dt} + \frac{\partial \mathbf{A}}{\partial t}\,\frac{dt}{dt} \\
&= v_x\,\frac{\partial}{\partial x}\,\mathbf{A} + v_y\,\frac{\partial}{\partial y}\,\mathbf{A} + v_z\,\frac{\partial}{\partial z}\,\mathbf{A} + 1\,\frac{\partial}{\partial t}\,\mathbf{A} \\
&= (\mathbf{v} \cdot \nabla)\mathbf{A} + \frac{\partial}{\partial t}\,\mathbf{A}
\end{aligned} \tag{9}$$

The field $\mathbf{A} = \mathbf{A}(\mathbf{r},\, t)$ is well defined here "everywhere" and at "all" times. The $\mathbf{v}(\mathbf{r},\, t)$ is defined *only* along a line in space at a series of times as m moves along. So, $d\mathbf{A}/dt$ is a "total" change in $\mathbf{A}$ as it *relates* to the movement of m, and it is a strange thing to really relate to intuitively, at least for me. With it, however, $\mathbf{F} = m\mathbf{a}$ becomes

$$\frac{d}{dt}\,(m\mathbf{v}) = e\left[-\nabla\left(\phi - \frac{1}{c}\,\mathbf{v} \cdot \mathbf{A}\right) - \frac{1}{c}\,\frac{d\mathbf{A}}{dt}\right] \tag{10}$$

Now we combine the two "total" derivatives on one side, and obtain $\mathbf{F} = m\mathbf{a}$ in the form:

$$\frac{d}{dt}\left(m\mathbf{v} + \frac{e}{c}\,\mathbf{A}\right) = -e\nabla\left(\phi - \frac{1}{c}\,\mathbf{v} \cdot \mathbf{A}\right) \tag{11}$$

This suggests that $\mathbf{P} = m\mathbf{v}$ may not be useful for conservation of momentum, but $\mathbf{p} \equiv \mathbf{P} + e\mathbf{A}/c$, the so-called conjugate "momentum", could take its place. It turns out that this $\mathbf{p}$ is the *only* $\mathbf{p}$ which survives into the quantum generalization, so it is far more important than ordinary momentum $\mathbf{P}$. Before we move on, it is tempting to at least mention that it is useful. In another context, to invent the Lagrangian, $L = L(\mathbf{v}, \mathbf{A}, \phi)$, a scalar function, through Eq. (11) in the form

$$\nabla_v L|_{r,t} \equiv m\mathbf{v} + \frac{e}{c}\mathbf{A} \equiv \mathbf{p} \qquad \nabla_r L|_{v,t} \equiv \nabla_r\left(-e\phi + \frac{e}{c}\mathbf{v}\cdot\mathbf{A}\right) \quad (12)$$

This is a strange thing, $L = L(\mathbf{v}, \mathbf{A}, \phi)$, with $\mathbf{A} = \mathbf{A}(\mathbf{r}, t)$ and $\phi = \phi(\mathbf{r}, t)$, so that $L = L(\mathbf{v}, \mathbf{r}, t)$ *indirectly*. You can verify that

$$L = \frac{1}{2}mv^2 - e\phi + \frac{e}{c}\mathbf{v}\cdot\mathbf{A} \quad (13)$$

meets the needs for a successful L. The value of L is its simplicity of form. No vectors! It has three scalar parts. When we were guessing at new physics laws, such as $e\mathbf{v}\times\mathbf{B} + e\mathbf{E}$ in the old days, for the way electricity pushes on charges e, we needed guidance to save time and focus the guesses. Guessing at L's is much simpler than guessing at $\mathbf{F}$ directly. In the quantum generalization, $\mathbf{F}$ fails to even be meaningful anymore. Yet L and H survive there, to give us good physics predictions, to test against actual measurements.

Our $\mathbf{F} = m\mathbf{a}$ equation now takes the incredible form

$$\frac{d}{dt}(\nabla_{\mathbf{v}}L) = \nabla_{\mathbf{r}}L,\ L = \frac{1}{2}mv^2 - e\phi + \frac{e}{c}\mathbf{v}\cdot\mathbf{A} \quad (14)$$

Here, $\nabla_{\mathbf{v}}$ means hold $\mathbf{r}$ and t constant and have L as a function of $\mathbf{v}$, $\mathbf{r}$, t first. If somehow, $\nabla_{\mathbf{r}}L = 0$, then $\nabla_{\mathbf{v}}L$ is a *constant of the motion* and sort of represents conservation of momentum. Equation (14) manages to cast $F = m\mathbf{a}$ in a form where we must guess at L, a scalar, instead of $\mathbf{F}$, a vector, as we said. (We are considering the general case here, so e is either positive or negative for the charged particle.)

We already know that $H = 1/2mv^2 + e\phi$ and $dH = 0$ so $H = H|_{t=0}$. Clearly then $H = L + 2e\phi - (e/c)\mathbf{v}\cdot\mathbf{A}$. But $\mathbf{p} \equiv m\mathbf{v} + (e/c)\mathbf{A}$ and, therefore,

$$\mathbf{p}\cdot\mathbf{v} = mv^2 + \frac{e}{c}\mathbf{v}\cdot\mathbf{A} \quad (15)$$

Therefore,

$$-L + \mathbf{p}\cdot\mathbf{v} = -\frac{1}{2}\,mv^2 + e\phi - \frac{e}{c}\,\mathbf{v}\cdot\mathbf{A} + mv^2 + \frac{e}{c}\,\mathbf{v}\cdot\mathbf{A}$$
$$= \frac{1}{2}\,mv^2 + e\phi = H \tag{16}$$

Therefore, $dH = 0 = -dL + d\mathbf{p}\cdot\mathbf{v} + \mathbf{p}\cdot d\mathbf{v}$ and dL is *not* simple, nor zero, along the path. The value of L for physics is its simplicity of form and ease of insertion to form the $\mathbf{F} = m\mathbf{a}$ equation. This produces a theory of how things get pushed or pulled due to interaction.

3. Poisson Bracket

We are almost finished now. We next invent the Poisson bracket as follows for functions $\alpha = \alpha(\mathbf{r},\,\mathbf{p})$ and $\beta = \beta(\mathbf{r},\,\mathbf{p})$ which are scalar functions of position, $\mathbf{r}$, conjugate momentum, $\mathbf{p}$, and possibly t.

$$(\alpha,\,\beta) \equiv (\nabla_r\alpha)\cdot(\nabla_p\beta) - (\nabla_r\beta)\cdot(\nabla_p\alpha) \tag{17}$$

(This is a crazy looking thing, to be sure.) Remember, ∇_r means here derivatives with respect to the components of $\mathbf{r}$ with $\mathbf{p}$ and t held constant, and $\nabla_{\mathbf{p}}$ means derivatives with respect to components of $\mathbf{p}$ with $\mathbf{r}$ and t held constant. How can this relate to our $\mathbf{F} = m\mathbf{a}$ equation? (It obviously does, or we would not bother with it.) We look at p_x and H, where H must now be expressed in $\mathbf{r}$ and $\mathbf{p}$. We have $\mathbf{p} \equiv m\mathbf{v} + (e/c)\mathbf{A}$. Therefore, any e sign has,

$$\mathbf{v} = \frac{1}{m}\,\mathbf{p} - \frac{e}{mc}\,\mathbf{A} \tag{18}$$

and

$$\mathbf{p}\cdot\mathbf{v} = \frac{p^2}{m} - \frac{e}{mc}\,\mathbf{p}\cdot\mathbf{A} \tag{19}$$

so that

$$H = \frac{1}{2}\,mv^2 + e\phi = \frac{1}{2}\,m\left(\frac{1}{m}\,\mathbf{p} - \frac{e}{mc}\,\mathbf{A}\right)\cdot\left(\frac{1}{m}\,\mathbf{p} - \frac{e}{mc}\,\mathbf{A}\right) + e\phi$$
$$= \frac{1}{2m}\left(p^2 - 2\frac{e}{c}\,\mathbf{p}\cdot\mathbf{A} + \frac{e^2}{c^2}\,A^2\right) + e\phi$$
$$= H(\mathbf{p},\,\mathbf{A}(\mathbf{r},\,t),\,\phi(\mathbf{r},\,t)) = H(\mathbf{p},\,\mathbf{r},\,t) \tag{20}$$

Now we can examine the Poisson bracket

$$
\begin{aligned}
(P_x,\, H) &= (\nabla_{\mathbf{r}},\, p_x)\cdot(\nabla_{\mathbf{p}}H) - (\nabla_{\mathbf{r}}H)\cdot(\nabla_{\mathbf{p}}p_x) \\
&= (\mathbf{0})\cdot(\nabla_{\mathbf{p}}H) - (\nabla_{\mathbf{r}}H|_{\mathbf{p},t})\cdot(\hat{\mathbf{i}}) = \left(-\left.\frac{\partial H}{\partial x}\right|_{y,z,\mathbf{p},t}\right) \\
&= \frac{-1}{2m}\left(0 - \frac{2e}{c}\,\mathbf{p}\cdot\frac{\partial \mathbf{A}}{\partial x} + \frac{e^2}{c^2}\frac{\partial}{\partial x}(A^2)\right) - e\,\frac{\partial \phi}{\partial x} \\
&= \frac{e}{mc}\,\mathbf{p}\cdot\frac{\partial \mathbf{A}}{\partial x} - \frac{e^2}{2mc^2}\left(2\mathbf{A}\cdot\frac{\partial \mathbf{A}}{\partial x}\right) - e\,\frac{\partial \phi}{\partial x} \qquad (21)
\end{aligned}
$$

This is supposed to equal dp_x/dt, which we know as

$$
\frac{dp_x}{dt} = -e\,\frac{\partial}{\partial x}\left(\phi - \frac{1}{c}\,\mathbf{v}\cdot\mathbf{A}\right)\bigg|_{y,z,t} = ? \;= -e\,\frac{\partial \phi}{\partial x} + \frac{e}{c}\,\mathbf{v}\cdot\frac{\partial \mathbf{A}}{\partial x} \qquad (22)
$$

from Eq. (11). We got the $\partial\phi/\partial x$ part right at least. The sneaky question is, "How do we deal with the partial derivatives?" They have different things held constant while we do $\partial/\partial x$. $\partial\phi/\partial t$ makes no reference to the particle directly. It is straightforward and clear. But what does

$$
\frac{\partial}{\partial x}(\mathbf{v}\cdot\mathbf{A}) = \frac{\partial}{\partial x}\left(\frac{dx}{dt}\,A_x(x,y,z,t) + \frac{dy}{dt}\,A_y + \frac{dz}{dt}\,A_z\right)\bigg|_{y,z,t} \qquad (23)
$$

really mean? Go back and look at Eq. (6). We skipped over the proof of this identity. Only by getting into the *details* can one see how $\nabla(\mathbf{v}\cdot\mathbf{A})$ gets constructed and, therefore, what it really means. (These things look so innocent on the surface.) Since this is getting tedious, we leave it here as an exercise for the bright student to *verify* rigorously that the = ? = in Eq. (22) is indeed an equality and

$$
(P_x,\, H) = \frac{dp_x}{dt} \qquad (24)
$$

Should $(p_x, H) = 0$, then P_x would be a *constant of the motion.* This is saying that

$$
\begin{aligned}
\nabla\left(\phi - \frac{1}{c}\,\mathbf{v}\cdot\mathbf{A}\right) &= 0 = \nabla\phi - \frac{1}{c}\,\nabla(\mathbf{v}\cdot\mathbf{A}) \\
&= \nabla\phi - \frac{1}{c}\,(\mathbf{v}\times(\nabla\times\mathbf{A}) - (\mathbf{v}\cdot\nabla)\mathbf{A} \\
&= -\mathbf{E} - \frac{1}{c}\frac{\partial \mathbf{A}}{\partial t} - \frac{1}{c}\,\mathbf{v}\times\mathbf{B} - (\mathbf{v}\cdot\nabla)\mathbf{A}
\end{aligned} \qquad (25)
$$

These terms are all clear and simple, in the last line. They are *not* independent, however. The next to the last line has only ϕ, $\mathbf{A}$, and $\mathbf{v}$, which are independent, and all the terms there are also well defined and clear. The conditions for $\mathbf{A}$ such that all these terms above vanish or the combination vanishes, can be rather subtle, besides the obvious such as $\mathbf{A} =$ constant or maybe $\mathbf{A} = \nabla\theta$ where θ is some scalar field ($\mathbf{B} = 0$).

Our final $\mathbf{F} = m\mathbf{a}$ equation version then is, for any e, m particle:

$$\begin{aligned} \frac{d}{dt}\mathbf{p} &= (p_x,\, H)\hat{\mathbf{i}} + (p_y,\, H)\hat{\mathbf{j}} + (p_z,\, H)\hat{\mathbf{k}} \\ \mathbf{p} &= m\mathbf{v} + \frac{e}{c}\mathbf{A} \\ H &= \frac{1}{2m}\left(\mathbf{p}\cdot\mathbf{p} - \frac{2e}{c}\mathbf{p}\cdot\mathbf{A} + \frac{e^2}{c^2}\mathbf{A}\cdot\mathbf{A}\right) + e\phi \end{aligned} \tag{26}$$

This is so complicated now that it is hard to remember what we are trying to solve for with this mess! Given $A_x(x,y,z,t)$, $A_y(x,y,z,t)$, $A_z(x,y,z,t)$, and $\phi(x,y,z,t)$ known "everywhere" at "all" times, and at $t = 0$ we have m at $\mathbf{r}_0$ with velocity $\mathbf{v}_0$, find $\mathbf{r}(t) = x(t)\hat{\mathbf{i}} + \cdots$ for the particle, from three coupled, second order differential equations. Once $\mathbf{r}(t)$ is found, take the derivatives and get $\mathbf{v}(t) = (dx(t)/dt)\hat{\mathbf{i}} + \cdots$. Then construct $\mathbf{p}(t) = m\mathbf{v}(t) + (e/c)\mathbf{A}(x(t), y(t), z(t), t)$, where $x(t)$ is the path solution in $\mathbf{r}(t)$, etc. Maybe p_x, or p_y, or p_z will be constant in time, but not generally. Calculate $H_0 = \frac{1}{2}mv_0^2 + -|e|\phi_0 = H(t)$, which is independent of $\mathbf{A}_0$ but depends on $|v_0|$ and $\mathbf{r}_0$ through $\phi(\mathbf{r},t)$, i.e., $\phi_0 \equiv \phi(\mathbf{r}_0,\ t=0)$.

Equation (26) gives us three coupled, first order differential equations for $p_x(x,y,z,t)$, $p_y(x,y,z,t)$, and $p_z(x,y,z,t)$, with $t = 0$ initial conditions $\mathbf{p}_0 = m\mathbf{v}_0 + (e/c)\mathbf{A}_0$ well defined. Note that these are independent of ϕ_0. But these Eq. (26) differential equations have $\nabla_\mathbf{p}$ and $\nabla_\mathbf{r}$ in them. The path of the particle $(x(t),\ y(t),\ z(t))$ is still unknown and must be a part of the solution for $\mathbf{p}$. We have then another equation set to consider besides Eq. (26).

We have $\mathbf{p}$ connected to $\mathbf{v}$ and $\mathbf{A}$, so we can write

$$\mathbf{v} = \frac{d\mathbf{r}}{dt} = \frac{1}{m}\mathbf{p} - \frac{e}{mc}\mathbf{A} \tag{27}$$

But can we put this in Poisson bracket form, like Eq. (26)? Consider

$$
\begin{aligned}
(x,\, H) &= (\nabla_r x)\cdot(\nabla_{\mathbf{p}} H) - (\nabla_r H)\cdot(\nabla_{\mathbf{p}} x) \\
&= \hat{\mathbf{i}}\cdot(\nabla_{\mathbf{p}} H) - (\nabla_r H)\cdot(\mathbf{0}) = \left.\frac{\partial H}{\partial p_x}\right|_{py,pz,\mathbf{r},t} \\
&= \frac{\partial}{\partial p_x}\left[\frac{1}{2m}\left((p_x^2+p_y^2+p_z^2) - \frac{2e}{c}\,(P_xA_x + P_yA_y + P_zA_z)\right.\right. \\
&\quad \left.\left.\left. + \frac{e^2}{c^2}\,(A_x^2+A_y^2+A_z^2)\right) + e\phi\right]\right|_{\mathbf{r},t,py,pz} \\
&= \frac{1}{2m}\left(2p_x - \frac{2e}{c}\,A_x + 0\right) + 0 \\
&= \frac{1}{m}\,P_x - \frac{e}{mc}\,A_x = v_x = \frac{dx}{dt}
\end{aligned}
\tag{28}
$$

Therefore, in summary:

$$
\begin{aligned}
\frac{d\mathbf{r}}{dt} &= (x,\, H)\hat{\mathbf{i}} + (y,\, H)\hat{\mathbf{j}} + (z,\, H)\hat{\mathbf{k}} \\
&= \nabla_{\mathbf{p}} H(\mathbf{r},\, \mathbf{p}) = \frac{1}{m}\,(\mathbf{p} - (e/c)\mathbf{A}) \\
\frac{d\mathbf{p}}{dt} &= (p_x,\, H)\hat{\mathbf{i}} + (p_y,\, H)\hat{\mathbf{j}} + (p_z,\, H)\hat{\mathbf{k}} \\
&= -\nabla_{\mathbf{r}} H(\mathbf{r},\, \mathbf{p}) = -e\nabla_{\mathbf{r}}\phi + (e/c)\left[\nabla_{\mathbf{r}}\left(\frac{d\mathbf{r}}{dt}\cdot\mathbf{A}\right)\right] \\
(\alpha,\, \beta) &= (\nabla_{\mathbf{r}}\alpha)\cdot(\nabla_{\mathbf{p}}\beta) - (\nabla_{\mathbf{r}}\beta)\cdot(\nabla_{\mathbf{p}}\alpha) \\
H &\equiv \frac{1}{2m}\left(\mathbf{p}\cdot\mathbf{p} - \frac{2e}{c}\,\mathbf{p}\cdot\mathbf{A} + \frac{e^2}{c^2}\,\mathbf{A}\cdot\mathbf{A}\right) + e\phi
\end{aligned}
\tag{29}
$$

This is totally equivalent to

$$
\begin{aligned}
&\frac{d}{dt}\,(m\mathbf{v}) = e\mathbf{E} + e\mathbf{v}\times\mathbf{B} \\
&\mathbf{B} \equiv \nabla_{\mathbf{r}}\times\mathbf{A}, \quad \mathbf{E} \equiv -\nabla_{\mathbf{r}}\phi - \frac{1}{c}\,\frac{\partial\mathbf{A}}{\partial t}
\end{aligned}
\tag{30}
$$

Equation (29) gives us 6 coupled, first order equations to solve instead of three coupled second order equations to solve for $\mathbf{r}(t)$ — the path of the particle m, moving under the influence of the external field $\phi(\mathbf{r},t)$, $\mathbf{A}(\mathbf{r},t)$. Also, Eq. (29) shows us rather simple forms for the right-hand sides, that do *not* involve the awkward

Poisson bracket structures. So what are these brackets really good for anyway? To answer that we must now turn to quantum issues. In preparation for that, let us look at

$$\begin{aligned}(x,\, p_x) &= (\nabla_{\mathbf{r}} x)\cdot(\nabla_{\mathbf{p}} p_x) - (\nabla_{\mathbf{r}} p_x)\cdot(\nabla_{\mathbf{p}} x)\\ &= (\hat{\mathbf{i}})\cdot(\hat{\mathbf{i}}) - (\mathbf{0})\cdot(\mathbf{0})\\ &= 1\\ (x,\, p_y) &= (\nabla_{\mathbf{r}} x)\cdot(\nabla_{\mathbf{p}} p_y) - (\nabla_{\mathbf{r}} p_y)\cdot(\nabla_{\mathbf{p}} x)\\ &= (\hat{\mathbf{i}})\cdot(\hat{\mathbf{j}}) - (\mathbf{0})\cdot(\mathbf{0})\\ &= 0 - 0 = 0\end{aligned} \tag{31}$$

Consider a new variable, composed from $\mathbf{r}$ and $\mathbf{p}$, such as angular momentum (but remember that $\mathbf{p}$ is *not* $m\mathbf{v}$). Define $\mathbf{L} \equiv \mathbf{r} \times \mathbf{p}$. We could then look at $(L_z,\, H)$, for example. Notice that $\mathbf{L}$ is defined such that it is *not directly* dependent on time, $(\partial \mathbf{L}/\partial t)|_{\mathbf{r},\,\mathbf{p}} = 0$. For all such time "absent" variables, $(L_z,\, H) = 0$ if and only if $L_z = L_{z0}$ at $t = 0$ and is a *constant of the motion* (theorem, no proof here). This gives the $(\ldots,\, H)$ bracket special significance, besides its role in the coupled equations of motion for $\mathbf{p}(t)$ and $\mathbf{r}(t)$ for the particle in a field. Also, $(H, H) = 0$, so $H = H_0$ which we knew already.

You could work out $(v_x,\, v_y)$ and see that this is $(e/mc)\,[\partial_x A_y - \partial_y A_x]$, which is not zero unless $\mathbf{A}$ is special. You could also verify that $(L_y,\, L_z) = L_x = -(L_z, L_y)$ and cyclic permutations. These are all just mathematical exercises with little significance for *classical* mechanics. As Corben and Stehle state,[1] "The chief interest in Poisson brackets lies in the fact that, if in classical mechanics the Poisson bracket of two dynamical variables vanishes, then according to quantum mechanics these variables may be simultaneously observed."

4. The Hydrogen Atom

For the hydrogen atom then, $L_z \neq 0$ would mean that L_x and L_y cannot both be known (since $(L_x, L_y) = L_z$). Because $(x, p_x) = 1 \neq 0$, we cannot know the position x and the conjugate momentum p_x at the same time. But $(x, p_y) = 0$ so both can be known. This is the first of many peculiar and classically unintuitive results for quantum mechanics. Classically, we can know $\mathbf{r}$ and $\mathbf{p}$ and thus *all* these components as precisely as our computer can calculate the solutions to the coupled equations.

For $(\ldots, H)$, we see that if this is classically zero for some variable, such as L_z, then we can know L_z and H, simultaneously. We know H *at all times*, since it is classically a constant of the motion. Therefore, we could know L_z at all times also, i.e., it would be another constant of the motion, both classically and quantum mechanically.

The classical Hamiltonian (energy), for a point mass with charge $-|e|$ attracted to a heavy nucleus with positive charge $Z|e|$ (Z a positive integer), is given by

$$\begin{aligned} H &= \frac{1}{2m}\left(\mathbf{p} + \frac{|e|}{c}\mathbf{A}\right)\cdot\left(\mathbf{p} + \frac{|e|}{c}\mathbf{A}\right) - \frac{Ze^2}{r} \\ &= \frac{1}{2m}\mathbf{p}\cdot\mathbf{p} - \frac{Ze^2}{r} + \frac{|e|}{mc}\mathbf{p}\cdot\mathbf{A} + \frac{e^2}{2mc^2}\mathbf{A}\cdot\mathbf{A} \end{aligned} \tag{32}$$

as should now be quite clear from our earlier work with H. The $\mathbf{p}\cdot\mathbf{A}$ term is small and the $\mathbf{A}\cdot\mathbf{A}$ term is very small (and often neglected). The $\mathbf{A}$ field here comes from the nucleus, which may have moving charge in it, producing magnetic effects at the orbiting particle. If we start our particle out at $t = 0$ with $\mathbf{r}_0$, $\mathbf{v}_0$, and $\mathbf{A}(\mathbf{r}, t)$ due to the nucleus assumed known at all t, then it moves in a complicated classical path but remains bound to the nucleus if $|\mathbf{v}_0|$ is not too large. Of course, it *radiates* as it is accelerated and this puts the "brakes on" but we have *totally ignored* that problem. It goes away at the quantum level for bound states, as does the concept of a path existing at all. Yet, H (as above) survives and tells us things about the bound states that can exist. The orbiting quantum electron acts like a point particle but it still has an internal structure of some kind such as to have a magnetic moment like a small bar magnet. This introduces more interaction terms in $\mathbf{F} = m\mathbf{a}$ besides $e\mathbf{E}$ and $e\mathbf{v} \times \mathbf{B}$. That was an exciting discovery and came later, after the "spinless" electron solutions were developed for the hydrogen atom. That story is told elsewhere in this book.

5. The Quantum Transition

All of the above classical mechanics variations of $\mathbf{F} = m\mathbf{a}$ were known, I suppose, by the time that de Broglie got the hunch that electrons may behave as waves, individually. He proposed

$mv = h/\lambda = \hbar k$ where $\hbar \equiv h/2\pi$ and $k \equiv 2\pi/\lambda$ (the wave number that goes with the wavelength λ). Bohr quantized the simple classical, elliptical orbits by the new guess $L_z = n\hbar$, $n = 1, 2, 3, \ldots, \infty$, and $n \approx \infty$ corresponds to very "classical" behavior. (The so-called correspondence principle.) The particles were far from the nucleus, for large n, and the allowed orbits were very closely spaced. One could not really tell the difference *there* (n large) between the classical description and the quantum description. But when other interactions were allowed, besides the simple Coulomb term Ze^2/r, his approach could not give accurate results. A whole new hierarchical level of discovery was then needed; a new set of equations to totally *replace* $\mathbf{F} = m\mathbf{a}$. The anchor upon which this new development hangs is the classical Hamiltonian, Eq. (32). Heisenberg made one set of guesses as to how to develop the needed new equations, based on jumping off from the Poisson brackets. Schrödinger made another set of guesses at the needed new development. They were both right, as it turned out and as Schrödinger later showed by building a bridge between the two approaches. They proved to be the extremes of a continuum of equivalent formulations, some of which are also useful.

There are two basic quantum problems, bound energy levels and scattering. One formulation is usually not very satisfactory for both types of problems only because it is *not easy*, not because it is incorrect. It may just require orders of magnitude more work than another approach.

I recall Feynman saying that he dazzled the mathematicians by his insights as to new theorems because he tried to relate the work to some *physical* model. If he could, then he had mental access to all the other formalisms that apply to the model and their insights. Those must give corresponding truths in this new mathematical context, since the real world is internally consistent and complicated. Solving one problem with one approach can be amazingly difficult, whereas that same approach makes another physical problem easy to solve.

This happened to Heisenberg when he tried to apply his analytical machinery to hydrogen. It was very difficult to calculate the energy levels. So difficult, in fact, that one author said there is no Heisenberg model for hydrogen! The Schrödinger approach proved very easy to apply to bound states and their allowed energy levels.

Heisenberg's approach was more consistent with the old classical outlook on the world. Clearly we have done enough to show that the solution to the hydrogen atom, at the classical level, involves learning how $\mathbf{r}$ and $\mathbf{p}$ *evolve in time* while H remains constant in time. $\mathbf{L} \equiv L_x\hat{\mathbf{i}} + L_y\hat{\mathbf{j}} + L_z\hat{\mathbf{k}}$ may evolve in time and certainly some of its components usually do. So Heisenberg's approach was to still think of these things as evolving in time, but he guessed that they also had new aspects, which did not show up at the crude, classical level. They became "operators" instead of ordinary numbers and this basically meant that they *no longer commute* in their multiplication. Matrices also don't commute, so he tried to think of them as matrix operators, which somehow evolve in time as their classical descriptions naturally suggest.

Heisenberg was in his early 20s when this was all being developed and he had more math to learn in order to finish the work. A similar thing happened to Einstein, in his 20s, with curved space gravity. He had to learn the details from curved space mathematical specialists, some of whom had *already speculated* that the whole universe might be radically curved on a large scale, but we would not notice it when looking at galaxies because the curving is so gradual. Einstein guessed at the curved space detail equations and the other experts then worked out the consequences of his equations (Schwarzschild, de Sitter, Friedmann, *etc.*) It is fun to note that Heisenberg knew little of matrices when he came to need them. He was at the time an assistant to Born. Born claims that since Heisenberg did not know what matrices were, he was unable to interpret his ideas clearly and submitted it to his superiors. In a letter[2] to Einstein, *years later*, 1948, Born wrote:

> "In the case of quantum mechanics, which you think so little of, the fulsome praise has descended entirely on Heisenberg and Schrödinger. Yet Heisenberg didn't know at all then what a matrix is (he had been my assistant therefore I can judge that)."

Born's major contribution seems to have been the interpretation of $\psi^*\psi$ as the probability of the electron being where its complex number wave function, ψ, has a certain value. One author says that the first scientists of quantum mechanics, Born, Heisenberg, and Jordan could not find a quantum mechanical approach to the

Kepler orbits in hydrogen and even doubted whether the unwieldy matrix formalism could handle the orbit problem.

Others have gone more into the comparisons between Heisenberg's 1925 development and Schrödinger's 1926 bridge between the two. Apparently, W. Pauli actually calculated the Schrödinger wave equation for hydrogen in 1925. He is best known for his exclusion principle and for adding coupling terms to the Hamiltonian which included the electron's dipole interaction with that of the nucleus, electron spin.

Basically, Heisenberg had square matrices as operators which developed in time and these operated on a *fixed* "thing" which was in the shadows. It was the wave function, essentially. Schrödinger developed, instead, fixed-in-time operators that operated on an evolving wave function, ψ. One can put some of the time development in the operators and some in the wave function as well. Whatever is *useful* at the time for the particular application, to solve real world problems, should be used.

6. Heisenberg's Basic Guess

We have seen that $(x, p_x) = 1$, $(x, p_y) = 0$, $(x, H) = dx/dt$, $(p_x, H) = dp_x/dt$ at the *classical, non-relativistic level* and this is equivalent to $F_x = ma_x = mdv_x/dt = d(mv_x)/dt \neq dp_x/dt$.

Heisenberg's basic guess was that the Poisson (α, β), for *classical* variables α and β, which are functions of $\mathbf{r}$, $\mathbf{p}$, and t, should be *replaced in quantum* by $-(i/\hbar)(\alpha\beta - \beta\alpha)$. Thus $(x, p_x) = 1$, which is classically true, is *replaced* by $-(i/\hbar)(xp_x - p_x x) = 1$ which then means that

$$xp_x = +i\hbar + p_x x \tag{33}$$

Thus, x and p_x cannot be ordinary numbers any longer, even if $x(t)$ and $p_x(t)$ are meaningful as changing operators with the flow of time. Since $\hbar$ is extremely small compared to x and p_x for, say, a charged speck of paper near a van de Graff generator ball, the fact that $xp_x \neq p_x x$ for such a speck of paper would not be noticed. Its quantum mechanical behavior would not be noticed at all. Inside the hydrogen atom, an electron has $x \sim 0.5$ A $=$ 0.05×10^{-9} meters (nm). There the non-commutativity is significant. The classical limit then involves $\hbar \to 0$, or more precisely $\hbar \to \ll (xp_x)$. Classically, $(x, p_y) = 0$ and this becomes, quantum mechanically, $(xp_y - p_yx)(-i/\hbar) = 0$, so that $xp_y = p_yx$. These "operators" commute. They can relate to classical values for the system such that both can be known at the *same time.* These quantum examples are not easy to relate to, however. Consider $(L_x, L_y) = L_z$, classically. This means quantum mechanically that $(-i/\hbar)(L_xL_y - L_yL_x) = L_z$ or $L_xL_y - L_yL_x = i\hbar L_z$ as operators. We cannot know L_x and L_y at the same time if there are classical values associated with these operators, for $L_z \neq 0$.

If you have had some matrix algebra, you can easily see that this relationship can be produced by suitable matrices. For example, $L_x \equiv (\hbar/2)\sigma_1$, $L_y \equiv (\hbar/2)\sigma_2$, $L_z \equiv (\hbar/2)\sigma_3$, $\sigma_1\sigma_2 \equiv i\sigma_3 \equiv -\sigma_2\sigma_1$ and cyclic, $\sigma_1\sigma_1 \equiv \sigma_2\sigma_2 \equiv \sigma_3\sigma_3 \equiv 1$, and $ii = -1$, will give the desired result. The $\{\sigma_1, \sigma_2, \sigma_3\}$ are called Pauli matrices. He used them in the mid-20s to *expand* H to include the electron dipole moment in hydrogen, and he got *more accurate* energy levels in the process. Experiment had shown that something more, theoretically, was needed in the Hamiltonian. Actually, Hamilton himself had long before invented $\{-i\sigma_1, -i\sigma_2, -i\sigma_3\}$, as what he called quaternion basis elements, about 1843. They were the key to *relativistic classical* physics and *relativistic quantum* but went unappreciated until Pauli reinvented them and then Dirac generalized them, in 1928, for relativistic electrons. (See the chapter on Paul Dirac).

Consider next $(x, H) = dx/dt$ which becomes $(-i/\hbar)(xH - Hx) = dx/dt$. This sort of gives the time development of the x-coordinate operator. Similarly, $(p_x, H) = dp_x/dt$ becomes $(-i/\hbar)(p_xH - Hp_x) = dp_x/dt$.

You can easily see the quandary that Heisenberg, Born, and Jordan were in at this point. We have the H for hydrogen. We know that $\mathbf{r}$ and $\mathbf{p}$ are operators and how their components commute or don't commute. But how do we do anything with this to get the *allowed* values of $H = H_0$ = constant, for the electron in "orbit" as a cloud about the nucleus??! Being bound to the nucleus, in some quantum state, means that the electron cloud "thins out" rapidly with large distances from the nucleus. This is a crucial, so-called boundary condition. It does not easily find expression for the hydrogen atom in Heisenberg's operators.

Schrödinger guessed *instead* that even though H *classically* gives the time evolution of the system and this means it gives $\mathbf{r}(t)$ and $\mathbf{p}(t)$, quantum mechanically he introduced a wave like function, ψ, in keeping with de Broglie's idea that the electron is a spread out wave. Then he guessed that $i\hbar(\partial/\partial t)\psi = H\psi(\mathbf{r}, t)$. The i here, $ii = -1$, requires the wave amplitude to be a *complex* number, rather than a real number as it is for light, sound, or water waves. The H here is the classical Hamiltonian of hydrogen, but with $\mathbf{p}$ and $\mathbf{r}$ obeying Heisenberg's non-commutivity relations. An operator system which works this way is $x \to x$ and $p_x \to -i\hbar(\partial/\partial x)$, when "passing" to the quantum world. These are not matrices as Heisenberg envisioned, but still are operators in a derivative sense. Notice that

$$\begin{aligned}
(xp_x - p_x x)f(x) &\equiv \left(x\left(-i\hbar\frac{\partial}{\partial x}\right) - \left(-i\hbar\frac{\partial}{\partial x}\right)x\right)f(x) \\
&\equiv x\left(-i\hbar\frac{\partial}{\partial x}\right)f(x) - \left(-i\hbar\frac{\partial}{\partial x}\,xf(x)\right) \\
&= -i\hbar x\,\frac{\partial f}{\partial x} + i\hbar\left[\frac{\partial x}{\partial x}\,f(x) + x\,\frac{\partial f(x)}{\partial x}\right] \\
&= -i\hbar x\,\frac{\partial f}{\partial x} + i\hbar f(x) + i\hbar x\,\frac{\partial f(x)}{\partial x} \\
&= i\hbar f(x) = (i\hbar)f(x), \qquad \text{for } any\ f(x) \qquad (34)
\end{aligned}$$

Therefore, $xp_x - p_x x = i\hbar$ as operators, as required by Heisenberg's approach. Notice that x and p_x do NOT evolve in time *here*! Only $\psi(x, t)$ would. The operators are time independent, unlike Heisenberg's approach.

We see H being indirectly associated with time development, in Schrödinger, but in a different way totally from Heisenberg. Which is better? Neither is, in *principle*; but Schrödinger's is in *practice* for the hydrogen atom. Which is better for our physics knowledge evolving on to new and even more radical physics, such as in relativistic quantum physics? Schrödinger is for sure. You also have seen here all of the convoluted classical things we had to go through to get to the Poisson brackets in order to even *start* the Heisenberg quantum development. No wonder then that quantum books skip Heisenberg completely for many chapters. They start with the simple, but radical guesses

$$\begin{aligned} &\mathcal{E} = H,\ \mathcal{E} \to i\hbar\frac{\partial}{\partial t},\ x \to x,\ p_x \to -i\hbar\frac{\partial}{\partial x} \\ &y \to y,\ p_y \to -i\hbar\frac{\partial}{\partial y}, \ldots,\ \mathcal{E}\psi = H\psi(\mathbf{r}, t) \end{aligned} \tag{35}$$

Here p_x is *not* particle momentum, but rather the *conjugate momentum* (classically); see Eq. (20). This subtle difference from $m\mathbf{v}$ only shows up when $\mathbf{A} \neq 0$.

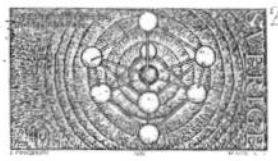

7. Into the Future

In the *relativistic* generalization of quantum mechanics, the free particle energy and momentum are used first to generate a guessed at, *free* particle wave equation. Then the couplings to other fields must be guessed at. Or, people instead guess directly at a Lagrangian that has parts for each free particle and a part that couples the free fields (particles). A machinery from classical mechanics is then used (variation of variables or of the path) to generate the coupled wave equations. Here, in relativity, time and space are put on an *equal* footing! Thus H and t are no longer favored, and all hell breaks lose in space-time *quantum* physics! Heisenberg was tied to Poisson brackets which in turn came from *non-relativistic* classical, where *time development is central.* Relativistic *classical* physics sort of exists under *limited* circumstances. Hitting two classical baseballs together at $v = 0.9c$ will actually produce quanta, however, with reasonable probability! Schrödinger originally considered

$$\mathcal{E}\psi = H\psi = \frac{\mathbf{p}\cdot\mathbf{p}}{2m}\psi \qquad [(\mathcal{E}/c)^2 - \mathbf{p}\cdot\mathbf{p}]\psi = (\mathcal{E}_0/c)^2\psi\,. \tag{36}$$

But even this is not sufficient because electrons are really particles. The ψ is only a *first approximation*, good for bound states and such. The full, best theory we have is quantum electrodynamics and that focuses on the particle concept. Feynman[3] has summarized the strange *guesses* that work there, to develop *successful* rules for calculating and predicting. The particle has a path but takes *all possible paths* and even goes backward in time and far into the future, as well as moving through all of space. This all "occurs" between the classical creation of the quanta and their arrival at classical detectors. Clearly this is a nice scattering machinery, but it may have problems with bound states. QED is general for electrical effects and, with an enormous computer, one could in *principle* calculate all electromagnetic effects in matter to at least 9-digit accuracy. In simple situations these calculations can really be done, and compare favorably with experiment to that incredible accuracy! All that stuff makes the hydrogen atom energy level problem seem like child's play by comparison. We have come a long way since the 1920s, when quantum was born. The hydrogen atom is the master prototype system even now. If we push the measurements far enough, the spread-out nature of the proton will begin to affect the energy level calculations. The details of the proton will need to be known and those details *do affect* the electron energy levels ever-so-slightly, in principle. Even the curved space, existing both in and around the proton, may show up eventually and affect the energy levels that we see, indirectly, through the emitted light quanta. There is at least one great level of mystery left to challenge you young people. You people stand now where Heisenberg and Schrödinger did then. The challenge is much tougher this time, and has stumped generations of their followers. How long before we solve the nucleus problem? What radical, new concepts must emerge, just as they were needed in the 1920s? Are our imaginations big enough?[4]

References

1. H. C. Corben and P. Stehle, *Classical Mechanics*, Wiley, NY, 1950, pp. 215–232.
2. M. Born, Albert Einstein, Hedwig and Max Born, Briefwechsel: 1916–1955, Nymphenburger, Munich, 1969.
3. R. P. Feynman, *QED*, Princeton University, Princeton, 1985.
4. J. D. Edmonds, Jr., "Klein–Gordon factorization within Dirac subalgebras: new mass terms and covariance extension", *Speculat. Sci. Techol.* **18** (1995) 216–228.

James D. Edmonds Jr. graduated from San Diego State University with high honors and distinction in physics. He worked as a physics lab technician during his last few years in college and that influenced him to apply to Cornell University in Engineering Physics. He finished his electron microscopy Ph.D. there, with minors in mathematics and theoretical physics, in 1967, and then went to Hughes Research Labs., in Malibu. There, Prof. R. P. Feynman lectured weekly for two hours and his inspiration, over two years, led Edmonds to soon switch to teaching at a small Methodist college in San Diego. He then taught for years in temporary positions and finally settled in a small college in Louisiana with the Cajuns.

Now, some 25 years after Hughes and 90 papers, he has finally put together some promising but radical ideas about the nature of mass and about the covariance of wave equations. He also learned about aerobatic airplanes, both flying and building, in this later years. He is now pulling all his insights together into a book, for undergraduates, on Dirac algebra and basic physics.

Chapter 8. Erwin Schrödinger, Wave Mechanics, and the Hydrogen Atom

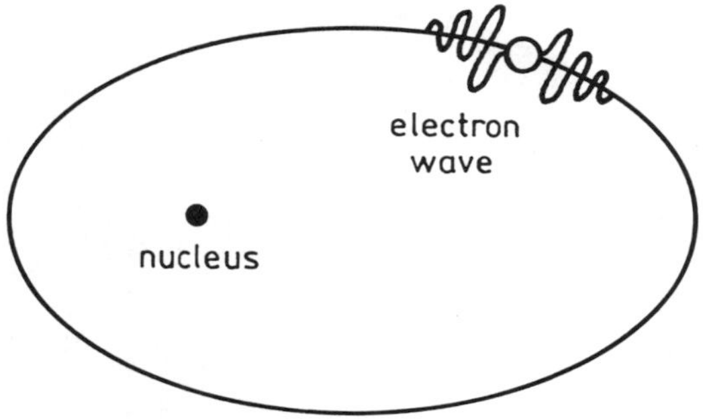

Owen L. de Lange

E. Schrödinger

Chapter 8

Erwin Schrödinger, Wave Mechanics, and the Hydrogen Atom

Owen L. de Lange

Physics Department,
University of Natal,
Private Bag X01 Scottsville,
Pietermaritzburg 3209, South Africa

Abstract

A brief account is given of some of the main events in Erwin Schrödinger's life, the circumstances leading to his discovery of wave mechanics in 1925–1926, and the application of this mechanics to a model of the hydrogen atom. Two methods of solving this model are outlined, namely, Sommerfeld's polynomial method and Schrödinger's shift operator method. The relation to Pauli's analysis of the quantum-mechanical hydrogen atom is also discussed.

1. Schrödinger's Life

> "His (Schrödinger's) private life seemed strange to bourgeois people like ourselves. But all this does not matter. He was a most lovable person, independent, amusing, temperamental, kind and generous, and he had a most perfect and efficient brain."
>
> *Max Born*

Erwin Schrödinger was one of the most important and interesting of the great twentieth century theoretical physicists. He was born in Vienna in 1887, the only child of Rudolf and Georgine Schrödinger. The family lived comfortably on the earnings of a linoleum business which Rudolf inherited from his family. Erwin later recalled that his father "assured for me an excellent upbringing and a carefree university education"; Rudolf was a cultured man with interests in chemistry, Italian painting and plant phylogenetics, and "to his growing son, he was a friend, teacher and inexhaustible conversation partner".[1] From his mother, aunts and several maids and nurses, Erwin received, and grew accustomed to, an abundance of doting feminine care.[1]

Until he was eleven, Schrödinger received his education at home from a private tutor and then, from 1898 to 1906, he attended the Akademisches Gymnasium in Vienna. Here he excelled at his studies, particularly mathematics and physics, but he also had strong interests in the Greek and Latin languages and in the Vienna theater. In 1906, he enrolled in the University of Vienna where he studied physics, mathematics, chemistry and astronomy.[2] He concentrated on physics and mathematics, and especially on the course of theoretical physics given by Professor Fritz Hasenöhrl, who was an excellent teacher. He continued to excel in these courses which were of a high standard and were later to stand him in good stead; it is considered that among theoretical physicists of his generation, Schrödinger had the finest mathematical ability, with the possible exception of Arnold Sommerfeld.[3] Schrödinger's first scientific research was an experimental project on the conductivity of insulators in moist air, which he submitted as his doctoral thesis; he was awarded the Doctor of Philosophy degree in May 1910.

Five months later he took leave of absence from the university to undergo military training as a "one-year volunteer" in the fortress artillery. He returned in October 1911 to an assistantship in experimental physics at the University of Vienna; this marks the start of his first period of scientific activity in experimental and theoretical physics. His research dealt with dielectric properties of matter, atmospheric electricity, X-ray diffraction and lattice dynamics.[2,3] On the basis of this work he obtained his Habilitation in January 1914 and became a Privatdozent for Physics at the University of Vienna. At about this time he fell in love with Felice Krauss, a young woman from a wealthy Austrian family. They wished to marry, and in view of his poor financial prospects as a physicist, he requested his father's permission to join the family linoleum business. Rudolf refused and insisted that his son continue with his academic career. The engagement was broken off and Felice subsequently married into a wealthy, old, imperial family. It has been suggested that this unhappy episode encouraged Erwin's later "negative view of bourgeois marriage".[3]

Schrödinger's first period of research was brought to an end by the advent of the First World War: he received his mobilization orders on 31 July 1914. There are two phases to his war-time military service. During the first phase, which lasted until approximately the middle of 1917, he served as an artillery officer at various places on the southeastern (Italian) front. Fortunately, he came through this unscathed by injury or illness, unlike the many promising young intellectuals who were killed in the general slaughter.[2] In the second phase he was stationed in Vienna where he taught meteorology to anti-aircraft officers until the end of the war in late 1918. Remarkably, he managed to continue his scientific research during the war: ten scientific papers were completed during this time, dealing with capillarity, specific heats, acoustics of the atmosphere, Brownian motion, statistical dynamics and general relativity.[2] He also filled several notebooks with discussions relating to statistical mechanics, general relativity and other theoretical topics.

The closing stages and the aftermath of the war brought turmoil and suffering to the people of Central Europe, and the Schrödinger family in Vienna was not spared this. Also, both Erwin and his mother suffered from ill health at this time, and the family experienced severe financial difficulties when their linoleum business was destroyed by the war. During the winter of 1918–19 "the Schrödingers often ate at a community soup kitchen".[4] Erwin's father died in December 1919 and his mother in September 1921. Despite these adverse circumstances he continued to devote himself to research at the University of Vienna between 1918 and 1920. The most important work from these years was on colour theory, but he also published the results of experiments on the coherence of light. It was during this time that his philosophical outlook was formed and he read widely on the European and Eastern philosophies, filling notebooks with his commentaries. In particular, he studied Schopenhauer and the Hindu scriptures. The latter he reworked "into his own words, and ultimately came to believe in them".[5] In 1920, two important changes occurred in his life; in March he married Annemarie Bertel, and the following month he moved to the University of Jena in Germany as an assistant to Max Wien. Although his marriage turned out to be rather unconventional and often stressful, it lasted until his death. The move to Germany also marked a lasting change and, apart from short visits, he did not return to his alma mater for another thirty six years.

Schrödinger's next three academic positions were all short-lived. After only six months in Jena he moved to the Technische Hochschule in Stuttgart for one semester as an associate professor. This was followed by a further semester as a professor of theoretical physics at the University of Breslau. Then in October 1921 he became a professor of theoretical physics in the University of Zurich. The stress of the preceding moves had taken their toll and he was first required to undertake a nine-month long recuperation for tuberculosis at Arosa, an Alpine resort. He remained at the University of Zurich for six years, and it was there that he made his greatest scientific discovery.

Schrödinger's main research during his first years in Zurich was on colour theory, statistical thermodynamics of ideal gases and quantum physics. Then late in 1925 there occurred a turning point in his life. In a letter to Albert Einstein, dated 3 November, he wrote: "A few days ago I read with the greatest interest the ingenious thesis of Louis de Broglie which I finally got hold of".[6] This remark refers to a doctoral thesis which a young French physicist presented to the University of Paris in 1924. In this thesis, de Broglie made the revolutionary suggestion that phase waves are connected with material particles; he had previously shown how one could use such an idea to explain Bohr's quantum condition for the angular momentum of the electron in a hydrogen atom.[6] Later in November, at the suggestion of a colleague, Pieter Debye, Schrödinger presented a colloquium on de Broglie's work to the Zurich physics department. In this talk he discussed and attempted to extend de Broglie's geometrical visualization of phase waves.[6,7] There was evidently some dissatisfaction with this approach, particularly from Debye who felt that one should deal instead with a wave equation. In a second colloquium a few weeks later, Schrödinger presented such an equation, and in this way the discovery of Schrödinger's wave equation for the hydrogen atom was anounced to the world.

The following months produced an outburst of scientific creativity from Schrödinger on a scale which has rarely been witnessed in the history of science. Between March and September 1926 he published in the prominent German journals, *Annalen der Physik* and *Die Natuurwissenschaften* a series of six monumental papers which presented a new fundamental theory of quantum physics.[8–13] This theory is based on the Schrödinger wave equation, and it is known as wave mechanics. Schrödinger also discussed in these papers certain applications which provided tests of wave mechanics; the new theory passed these and all subsequent tests.

The mid-1920s were hectic years in physics, and several theoreticians were active in the search for a new and acceptable quantum mechanics. In the second half of 1925 the first successful modern theory of quantum mechanics was presented by Werner Heisenberg, Max Born, and Pascual Jordan in a series of papers published in the *Zeitschrift für Physik*.[14–16] This theory, which is now known as matrix mechanics, appeared to be very different from wave mechanics, the former being essentially a theory of a discontinuum whereas "Schrödinger's concept of a wavefunction filling all space evoked a continuum even more pervasive than that of classical mechanics".[7] Yet when both theories were applied to the same problem they gave the same result. In Schrödinger's third paper on wave mechanics, entitled "On the relationship between the quantum mechanics of Heisenberg, Born, and Jordan, and mine",[10] it was shown that the two theories are, in fact, mathematically equivalent. Subsequently, Paul Dirac developed an abstract general formulation of quantum mechanics which includes the matrix and wave theories as particular cases.[17]

News of wave mechanics spread rapidly, not only through the publications themselves, but also via Schrödinger's extensive correspondence with other physicists such as Wilhelm Wien (the editor of *Annalen der Physik*), Alfred Landé, Arnold Sommerfeld, Hendrik Lorentz and Max Planck. It soon became clear that wave mechanics provided a convenient and powerful method for understanding and predicting physical phenomena, and it quickly emerged as the preferred version of the theory. Indicative of its importance is Dirac's remark that, "The underlying physical laws necessary for the mathematical theory of a large part of physics and the whole of chemistry are thus completely known ...."[18] Not surprisingly,

Schrödinger's discovery earned for him a place among the immortals of science.

During the second half of 1926, Schrödinger lectured on wave mechanics in Zurich, Stuttgart, Berlin, Munich, and Copenhagen, where he had long discussions with Niels Bohr on the physical interpretation of his theory. This was followed by a three-month long lecture tour of the United States. In August 1927, he moved to the University of Berlin as successor to Max Planck, who had recently retired. Although the decision to leave Zurich was a difficult one, the Schrödingers were soon pleased to be in Berlin with its rich scientific and artistic life, and where the university faculty included renowned scientists, such as Einstein, with whom Schrödinger formed a close friendship. Schrödinger was now forty and he occupied the most prestigious chair in European physics. During the next few years he travelled extensively throughout Europe, both on lecture tours and on holidays. His stay in Berlin was terminated by the ominous political developments which followed the appointment of Adolf Hitler as Chancellor of Germany in January 1933.

Although he was not a Jew, Schrödinger left Germany in July 1933 because, as he put it, "I could not endure being bothered by politics".[19] The Schrödingers journeyed via Zurich and the Tyrol (where they vacationed) to Paris, Brussels and finally Oxford, arriving there on 4 November, 1933. Schrödinger had previously been in negotiations with Frederick Lindemann, a professor of physics at Oxford University, who had arranged for him a fellowship at Magdalen College. On the 9 November he was informed that he was to receive, jointly with Dirac, the Nobel Prize in physics for 1933. (The other founder of modern quantum theory, Heisenberg, was simultaneously awarded the prize for 1932.) Schrödinger was cited for "establishing a new system of mechanics which also holds good for motions within the atoms and molecules" and for the "discovery of new fruitful forms of atomic physics and the applications of these".[20] During the three years he spent in Oxford, Schrödinger continued to work on the interpretations of quantum mechanics. It must have been a worrying time for him: while on a visit to Princeton, where he was offered (and declined) the professorship of mathematical physics, he wrote to Lindemann that in spite of a "so-called scientific reputation, I am actually without what a man of my age and métier considers a *Lebenstellung*" and "I am afraid that my wife could neither live upon the German pension nor on the '*Schrödingergleichung*' ".[21] The Schrödingers found it difficult to adapt to life in Oxford. After two visits to Spain, Schrödinger considered accepting a professorship in Madrid, but this was prevented by the outbreak of the Spanish civil war. He then considered a position at the University of Edinburgh, but this also came to naught.[21] Eventually, in October 1936 he returned to Austria, after an absence of fifteen years, to a professorship at the University of Graz.

Schrödinger later came to regard his decision to return to Austria as a serious mistake. Certainly, after the *Anschluss* in March 1938 his position became increasingly difficult: on the one hand he was regarded by high officials in the German government as a "fanatical opponent" of National Socialism and, on the other, his attempt to appease these officials by publishing a contentious letter in the Graz *Tagespost* only served to harm his relations with supporters abroad, particularly at Oxford.[22] In August 1938, he was dismissed from the University of Graz for "political unreliability" and soon after, on 14 September, the Schrödingers' position had become so precarious that they had to flee Austria. They travelled by train to Rome and then Geneva. Eventually, at the beginning of October, they arrived once again in Oxford where, it turned out, they were no longer welcome, partly for reasons to do with the letter Schrödinger had written in Graz. However, he was unrepentant, remarking that with the Nazis "you take care".[22] After two months in Oxford he accepted a one-year visiting professorship at the University of Gent, commencing in December. Meanwhile he was involved in negotiations with Eamon de Valera, the prime minister of Ireland, who had a passionate interest in mathematics and wished to establish an Institute for Advanced Studies in Dublin. So it happened that in October 1939, after the outbreak of the Second World War, the Schrödingers arrived in Dublin where he had secured a temporary professorship. He was grateful to the University of Gent for the hospitality it had offered and wrote to the rector "to express to you how deeply indebted I feel to your fatherland for the refuge and help that it unreservedly granted us in critical times".[22]

On 1 June 1940, the Dublin Institute for Advanced Studies was founded and Schrödinger was appointed a senior professor; he was to remain at the Institute for the next sixteen years. He was thankful to have found refuge in Ireland. In a letter to Born he wrote: "To be reinstated to absolute security ... at 53 by a foreign government, in my case fills you with — well, infinite gratitude to that country".[23] In the early years of the war he sometimes felt despair as, for example, when he wrote to Hermann Weyl in April 1941 that, "I cannot see what should change the present state. I have believed long enough. I can no longer. It is all nothing"; but two months later, with the German invasion of Russia, he wrote "the banner is turned around".[23] His main research interests in Dublin included quantum mechanics, nonlinear electromagnetic theory and unified field theory, and he also pursued his interests in philosophy and religion. He wrote a book on Statistical Thermodynamics and a book entitled *What is Life?*, which had an important impact on the development of molecular biology.[23]

After the war, Schrödinger expressed an interest in returning to live and work in Austria, but this was prevented for more than a decade by the complicated political situation in that country. Finally, in July 1955, he was appointed professor of theoretical physics at his alma mater, the University of Vienna, a position which he assumed in April 1956. He retired in 1958 and became Professor Emeritus in September of that year. His health had not been good for some years as he was suffering from atherosclerosis aggravated by chronic respiratory problems. He died in Vienna on 4 January 1961 and was buried in Alpbach. On his gravestone is carved the symbol "Ψ" denoting the wavefunction which he discovered.

Walter Moore, in a recent biography of Schrödinger, wrote that he "had the most complex personality of all the great creators of modern physics. He was a passionate foe of injustice yet he regarded any political activity as contemptible. He despised pomp and circumstance yet he took a childlike pleasure in honors and medals. He was devoted to the vedantic concept that all men are members of one another, yet he eschewed cooperative work of any kind. His mind was dedicated to precise reasoning, yet his temperament was as volatile as a prima donna's. He claimed to be an atheist, but he always used religious symbolism and believed his scientific work was an approach to the godhead".[24]

2. Wave Mechanics and the Hydrogen Atom

> "The discovery of the Schrödinger wave equation was a great event for physics. It may well be judged as having been an even greater event for chemistry and biology, which have progressed so rapidly during the last 60 years, with much of the progress owed to the new ideas provided by quantum mechanics."
>
> *Linus Pauling*

We first give a brief description of those parts of de Broglie's theory of phase waves[25] which Schrödinger used in the early stage of his development of wave mechanics. De Broglie's work was motivated by a desire to extend the wave-particle duality of photons to material particles.[26] He started with the energy-frequency relation of Planck,

$$E = h\nu \tag{1}$$

and the energy-mass relation of Einstein,

$$E = mc^2 \,. \tag{2}$$

(Here h is Planck's constant and c is the speed of light *in vacuo.*) By equating these expressions in the rest frame of the particle, de Broglie was encouraged to postulate that in this frame there is a periodic phenomenon with frequency

$$\nu_0 = m_0c^2/h \,, \tag{3}$$

where m_0 is the rest mass of the particle. He then considered a fixed observer O, relative to whom the particle moves with velocity v. According to the phenomenon of time dilation in the theory of relativity, O will detect a frequency

$$\nu_1 = \nu_0(1-\beta^2)^{\frac{1}{2}} \tag{4}$$

where $\beta = v/c$. But the mass of the particle relative to O is

$$m = m_0(1-\beta^2)^{-\frac{1}{2}} \,, \tag{5}$$

and this defines another frequency

$$\nu = mc^2/h = (m_0c^2/h)(1-\beta^2)^{-\frac{1}{2}} \,, \tag{6}$$

connected with the total energy (2) of the particle.

Next, de Broglie associated a scalar travelling wave (a phase wave) of frequency ν with the particle, and required that this wave should keep in phase with the internal oscillation of frequency ν_1: he showed that the condition for this "harmony of phases" is[25,27]

$$u = c^2/v\,, \tag{7}$$

where u is the phase velocity of the travelling wave. Corresponding to this velocity there is a wavelength

$$\lambda = u/\nu = h(1-\beta^2)^{\frac{1}{2}}/m_0 v = h/p\,, \tag{8}$$

where p is the relativistic momentum of the particle. (Equation (8) is the famous de Broglie relation connecting the momentum of a particle to its wavelength; although it is implicit in de Broglie's work, he used frequency and phase velocity in his presentation.) De Broglie also generalized the above results, which are for a free particle, to a charged particle in an electromagnetic field. In particular, for a field with scalar potential ϕ, he showed that Eqs. (6) and (7) become[25,27]

$$\nu = (1/h)[m_0 c^2(1-\beta^2)^{-\frac{1}{2}} + e\phi] \tag{9}$$

and

$$u = \frac{c^2}{v} + \frac{e\phi}{m_0 v}(1-\beta^2)^{\frac{1}{2}}\,, \tag{10}$$

where e is the charge of the particle.

Next we discuss Schrödinger's development of wave mechanics. As the first testing-ground for his theory of matter waves, Schrödinger used the hydrogen atom. Thus in applying Eqs. (9) and (10), he took $e\phi = -e^2/r$ where e is the electronic charge and $r = (x^2+y^2+z^2)^{\frac{1}{2}}$ is the electron-proton distance. He used (9) to eliminate the speed of the electron v from (10), obtaining

$$u = c\,\frac{h\nu}{[(h\nu + e^2/r)^2 - m_0^2c^4]^{1/2}}\,. \tag{11}$$

Then he substituted (11) in the time-independent wave equation[a]

$$\nabla^2\psi + (4\pi^2\nu^2/u^2)\psi = 0 \tag{12}$$

[a]Equation (12) follows from the usual equation of wave motion, $\nabla^2\psi - (1/u^2)(\partial^2\psi/\partial t^2) = 0$, if one assumes the harmonic time-dependence $\psi \sim \exp(i2\pi\nu t)$.

where

$$\nabla^2 \equiv \frac{\partial^2}{\partial x^2} + \frac{\partial^2}{\partial y^2} + \frac{\partial^2}{\partial z^2} \tag{13}$$

is the Laplacian and $\psi(\mathbf{r})$ is the phase wavefunction. In this way he found the wave equation[6]

$$\nabla^2\psi + \frac{4\pi^2}{h^2c^2}\left[\left(E + m_0c^2 + \frac{e^2}{r}\right)^2 - m_0^2c^4\right]\psi = 0\,, \tag{14}$$

where $E = h\nu - m_0c^2$ is the energy of the electron relative to its rest–mass energy. He also considered the non-relativistic limit: setting $(1-\beta^2)^{\frac{1}{2}} \approx 1 - \frac{1}{2}\beta^2$ in (9) and (10), and repeating the above steps (or taking $E + e^2/r \ll m_0c^2$ in (14)), one has

$$\nabla^2\psi + \frac{8\pi^2 m_0}{h^2}\left(E + \frac{e^2}{r}\right)\psi = 0\,. \tag{15}$$

Equation (15) is known as the (time-independent) Schrödinger wave equation for the hydrogen atom. The above calculations appear in Schrödinger's notebooks, and they are one of the earliest (perhaps the first) derivations of (14) and (15).[6] In his first paper in the series "Quantisierung als Eigenwertproblem" he gave a different derivation of (15) which was based on a variational principle[8] and made no reference to de Broglie's waves.[b]

He now imposed conditions on the wavefunction $\psi(\mathbf{r})$, namely, that it should be "single-valued, finite, and continuously twice differentiable over the whole configuration space".[8] He then asked (in his notebooks and in Ref. 8) the crucial question: can (14) and (15) yield correct results for the energy E? To understand the subsequent analysis it is helpful to recall the results of two previous models for the hydrogen atom. First, according to the non-relativistic Bohr model the energy levels are

$$E = -\frac{2\pi^2 m_0 e^4}{h^2 n^2} \tag{16}$$

where $n = 1, 2, \ldots$ is the principal quantum number. Second, the relativistic calculation of Sommerfeld based on the old quantum

[b] In his second paper of the series, Schrödinger described his derivation in Ref. 8 as "incomprehensible" and gave yet another derivation based on the analogy between mechanics and optics.[9]

mechanics yielded[28]

$$E = m_0c^2 \left\{1 + \frac{\alpha^2}{[N + \sqrt{k^2 - \alpha^2}]^2}\right\}^{-\frac{1}{2}} - m_0c^2 \tag{17}$$

where $N = 0, 1, \ldots$ is the radial quantum number, $k = 1, 2, \ldots$ is the azimuthal quantum number, and $\alpha = 2\pi e^2/hc$ is the fine structure constant. Equation (17) gives good agreement with the observed fine structure of the hydrogen spectrum.

In his published work, Schrödinger first considered the nonrelativistic equation (15).[8] A method of solving such a problem was familiar to him from his student days.[6] Because of the spherical symmetry of (15) he chose spherical polar coordinates r, θ (the colatitude) and ϕ (the longitude). Then he assumed a separation of variables, that is, ψ is taken to be a product of a function of r, a function of θ and a function of ϕ. The angular dependence is easily shown to be the spherical harmonic $Y_{\ell m}(\theta, \phi)$, where $\ell = 0, 1, \ldots$ and m is restricted to the $2\ell + 1$ values from $-\ell$ to ℓ. With

$$\psi(\mathbf{r}) = \chi(r)Y_{\ell m}(\theta, \phi) \tag{18}$$

in (15), the equation for the radial part $\chi(r)$ is readily found to be

$$\frac{d^2\chi}{dr^2} + \frac{2}{r}\frac{d\chi}{dr} + \left(-A + \frac{2B}{r} - \frac{C}{r^2}\right)\chi = 0\,, \tag{19}$$

where

$$A = -\frac{8\pi^2 m_0 E}{h^2}, \quad B = \frac{4\pi^2 m_0 e^2}{h^2}, \quad C = \ell(\ell + 1) \tag{20}$$

are constants. Now came the most challenging part of the problem: to find solutions to (19) which satisfy the above conditions. To do this, Schrödinger used a result from the mathematical literature, namely an integral representation for the *general* solution to (19).[6,8] By a careful analysis of the behavior of this solution at $r = 0$ and $r = \infty$, he was able to establish the following results: (i) For $E < 0$ (that is, for bound states of the electron, corresponding to elliptical orbits in the old quantum theory) acceptable solutions to (19) exist only if E is given by (16) and if[c]

$$\ell = 0, 1, \ldots, n - 1\,. \tag{21}$$

[c]Here, and elsewhere, we use standard notation rather than Schrödinger's notation, where n and ℓ are interchanged.

(ii) For $E > 0$ (which means the atom is dissociated, corresponding to hyperbolic orbits in the old quantum theory), (19) has acceptable solutions for all positive E. He also wrote down explicit expressions for the radial wavefunctions $\chi_{n\ell}(r)$. The general solution to (19) and its analysis are rather cumbersome, and are not given here; for these the reader can refer to Refs. 6, 8 and 29. Later on we will describe two alternative methods of analysing the Schrödinger model of the hydrogen atom, namely, Sommerfeld's polynomial method and the shift operator method.

Schrödinger emphasized the significance of his results by writing "... that the usual rules for quantization can be replaced by another requirement, in which the mention of "whole numbers" no longer occurs. Instead the integers occur in the same natural way as the integers specifying the number of nodes of a vibrating string".[8] His first paper on the hydrogen atom announced the advent of wave mechanics, and it is generally regarded as one of the greatest achievements in twentieth century physics. Even in a purely mathematical context, his achievement of devising a partial differential equation whose eigenvalue spectrum has both discrete and continuous parts, was a considerable one which had evaded a mathematician of the stature of David Hilbert.[6]

The most popular textbook method of solving for the bound states of tractable wave-mechanical problems is Sommerfeld's polynomial method.[30,31] Applying this method to the radial equation (19), one seeks a solution of the form

$$\chi = e^{-\rho/2}\rho^{\ell}P_N(\rho)\,, \tag{22}$$

where $\rho = 2\sqrt{A}r$ and

$$P_N(\rho) = \sum_{s=0}^{N} a_s\rho^s \tag{23}$$

is a polynomial of degree $N(=0,1,\dots)$. (In (22), the exponential factor is obtained from the asymptotic form of χ and the second factor from solving the characteristic equation[30]; an explanation for the polynomial factor will be given below.) For (22) to be a solution to (19) one readily finds that

$$B/\sqrt{A} = N + \ell + 1 = n\,. \tag{24}$$

The non-negative integer N is known as the radial quantum number. Using (20) in (24) one obtains (16). By construction, the ansatz (22) is sufficient to obtain an acceptable solution; it may be shown that it is also necessary.[30,31]

The analysis of the relativistic equation (14) is similar to the non-relativistic problem: the radial equation also has the form (19) but with

$$A = -\frac{4\pi^2}{h^2c^2}\,[(E+m_0c^2)^2 - m_0^2c^4], \qquad B = \frac{4\pi^2e^2(E+m_0c^2)}{h^2c^2},$$
$$C = \ell(\ell+1) - \alpha^2\,. \tag{25}$$

Employing the polynomial method, one starts with

$$\chi = e^{-\rho/2}\rho^{\gamma}P_N(\rho)\,, \tag{26}$$

where γ is given as a root of the characteristic equation by[30]

$$\gamma = \sqrt{(\ell+\tfrac{1}{2})^2 - \alpha^2} - \tfrac{1}{2}\,. \tag{27}$$

For (26) to be a solution to (19) it is necessary that

$$B/\sqrt{A} = N + \gamma + 1\,. \tag{28}$$

From (28) and (25) one obtains the energy eigenvalues for bound states

$$E = m_0c^2\left\{1 + \frac{\alpha^2}{[N+\frac{1}{2}+\sqrt{(\ell+\frac{1}{2})^2-\alpha^2}]^2}\right\}^{-\frac{1}{2}} - m_0c^2\,. \tag{29}$$

There are several interesting historical and physical facts about the relativistic wave equation (14) and its eigenvalues (29). In a notebook, Schrödinger obtained and solved (14) before he treated the non-relativistic hydrogen atom.[6] By comparing a result like (28) (which he derived from an analysis of the general solution to (19)) with a similar expression in Sommerfeld's theory[28] he concluded that "... if the relativistic Kepler problem is worked out in detail, it is found to lead strangely to *half-integral partial* quanta (radial and azimuthal)".[8] One can understand this remark by comparing (17) and (29): in place of the integer k one has $\ell + \frac{1}{2}$, and instead of N one has $N + \frac{1}{2}$. These differences are sufficient to spoil completely the agreement with the measured fine structure of the hydrogen spectrum.[30] This failure so disappointed Schrödinger that he abandoned his new wave theory for a while before returning to, and publishing, the successful non-relativistic version of the theory.[6] Later, the reason for this failure of Schrödinger's relativistic theory became clear: the electron has spin angular momentum equal to $\frac{1}{2}\hbar$ (where $\hbar$ is $h/2\pi$), and it is therefore described by a different relativistic wave equation, the Dirac equation (see Chapter 9). The Dirac equation for a hydrogen atom leads to exactly the Sommerfeld result (17) of the old quantum theory, a quirk known as the "Sommerfeld puzzle" and which has been explained in the literature.[32] By contrast, the relativistic equation (14) applies to a spinless charged particle in the Coulomb field of a nucleus. More than fifty years after Schrödinger's work, the difficult task was undertaken of studying such a system experimentally. Using pionic atoms like π^-Ti and π^-Fe it was found that the measured spectra are in good agreement with (29) (with α replaced by $Z^2\alpha$ to take account of the atomic number of the nucleus).[33,34] Schrödinger could not know that his first wave-mechanical result, far from being incorrect, was a prediction for a physical system which had not yet been produced in the laboratory. It is therefore ironic that he delayed the publication of his seemingly incorrect relativistic equation.[13] Consequently, the relativistic wave equation for a spinless particle was also presented, almost simultaneously, by several other authors[30,35,36] and it is now known as the Klein–Gordon equation; (14) is the time-independent Klein–Gordon equation for the Coulomb (or Kepler) problem.

In his series of papers[8–13] Schrödinger developed wave mechanics into the form in which it is used today. Thus he came to realize that, for a general time-varying system, the wavefunction is a complex, time-dependent function $\Psi(\mathbf{r}, t)$ which satisfies the equation

$$H\Psi = i\hbar\frac{\partial\Psi}{\partial t}\,. \tag{30}$$

Here H is the Hamiltonian operator: for a non-relativistic particle in a time-independent potential $V(\mathbf{r})$ it is given by

$$H = -\frac{\hbar^2}{2m_0}\nabla^2 + V(\mathbf{r})\,. \tag{31}$$

Equation (30) is the famous time-dependent Schrödinger wave equation. For stationary states of the Hamiltonian (31)

$$\Psi(\mathbf{r},t) = \psi(\mathbf{r})e^{-\frac{i}{\hbar}Et}\,, \tag{32}$$

where the energy E is a constant, and (30) reduces to the time-independent Schrödinger wave equation

$$H\psi = E\psi\,. \tag{33}$$

If

$$V = -k/r \tag{34}$$

(k is a constant) in (33), then one is dealing with the non-relativistic Coulomb (or Kepler) problem. The non-relativistic equation for the hydrogen atom (15) corresponds to the choice $k = e^2$.

Schrödinger also discussed the difficult question of the physical meaning of the wavefunction.[8,13] When compared with the immediate success of his wave equation, his attempts to achieve an understanding of the wavefunction appear halting and unconvincing. His ideas on the interpretation of the wavefunction underwent changes while he developed wave mechanics, and "Instead of publishing just one final result, he revealed the whole process of his search, the picture of his long wandering in the darkness".[37] What emerged was an electrodynamic interpretation of the wavefunction of an electron. According to this interpretation, and for the simplest case of a hydrogen-like atom, $\psi(\mathbf{r})$ determines the electric charge and current densities of the electron:

$$\rho(\mathbf{r}) = e\psi^*(\mathbf{r})\psi(\mathbf{r}) \tag{35}$$

and

$$\mathbf{j}(\mathbf{r}) = \frac{ie\hbar}{2m_0}\left(\psi(\mathbf{r})\boldsymbol{\nabla}\psi^*(\mathbf{r}) - \psi^*\boldsymbol{\nabla}\psi(\mathbf{r})\right). \tag{36}$$

Thus in this picture the electron is not a point charge but has dimensions of the same order as those of the atom. Schrödinger extended this interpretation to multi-electron atoms, and in the process touched on the concept of a density matrix.[13,37] He used his picture and classical electrodynamics to offer an explanation as to why an electron moving around a nucleus does not lose energy by radiation.[13,37] Schrödinger's interpretation of the wavefunction was not accepted by most other physicists. Soon a paper by Born appeared which contained the famous probability interpretation, according to which $|\psi(\mathbf{r})|^2 dV$ is the probability of finding a particle in a volume dV.[38] This came to be the generally accepted interpretation, although Schrödinger's electrodynamic picture continues to attract attention.[37,39]

Schrödinger developed in detail the perturbation theory of wave mechanics and applied this to the Stark effect in atomic hydrogen.[11] (That is, to the effect of a uniform electric field on the hydrogen spectrum.) For a hydrogen atom in a uniform electric field of strength F in the z-direction, the wave equation (33) is

$$\nabla^2\psi + \frac{8\pi^2 m_0}{h^2}\left(E + \frac{e^2}{r} - eFz\right)\psi = 0\,. \tag{37}$$

Schrödinger used parabolic coordinates to treat this problem, and showed that first-order perturbation theory gives the result

$$E = -\frac{2\pi^2 m_0 e^4}{h^2 n^2} - \frac{3h^2 Fn(k_2 - k_1)}{8\pi^2 m_0 e}\,, \tag{38}$$

where k_1 and k_2 are non-negative integers ("parabolic quantum numbers").[11] He pointed out that (38) "is the well-known formula of Epstein for the term values in the Stark effect of the hydrogen spectrum".[11] He also calculated the intensities of the hydrogen lines. With this, the major part of Schrödinger's work on the wave mechanics of the hydrogen atom was complete. In the following decades there has been a continuing interest in the quantum-mechanical Coulomb problem, and in the rest of this chapter we will discuss a few of these developments.

3. Further Developments[d]

> "The general method described here could be called that of adjoint first order operators. I think it gives the most direct insight, hitherto available, into the way in which the line spectrum depends on the structure of the equation."
>
> *Erwin Schrödinger*

In 1940 and 1941, shortly after his arrival in Dublin, Schrödinger published a series of papers on a novel method of solving certain

[d]Much of the material in this section is more advanced than that in the preceding sections, and it may be omitted in a first reading of this book.

wave-mechanical problems, including the Coulomb problem.[40–42] He wrote that "The attractive feature of the method to be described here is that it avoids cumbersome transformations, recourse to the ready-made equipment of the mathematical warehouse or expansion into power-series. It yields all the eigenfunctions of the line spectrum by *one* quite elementary quadrature".[40] The essence of this method is the factorization of a second-order operator into two first-order operators, with the latter two operators usually being mutually adjoint. This technique has become known as the shift operator method or the factorization method, and it has been extensively developed, particularly in the classic paper by Infeld and Hull.[43] The simplest application, which can be found in most textbooks on quantum mechanics, is to the one-dimensional harmonic oscillator. Here we will describe an application to the nonrelativistic Coulomb problem.

For this purpose, it is convenient to rewrite the radial equation (19) (with $e^2 = k$) as

$$H_\ell \chi_{E\ell} = E\chi_{E\ell}\,, \tag{39}$$

where

$$H_\ell = -\frac{\hbar^2}{2m_0}\left(\frac{d}{dr} + \frac{1}{r}\right)^2 + \frac{\hbar^2 \ell(\ell+1)}{2m_0 r^2} - \frac{k}{r} \tag{40}$$

is the radial Hamiltonian operator. The quadratic form (40) can be factorised

$$R^-_{\ell+1} R^+_\ell = 2m_o H_\ell + (m_0 k/\hbar)^2 (\ell+1)^{-2}\,, \tag{41}$$

$$R^+_\ell R^-_{\ell+1} = 2m_o H_{\ell+1} + (m_0 k/\hbar)^2 (\ell+1)^{-2}\,, \tag{42}$$

where

$$R^+_\ell = \hbar\frac{d}{dr} - \frac{\hbar\ell}{r} + \frac{m_0 k}{\hbar(\ell+1)}\,, \tag{43}$$

$$R^-_\ell = -\hbar\frac{d}{dr} - \frac{\hbar(\ell+1)}{r} + \frac{m_0 k}{\hbar\ell}\,. \tag{44}$$

It is easy to show, using (41), (42) and (39) that $R^\pm_\ell$ perform the transformations (shift operations)[44]

$$R^\pm_\ell \chi_{E\ell} = \alpha^\pm \chi_{E,\ell\pm 1} \tag{45}$$

where

$$|\alpha^{\pm}|^2 = 2m_0E + (m_0k/\hbar)^2(\ell + \tfrac{1}{2} \pm \tfrac{1}{2})^{-2} . \qquad (46)$$

(In (45) and (46) the upper signs go together, as do the lower signs.) The operators $R_\ell^{\pm}$ are known as shift operators (raising and lowering operators) for the quantum number ℓ.

These operators preserve the required quadratic integrability (normalizability) of $\chi_{E\ell}$; that is, if $\chi_{E\ell}$ is normalized, i.e.

$$\int_0^\infty |\chi_{E\ell}|^2 r^2 dr = 1 ,$$

so are $\chi_{E,\ell\pm1}$ in (45). It follows that the right-hand side of (46) must be non-negative. This is the crucial condition, from which it follows that for bound states ($E < 0$) of an attractive Coulomb potential the raising operations for ℓ in (45) *cannot continue indefinitely*, but must terminate at some value $\ell_{\max}$, such that

$$2m_0E + (m_0k/\hbar)^2(\ell_{\max} + 1)^{-2} = 0 . \qquad (47)$$

If we let $\ell_{\max} + 1 = n$, then $n = 1, 2, \ldots$ and $\ell = 0, 1, \ldots, n-1$. Thus n is the usual principal quantum number, and (47) gives the Bohr formula for E. Equations (45)–(47) yield the annihilation condition

$$R_{n-1}^{+}\chi_{n,n-1} = 0 \qquad (48)$$

where

$$n = (-m_0k^2/2\hbar^2E)^{\frac{1}{2}} . \qquad (49)$$

Equation (48) is a first-order differential equation which can easily be solved for $\chi_{n,n-1}$. In particular, for the ground state χ_{10}, (48) is

$$\left(\frac{d}{dr} + \frac{m_0k}{\hbar^2}\right)\chi_{10} = 0 . \qquad (50)$$

The normalized solution is

$$\chi_{10} = 2(m_0k/\hbar^2)^{\frac{3}{2}} \exp(-m_0kr/\hbar^2) . \qquad (51)$$

The operators $R_\ell^{\pm}$ for the hydrogen atom were deduced by Schrödinger, who also showed how the factorization method could be used treat perturbations and illustrated this with an application to the Stark effect in hydrogen.[40]

One can, by factorization, also construct shift operators, $Q_n^{\pm}$ which change the energy of bound states of the Coulomb problem, i.e. the value of n in (49),[40,44]

$$Q_n^{\pm}\chi_{n\ell} = \beta^{\pm}\chi_{n\pm1,\ell} \tag{52}$$

where $\beta^{\pm}$ are independent of r. Using the raising operations in (45) and (52) one can obtain all the wavefunctions $\chi_{n\ell}$ from the ground state (51).[45] These calculations show that it is the application of the energy shift operators Q_n^{+} that produces the polynomial factor in (22); the degree of the polynomial (the radial quantum number N) is just the number of times that these raising operators must be applied to $\chi_{\ell+1,\ell}$ to generate $\chi_{n\ell}$ with $n = N + \ell + 1$.[45]

It is helpful for what follows to place the Schrödinger model of the Coulomb problem in the context of the abstract (Dirac) formalism of quantum mechanics.[46] In Dirac's language, Schrödinger devised a model for which H, $\mathbf{L}^2$ and L_z is a possible choice for a complete set of commuting observables,[e] and for which the coordinate representation of wave mechanics is used. In the Dirac formalism, $|n\ell m\rangle$ denotes the ket for the system and the eigenvalue equations are[f]

$$H|n\ell m\rangle = E|n\ell m\rangle\,, \tag{53}$$

$$\mathbf{L}^2|n\ell m\rangle = \hbar^2\ell(\ell+1)|n\ell m\rangle\,, \tag{54}$$

$$L_z|n\ell m\rangle = \hbar m|n\ell m\rangle\,. \tag{55}$$

For bound states, n is defined in terms of E by (49). The operators $\mathbf{r}$ and $\mathbf{p}$ which appear in (53)–(55) are abstract operators satisfying the canonical commutation relations

$$[r_i, r_j] = [p_i, p_j] = 0, \quad [r_i, p_j] = i\hbar\delta_{ij}\,, \tag{56}$$

where i and $j = 1, 2, 3$ and δ_{ij} is the Kronecker delta function. The coordinate-space wavefunctions are the quantities

$$\psi_{n\ell m}(\mathbf{r}) = \langle\mathbf{r}|n\ell m\rangle\,, \tag{57}$$

[e]Here $H = \mathbf{p}^2/(2m_0) - k/r$ is the Hamiltonian and $\mathbf{L} = \mathbf{r}\times\mathbf{p}$ is the orbital angular momentum operator.
[f]Here $m = -\ell, -\ell+1, \ldots, \ell$ and ℓ is a non-negative integer or half-integer: the half-integer values can be excluded with further analysis.[47]

where, using Dirac's notation and terminology, $\langle \mathbf{r}|$ is an eigenbra of the position operator $\mathbf{r}$ for a spinless free particle.

One can also choose to work in the momentum representation with a momentum-space wavefunction

$$\bar{\psi}_{n\ell m}(\mathbf{p}) = \langle \mathbf{p}|n\ell m\rangle\,. \tag{58}$$

In particular, the non-relativistic Coulomb problem in this representation can be treated by the same methods used for the coordinate representation. Thus one can use (53), (56) and (58) to deduce a wave equation for $\bar{\psi}_{n\ell m}$.[48] After the separation of variables

$$\bar{\psi}_{n\ell m}(\mathbf{p}) = \bar{\chi}_{n\ell}(p) Y_{\ell m}(\theta_p, \phi_p)\,, \tag{59}$$

where $p = (p_x^2 + p_y^2 + p_z^2)^{\frac{1}{2}}$, one obtains for the radial wave equation[48]

$$\begin{aligned} -\left[\left(\frac{d}{dp} + \frac{1}{p}\right)(p^2 - 2m_0 E)\right]^2 \bar{\chi}_{n\ell} + \frac{\ell(\ell+1)}{p^2}(p^2 - 2m_0 E)^2 \bar{\chi}_{n\ell} \\ = 4(m_0 k/\hbar)^2 \bar{\chi}_{n\ell}\,. \end{aligned} \tag{60}$$

This equation can be analysed by the same methods used for (19). For example, one can, by factorization, derive the shift operators[48]

$$P_{E\ell}^{\pm} \bar{\chi}_{n\ell} = \gamma^{\pm} \bar{\chi}_{n,\ell+1} \tag{61}$$

where

$$P_{E\ell}^{+} = \left(\frac{d}{dp} + \frac{1}{p}\right)(p^2 - 2m_0 E) + \frac{\ell+1}{p}(p^2 + 2m_0 E)\,, \tag{62}$$

$$P_{E\ell}^{-} = -\left(\frac{d}{dp} + \frac{1}{p}\right)(p^2 - 2m_0 E) + \frac{\ell}{p}(p^2 + 2m_0 E)\,, \tag{63}$$

and

$$|\gamma^{\pm}|^2 = 4(m_0 k/\hbar)^2 + 8 m_0 E(\ell + \tfrac{1}{2} \pm \tfrac{1}{2})^2\,. \tag{64}$$

The shift operations (61) are the momentum-representation counterparts of the coordinate-representation shift operations (45): the analysis of (61) is similar to that of (45), and for bound states the non-negativity of (64) yields again the Bohr formula for E. One can also derive shift operations for E (i.e. for n). All the radial momentum-space wavefunctions for bound states $\bar{\chi}_{n\ell}$ can be generated from the ground state $\bar{\chi}_{10}$ using the shift operators for n

and ℓ.[48] As is well known, the $\bar{\chi}_{n\ell}(p)$ are Fourier transforms of the $\chi_{n\ell}(r)$.[49]

At about the same time that Schrödinger used wave mechanics to solve the non-relativistic hydrogen atom problem, Wolfgang Pauli solved the same problem using Heisenberg's matrix mechanics.[g,50] We conclude this chapter with a brief discussion of the relationship between these two approaches. Pauli started with the property that the classical Coulomb problem possesses, in addition to the angular momentum **L**, other vector constants of the motion. In particular, he considered the Laplace–Runge–Lenz vector (or second Laplace vector)[51]

$$\mathbf{A}_c = (m_0 k)^{-1}\mathbf{L} \times \mathbf{p} + \hat{\mathbf{r}} \tag{65}$$

where $\hat{\mathbf{r}} = r^{-1}\mathbf{r}$, and $\mathbf{r}$ and $\mathbf{p}$ are the classical position and momentum vectors of the particle. The vector (65) has a simple physical significance: the magnitude A_c is equal to the eccentricity of the conic section describing the trajectory, and the direction of $\mathbf{A}_c$ is from a focus and away from the point of closest approach.[47,51] From (65), Pauli constructed a vector operator

$$\mathbf{A} = (2m_0 k)^{-1}(\mathbf{L} \times \mathbf{p} - \mathbf{p} \times \mathbf{L}) + \hat{\mathbf{r}}\,, \tag{66}$$

where $\mathbf{r}$ and $\mathbf{p}$ are operators satisfying the commutation relations (56). He showed that **A** is also a constant of the motion, and derived a relation between **A** and H. Then, using properties of **A**, he determined the eigenvalues of H, obtaining the usual Bohr formula.[50]

From the Pauli operator (66) one can construct shift operators as follows: Let

$$\mathbf{C}^{\pm} = \mathbf{A} \pm i\mathbf{B} \tag{67}$$

where

$$\mathbf{B} = (\mathbf{A} \times \mathbf{L} - \tfrac{1}{2} i\hbar \mathbf{A})(\mathbf{L}^2 + \tfrac{1}{4}\hbar^2)^{-\frac{1}{2}}\,. \tag{68}$$

Then it can be shown that[47]

$$C^{\pm}_{\mu}|n\ell m\rangle = \gamma^{\pm}_{\mu}|n, \ell \pm 1, m + \mu\rangle \tag{69}$$

[g] The manuscripts by Pauli and Schrödinger, respectively, were received by the journals on 17 January 1926 and 27 January 1926; both were published in March 1926.

for $\mu = 0, \pm 1$. Here $\gamma_\mu^\pm$ are independent of $\mathbf{r}$ and $\mathbf{p}$, and $C_0^+ = C_z^+$, $C_{\pm 1}^+ = C_x^+ \pm i C_y^\pm$, etc. Thus $\mathbf{C}^\pm$ provides shift operators for angular momentum in the kets $|n\ell m\rangle$: they are abstract (representation independent) operators that transform between degenerate kets of the Coulomb problem.[h]

It is natural to ask: what is the connection between the shift operators $C_\mu^\pm$ and the Schrödinger-type shift operators (43), (44) and (62), (63) that act on radial wavefunctions? The answer is provided by an interesting property of $\mathbf{C}^\pm$ known as bi-factorizability: the operators $\mathbf{C}^\pm$ can be factorized into a product of a vector operator and a scalar operator, and this can be done in two different ways.[47] That is,

$$\mathbf{C}^\pm = \mathbf{U}^\pm R^\pm \quad \text{and} \quad \mathbf{C}^\pm = \mathbf{V}^\pm P^\pm . \tag{70}$$

Explicit expressions for the vector operators $\mathbf{U}^\pm$ and $\mathbf{V}^\pm$ are given in Ref. 47: they are shift operators for ℓ and m in the eigenkets $|\ell m\rangle$ of a complete set of commuting operators $\mathbf{L}^2$ and L_z. For our purposes, the interesting operators are the scalars $R^\pm$ and $P^\pm$. They are given by[47]

$$\begin{aligned} R^\pm = {} & \pm i\hbar^{-1}(\hat{\mathbf{r}} \cdot \mathbf{p} - i\hbar r^{-1})\left(\sqrt{\mathbf{L}^2 + \tfrac{1}{4}\hbar^2} \pm \tfrac{1}{2}\hbar\right) \\ & - r^{-1}\left(\sqrt{\mathbf{L}^2 + \tfrac{1}{4}\hbar^2} \pm \tfrac{1}{2}\hbar\right)^2 + m_0 k/\hbar \end{aligned} \tag{71}$$

and

$$\begin{aligned} P^\pm = {} & \mp i\hbar^{-1}(\hat{\mathbf{p}} \cdot \mathbf{r} + i\hbar p^{-1})(p^2 - 2m_0 H) \\ & + \hbar^{-1} p^{-1}(p^2 + 2m_0 H)\left(\sqrt{\mathbf{L}^2 + \tfrac{1}{4}\hbar^2} \pm \tfrac{1}{2}\hbar\right). \end{aligned} \tag{72}$$

In the coordinate representation the operator $\hat{\mathbf{r}} \cdot \mathbf{p}$ in (71) can be replaced by $-i\hbar d/dr$. Also, for the purpose of operating on $|n\ell m\rangle$, the term $\sqrt{\mathbf{L}^2 + \frac{1}{4}\hbar^2}$ in (71) can be replaced by its eigenvalue $\hbar(\ell + \frac{1}{2})$. In this way, (71) reduces to $(\ell + 1)$ times (43) and ℓ times (44). Similarly, (72) leads to (62) and (63). Thus, in this sense, the Schrödinger-type differential operators are contained in the abstract Pauli vector operator.

[h] It is the existence of the extra constant of the motion $\mathbf{A}$ that accounts for the extra degeneracy (degeneracy in ℓ) of the non-relativistic Coulomb problem.

Finally, we note that in the classical limit ($\hbar \to 0$ and $\mathbf{r}$ and $\mathbf{p}$ replaced by c-numbers), (67) reduces to $\mathbf{A}_c \pm i\mathbf{B}_c$ where $\mathbf{A}_c$ is given by (65), and

$$\mathbf{B}_c = \mathbf{A}_c \times \hat{\mathbf{L}} \tag{73}$$

is the Hamilton vector (or first Laplace vector).[51] The vectors (65) and (73) have been known since the earliest days of classical mechanics,[47,51,52] and it is interesting that their operator counterparts are contained in some of the first work on quantum mechanics.

Acknowledgements

I would like to thank Professor R. E. Raab for helpful comments. In writing this short account of Schrödinger and his work on the hydrogen atom, I have found Moore's biography of Schrödinger[1] and Mehra and Rechenberg's two volumes on Schrödinger[2,6] invaluable.

References

1. W. Moore, *Schrödinger*, Cambridge University Press, Cambridge, 1989, chap. 1.
2. J. Mehra and H. Rechenberg, *The Historical Development of Quantum Theory*, vol. 5, part 1, Springer-Verlag, New York, 1987, chap. 1.
3. Ref. 1, chap. 2.
4. Ref. 1, chap. 3.
5. Ref. 1, chap. 4.
6. J. Mehra and H. Rechenberg, *The Historical Development of Quantum Theory*, vol. 5, part 2, Springer-Verlag, New York, 1987, chap. 3.
7. Ref. 1, chap. 6.
8. E. Schrödinger, "Quantisierung als Eigenwertproblem. (Erste Mitteilung)", *Ann. Phys.* **79** (1926) 361–376. English translation: "Quantisation as a problem of proper values. Part 1", in *Collected Papers on Wave Mechanics*, 2nd edn., translated by J. F. Shearer, Blackie, London, 1927, pp. 1–12.
9. E. Schrödinger, "Quantisierung als Eigenwertproblem. (Zweite Mitteilung)", *Ann. Phys.* **79** (1926) 489–527. English translation: "Quantisation as a problem of proper values. Part 2", in *Collected Papers on Wave Mechanics*, ibid., pp. 13–40.
10. E. Schrödinger, "Über das Verhältnis der Heisenberg–Born–Jordanschen Quantenmechanik zu der meinen", *Ann. Phy.* **79** (1926) 734–756. English translation: "On the relation between the quantum mechanics of Heisenberg, Born, and Jordan, and mine", in *Collected Papers on Wave Mechanics*, ibid., pp. 45–61.
11. E. Schrödinger, "Quantisierung als Eigenwertproblem. (Dritte Mitteilung: Storungstheorie mit Anwendung auf den Starkeffekt der

Balmerlinien)", *Ann. Phys.* **80** (1926) 437–490. English translation: "Quantisation as a problem of proper values. Part 3: Perturbation theory, with applications to the Stark effect of Balmer lines", in *Collected Papers on Wave Mechanics*, ibid., pp. 62–101.
12. E. Schrödinger, "Der stetige Übergang von der Mikro-zur Makromechanik", *Natuurwissenschaften* **14** (1926) 664–666. English translation: "The continuous transition from micro- to macromechanics", in *Collected Papers on Wave Mechanics*, ibid., pp. 41–44.
13. E. Schrödinger, "Quantisierung als Eigenwertproblem. (Vierte Mitteilung)", *Ann. Phys.* **81** (1926) 109–139. English translation: "Quantisation as a problem of proper values. Part 4", in *Collected Papers on Wave Mechanics*, ibid., pp. 102–123.
14. W. Heisenberg, "Über quantenmechanische Umdeutung kinematischer und mechanischer Beziehungen", *Z. Phys.* **33** (1925) 879–893. English translation: "Quantum theoretical re-interpretation of kinematic and mechanical relations", in *Sources of Quantum Mechanics*, B. L. van der Waerden, Dover, New York, 1968, pp. 261–276.
15. M. Born and P. Jordan, "Zur Quantenmechanik", *Z. Phys.* **34** (1925) 858–888.
16. M. Born, W. Heisenberg, and P. Jordan, "Zur Quantenmechanik. II", *Z. Phys.* **36** (1926) 557–615.
17. See, for example, P. A. M. Dirac, *The Principles of Quantum Mechanics*, Clarendon Press, Oxford, 1930.
18. P. A. M. Dirac, "Quantum mechanics of many-electron systems", *Proc. R. Soc.* **A123** (1929) 714–733.
19. Ref. 1, chap. 7.
20. *Nobel Lectures in Physics*, 1922–1941, Elsevier, Amsterdam, 1965, p. 288.
21. Ref. 1, chap. 8.
22. Ref. 1, chap. 9.
23. Ref. 1, chap. 10.
24. Ref. 1, p. 4.
25. L. de Broglie, "Recherches Sur La Theorie Des Quanta", Ph.D. dissertation, Paris, 1924. Reprinted by Masson et Cie., Paris, 1963.
26. J. Mehra and H. Rechenberg, *The Historical Development of Quantum Theory*, vol. 2, part 1, Springer-Verlag, New York, 1982, chap. 5.
27. See also, for example, J. M. Espinosa, "Physical properties of de Broglie's phase waves", *Am. J. Phys.* **50** (1982) 357–362.
28. A. Sommerfeld, "Zur Quantentheorie der Spektrallinien", *Ann. Phys.* **51** (1916) 1–94.
29. B. H. Armstrong and E. A. Power, "Acceptable solutions and boundary conditions for the Schrödinger and Dirac equations", *Am. J. Phys.* **31** (1963) 262–268.
30. A. Sommerfeld, *Wave Mechanics*, Methuen, London, 1930.

31. A. Rubinowicz, *Sommerfeldsche Polynommethode*, Springer, Berlin, 1972.
32. L. C. Biedenharn, "The 'Sommerfeld Puzzle' revisited and resolved", *Found. Phys.* **13** (1983) 13–33.
33. L. Delker *et al.*, "Experimental verification of the relativistic fine-structure term of the Klein–Gordon equation in pionic titanium atoms", *Phys. Rev. Lett.* **42** (1979) 89–92.
34. K. C. Wang *et al.*, "Experimental determination of the relativistic fine-structure splitting in pionic Ti and Fe atoms", *Phys. Rev.* **A22** (1980) 1072–1077.
35. H. Kragh, "Equation with many fathers. The Klein–Gordon equation in 1926", *Am. J. Phys.* **52** (1984) 1024–1033.
36. Ref. 6, chap. 4.
37. A. D. Vlasov, "The Schrödinger atom", *Physics-Uspekhi* **36** (1993) 94–99.
38. M. Born, "Quantenmechanik der Stoßvorgänge", *Z. Phys.* **38** (1926) 803–827.
39. J. Dorling, "Schrödinger's original interpretation of the Schrödinger equation: a rescue attempt" in *Schrödinger, Centenary Celebration of a Polymath*, C. W. Kilmister, ed., Cambridge University Press, Cambridge, 1987, pp. 16–40.
40. E. Schrödinger, "A method of determining quantum-mechanical eigenvalues and eigenfunctions", *Proc. R. Irish Acad.* **A46** (1940) 9–16.
41. E. Schrödinger, "Further studies on solving eigenvalue problems by factorization", ibid. **46** (1940) 183–206.
42. E. Schrödinger, "The factorization of the hypergeometric equation", ibid. **47** (1941) 53–54.
43. L. Infeld and T. E. Hull, "The factorization method", *Rev. Mod. Phys.* **23** (1951) 21–68.
44. O. L. de Lange and R. E. Raab, *Operator Methods in Quantum Mechanics*, Clarendon Press, Oxford, 1991, chap. 5.
45. Ref. 44, chap. 6.
46. P. A. M. Dirac, *Quantum Mechanics*, 4th edn., Clarendon Press, Oxford, 1958.
47. Ref. 44, chap. 9.
48. See, for example, Ref. 44, chap. 7 and references therein.
49. See, for example, P. M. Morse and H. Feshbach, *Methods of Theoretical Physics*, Part II, McGraw-Hill, New York, 1953, chap. 12.
50. W. Pauli, "Über das Wasserstoffspektrum vom Standpunkt der neuen Quantenmechanik", *Z. Phys.* **36** (1926) 336–363. English translation: "On the hydrogen spectrum from the standpoint of the new quantum mechanics", in *Sources of Quantum Mechanics*, B. L. van der Waerden, Dover, New York, 1968, pp. 387–415.

51. See, for example, H. Goldstein, "More on the prehistory of the Laplace or Runge-Lenz vector", *Am. J. Phys.* **44** (1976) 1123–1124 and references therein.
52. O. Volk, "Miscellanea from the history of celestial mechanics", *Celestial Mechanics* **14** (1976) 365–382.

Owen L. de Lange is a professor of Physics in the University of Natal, South Africa. As a research student he worked under Professor F. R. N. Nabarro on solid state physics and under Professor F. A. Otter on superconductivity. His research interests include conventional and high-temperature superconductivity, dislocation interactions and work hardening in metals, and algebraic methods in quantum mechanics. He is the author (together with Professor R. E. Raab) of a monograph, *Operator Methods in Quantum Mechanics.* He is also interested in the history of physics.

Chapter 9. Dirac and the Emergence of Quantum Mechanics

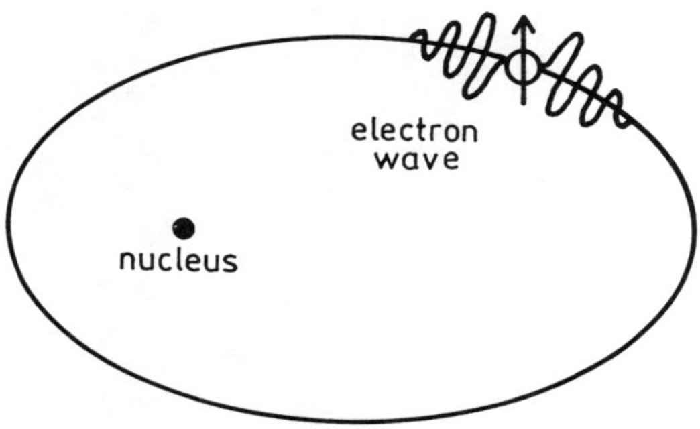

Vilas Y. Rajopadhye

P. A. M. Dirac

Chapter 9

Dirac and the Emergence of Quantum Mechanics

Vilas Y. Rajopadhye
Laxmi Park, Navi Peth,
Pune 411 030, India

Abstract

Discovering the non-commutativity of conjugate variables as the most significant feature of Heisenberg's work, Dirac provided a mathematical structure to quantum theory. He then used that formalism to derive Bohr's formula for the hydrogen spectrum and also to develop a quantum theory of radiation wherein he derived the A and B coefficients in terms of which Planck's black-body radiation formula had been obtained by Einstein.

Dirac's greatest achievement was the relativistic equation of the electron, eliminating thereby the ersatz nature of Goudsmit's concept of the electron spin. This equation was applied to predict the energy levels of the hydrogen atom in agreement with Sommerfeld's fine structure formula. This equation also led to the concept of the anti-particle.

1. Childhood and Schooling

Paul Adrien Maurice Dirac, who is recognized as one of the greatest physicists of the century, was born in Bristol on 8th August 1902. His father, Charles Dirac was a Swiss citizen, who revolted against his parents after graduating from the University of Geneva, left home and migrated to England in 1890, where he made his living by teaching the French language. Six years later, he moved to the Technical College in Bristol which later became the engineering faculty of the University of Bristol. There he met his future wife who was at the time working in the college library. Paul was his second son, and was registered at birth as Swiss, though his mother was English. He acquired British nationality later in 1919 when his father became a British citizen.

Dirac stated later that he had no social life as a child.[1] As he was more intelligent than his elder brother, his father wanted to educate him better, and consequently made a rule that Paul should talk to him only in French, because he thought that it would be good for him to learn French that way. But unable to express himself well in French, young Paul preferred to remain silent rather than speak in English and earn the displeasure of his father at the dinner table. Having been forced to remain silent may have been a traumatic experience that made him laconic.

Paul graduated in electrical engineering from the University of Bristol with first class honors in 1921, but he was not considered to be a promising engineer. Before graduation, as per the regulations of the University, he worked for a term with the British Thompson House, a well-known engineering firm at Rugby; but their report was not favorable. As a result he did not find an engineering job. That kind of work also did not hold his interest. To occupy his time, he was encouraged by a former professor to start some practical projects in electrical engineering. At the same time, his teachers in mathematics, who were unhappy that Dirac went over to engineering rather than taking mathematics, proposed that he should follow mathematics lectures unofficially, without paying any fees and covering the graduation course in two years. This was personally attractive to him and financially suitable, as he could continue to live at home. He accepted the offer and graduated with first class honors in mathematics in 1923. He also won the 1851 Exhibition Senior Studentship at Cambridge and started his research work at St. John's College.

In Cambridge, theoretical physics was taught by the mathematics department. Dirac desired to work on relativity and electromagnetism with Ebenezer Cunningham, who had earlier visited Bristol and so was known to him. However, Cunningham had by then decided not to take any more students and Ralph H. Fowler was appointed as Dirac's supervisor. Fowler, a first-rate, all-round theoretical physicist, was more interested in new developments of physics such as quantum theory, spectroscopy, statistical mechanics and their application to astrophysical problems. He was also in touch with the experimental work in physics at Cavendish Laboratory, not only because he was Ernest Rutherford's son-in-law, but also due to his own interest in the latest experimental results. When Dirac came to Cambridge, he was more interested in relativity and electrodynamics which he had studied in the mathematics course at Bristol. He had not studied any physics earlier, but as Fowler's research student, he had to learn modern theoretical ideas in physics through lectures and self-study, and he discovered that they were far from uninteresting. Earlier he knew nothing about atoms, which were then believed to exist only in the imagination of a few scientists. But he was surprised to see that the equations of electromagnetic theory were being used to study the atom. He worked hard to improve his knowledge quickly to the level of Cambridge graduates.

Within a year he mastered Sommerfeld's *Atomic Structure and Spectral Lines* as well as Whittaker's Analytical Dynamics, which taught him the methods of Hamiltonian dynamics and transformation theory. He also attended Arthur S. Eddington's lectures on special and general relativity and tensor analysis, and mastered Eddington's book, *The Mathematical Theory of Relativity.* Dirac used the opportunity to discuss relativity with Eddington, about whom he later said, "It was really a wonderful experience to meet the man who was the fountainhead of relativity so far as England was concerned."

Dirac also attended Cunningham's course on electromagnetic theory and relativity. One day, Cunningham worked out a long calculation on the electric and magnetic fields of the radiation, and said to the class, "This is an extraordinarily simple result in the end, but why?" A week later, Dirac, who had been in Cambridge only for about a year, gave a piece of paper on which the answer was written out.

During those days Dirac became acquainted with many classmates, who later became famous physicists. John C. Slater, eventually a professor at M.I.T., had come to Cambridge on a post-doctoral fellowship, and Georges E. Lemaître, a future cosmologist, also attended the same course. Lemaître was already a Ph.D. in mathematics but, after three years of scriptural studies, he became a priest and is thus now known as Abbe Lemaître. He had mastered the general theory of relativity before he went to Cambridge on a

Belgian Government scholarship to work with Eddington. It was many years later that they recalled their sitting in the same class. Llewellyn H. Thomas — who also sat through the courses of Fowler, Cunningham, and Eddington — remarked that Dirac was a quiet man who made no impression till he published his first paper on quantum mechanics. "He is a man of few words," said Thomas[2] in 1962. "If you ask him a question, he'd say, oh, that's very difficult. Then a week later he'd come back with the answer completely worked out."

Dirac's scientific perspectives gradually became wider. As he came in contact with people, he became less shy and less introverted. He also started attending weekly colloquia and tea parties at the home of Henry Baker, a professor of geometry, and even gave a seminar at one of them.

2. Dirac's Early Work

After about six month's stay in Cambridge, Dirac submitted his first paper[3] to the Cambridge Philosophical Society. The paper dealt with a problem in statistical mechanics which arose out of Fowler's course. With this work accomplished, Dirac became confident that he could make a significant contribution to the advancement of science. But he did not have any definite ideas about the area in which to concentrate his efforts. So, his early papers were written on diverse topics: relativity, old quantum theory and statistical mechanics. His ability to solve difficult theoretical problems became known even outside Cambridge. Fowler spoke about him to Charles G. Darwin, professor at Edinburgh, who in turn asked Dirac to solve a problem, which was worrying him for some time: Prove that in quantizing a dynamical system, the result is independent of the coordinate system.

Very early in his career Dirac developed a style characterized by conciseness. Conceptual clarity, directness, technical accuracy and logical presentation were the virtues he developed very quickly. Before writing a manuscript, he would draw up the whole work in front of his mind. Only then would he commit it to paper in his meticulous handwriting, the first draft needing hardly any change.

In his first paper, Dirac discussed the disassociation equilibrium of molecular hydrogen into its constituent atoms. Therein, he calculated the spatial shift in equilibrium when a temperature gradient is applied to the ends of a tube containing the gas. During the next fourteen months, he worked hard and published several papers in rapid succession on various topics, mainly arising from the courses he was taking. In his work he placed an emphasis on general principles, but it mostly consisted of improvements in the work of others.

Later on he decided to concentrate on quantum theory, and so kept in touch with the current literature to discover problems for his work. The Bohr–Sommerfeld quantization was then quite successful in handling systems with many degrees of freedom. In that area, the Hamilton–Jacobi theory played a very important role and Dirac had become very proficient in it. Bohr's correspondence principle and the adiabatic invariance principle of Ehrenfest were the bedrocks of most of the contemporary theoretical work.

In his first paper[4] in this area, Dirac extended and simplified Schrödinger's work on the relation between the Doppler principle and the Bohr formula, making use of a relativistic generalization. He then discovered[5] very simple conditions, consistent with the selection rules, under which the quantum integrals are adiabatically invariant. He also later studied[6] the effect on an atomic system of an adiabatically varying magnetic field, taking into account the small electrical field the process would generate. Through these types of practice problems, Dirac was trying to obtain a rigorous mathematical basis for the quantum theory.

Dirac's work from 1925 to 1928 brought him worldwide recognition. He wrote many papers during this period, the very first one[7] of which elevated his status from a student to a serious scientist. In order to understand the significance and impact of this work on the development of physics, it is necessary to recapitulate the basic ideas of classical physics and the facts concerning the successes and failures of the quantum theory, which is a twentieth century phenomenon.

3. Classical Mechanics and Electromagnetism

By the end of the nineteenth century, the laws of physics were regarded as well-established with respect to two main branches. The first main branch was classical mechanics. The laws of particle dynamics were set up by Newton in the seventeenth century, but were later recast into a more useful form by Hamilton. This form uses a function $H(q,p)$, called the Hamiltonian, which equals the total energy expressed in terms of two conjugate variables: the position coordinate q and the linear momentum p.

For a single particle in one-dimensional motion, the Hamiltonian equations take the form

$$\dot{q} = \frac{\partial H(q,p)}{\partial p}, \tag{1}$$

$$\dot{p} = -\frac{\partial H(q,p)}{\partial q} \tag{2}$$

Here and hereafter, the dot above a quantity signifies the derivative of that quantity with respect to time. The Hamiltonian for a single particle, of mass m_o and immersed in a potential $V(q)$, is given by

$$H(q,p) = \frac{p^2}{2m_o} + V(q).$$

Using (1) and (2) we obtain $\dot{q} = \frac{p}{m_o}$ and $\dot{p} = -\frac{dV}{dq}$. These two equations can be solved to obtain the trajectory of the particle in terms of the initial position and velocity.

In Hamilton's formalism the variables p and q stand on the same footing, in spite of their distinguishing names. This permits the possibility of their transformation into a new set of variables, which lead to new equations of motion that, however, retain the same form. This made it possible to develop new forms of mechanics, one of which is the Hamilton–Jacobi theory where use is made of action-angle variables. These are particularly useful in the study of problems involving periodic motion. The action variable $J = \oint p \cdot dq$ is of the nature of angular momentum. Its conjugate variable is an angle ω, which has the property of changing by unity in one complete period. The equation of motion in terms of J and ω is given by $\dot{\omega} = \frac{\partial H}{\partial J} = \nu$, where ν is the frequency of the system.

The second main branch of classical physics was electromagnetism. Its laws are described by Maxwell's equations in terms of the scalar potential ϕ and the vector potential A. The propagation of electromagnetic waves (of which visible light is an example) transmitting energy and momentum, was predicted by James C. Maxwell and experimentally confirmed by Heinrich Hertz.

For a system of a charged particle in the presence of an electromagnetic field, the potential energy V is modified to $V + e\phi$ and the momentum to $p - \frac{e}{c} A$, where e is the charge of the particle and c is the speed of light in vacuum. Then the Hamiltonian is given as

$$H = \frac{p^2}{2m_o} + V(q) + e\phi - \frac{e}{c}\,\dot{q} \cdot \mathbf{A} \tag{3}$$

$\dot{q}$ being the velocity of the particle. The last term on the right side of (3) is called the interaction energy, and is often expressed as $-\frac{e}{m_o c}\,\mathbf{p} \cdot \mathbf{A}$.

In the combined system of a charged particle and an electromagnetic field, energy transfers continually take place. When the energy of the field is transferred to the particle as its kinetic energy, the latter is accelerated. As a result, the particle generates electromagnetic radiation. Thus the particle's mechanical energy is converted into the energy of the electromagnetic field, and the cycle repeats itself.

At the beginning of the twentieth century, the equations of classical mechanics and electromagnetism had been confirmed by a large body of experimental observations and had resulted in several practical applications. In particular, they were also known to be applicable to free electrons that were discovered by Joseph J. Thomson in 1895 and for which he was awarded the Nobel Prize in 1906.

4. Quantum Theory

However, Maxwell's equations did not satisfactorily account for the exchange of energy between radiation and atoms. The frequency spectrum of radiation intensity from atoms was not consistent with the theoretical results. It was Max Planck who, in 1900, proposed the quantum hypothesis. The incorporation of the quantum hypothesis into the classical theory alone was found to produce the correct intensity-frequency curve for the black-body radiation.

According to Planck's hypothesis, a material oscillator of atomic dimensions gains or loses energy in discrete amounts of magnitude $h\nu$, where h is the Planck constant and ν is the frequency of the oscillator. The energy of the oscillator can then be written as $E = nh\nu$, where n is an integer. Often, it is more convenient to use the Planck–Dirac constant $\hbar = h/2\pi$ and the angular frequency $\omega = 2\pi\nu$. Thus,

$$E_n = n\hbar\omega \tag{4}$$

is the energy of the nth state, and the lowest energy of the oscillator is $\hbar\omega$. Suppose the oscillator's mass is m_o and the amplitude of oscillation in the lowest energy state is q_o. Then, $m_o q_o^2 \omega^2 = \hbar\omega$, which determines the amplitude of the lowest energy state.

Five years later, Albert Einstein showed by independent reasoning that the energy of radiation is also quantized. One quantum of radiation of frequency ν has energy

$$E = h\nu \tag{5}$$

and linear momentum

$$p = h/\lambda\,, \tag{6}$$

where λ is the wavelength. These equations explained the phenomenon of photo-electricity, but were in direct contradiction with classical electromagnetism. Another three years went by, when Einstein applied the concept of quantum of energy to vibrating atoms of a solid to explain the variation of specific heat with temperature.

The most important application of quantum theory was made by Niels Bohr for the explanation of the stability and the spectrum of the hydrogen atom. He used Rutherford's model of a nuclear atom. Thomson had shown that Rutherford's model would result in an unstable hydrogen atom: The electron would lose all its potential energy in a flash (within 10^{-11} s), by spiralling into the nucleus and thus be destroyed.

Bohr however worked out many results from his model, all of which were experimentally confirmed. For the purpose of answering the objections of Thomson, he made an *ad hoc* assumption about stationary orbits, which are characterized by the condition that angular momentum of the electron must be an integral multiple of $\hbar$. Thus for the lowest energy state or the ground state,

$$m_o \omega^2 r_o = \hbar$$

and the angular momentum is thus quantized. Using this condition, and assuming the hydrogen nucleus to apply a Coulombic potential on the orbiting electron, Bohr found the radius of the smallest orbit to be

$$r_o = \frac{\hbar^2}{m_o e^2} \, .$$

This came to be of the correct magnitude, as confirmed from several experiments. Bohr went on to explain the hydrogen spectrum correctly and calculate the so-called Rydberg constant in terms of e, h and c. He also predicted the variation of the Rydberg constant with the nuclear mass in one-electron systems.

Subsequent efforts to explain the spectra of other atoms using Bohr's model met with only partial success. Progress in quantum theory consisted of the discovery of various quantum rules and selection principles, coupled with *ad hoc* assumptions, theorems and computational devices. The most ingenious architects of these extensions were Bohr in Copenhagen and Arnold Sommerfeld in Munich, of whom accounts are given in the earlier chapters of this book.

The search for a sound theoretical basis for all the rules of the quantization of energy and angular momentum went on for a long time, but without success. The first breakthrough was provided by Werner Heisenberg, who was a student of Sommerfeld in Munich. He analyzed the spectroscopic data of alkali atoms and their Zeeman spectra, and arrived at the conclusion that a proper explanation required the inclusion of half-integer quantum numbers. Wolfgang Pauli, a close friend, ridiculed the idea[8] saying that "Today you need half quantum numbers; tomorrow you would advocate half of these, and eventually land up into a continuum."

5. Heisenberg's Approach

Heisenberg was a contemporary of Dirac, senior by only a few months. After getting a Ph.D. from Munich, he became a research assistant to Max Born in Göttingen but later moved to Copenhagen to work with Bohr. In collaboration with Hendrik A. Kramers, Heisenberg developed the quantum theory of optical dispersion.

Heisenberg sought a proper relationship between the Bohr formula and the interaction term of the Hamiltonian in (3). He wanted to do away with the concept of orbits and other non-measurable parameters, and wanted to construct a theoretical framework based only on observables such as frequencies and the intensities of the spectral lines.

Heisenberg started with the Fourier expansion of the coordinate $x(t)$ as the sum

$$q(t) = \sum_k q_k e^{ik\omega t}$$

where k are integers, q_k are the Fourier coefficients, ω is the circular frequency, t denotes time, $i = \sqrt{-1}$, and $e = 2.71828...$ is the base of natural logarithms. He realized that the frequencies of the spectral lines are not the fundamental frequency ν and its harmonics. The circular frequencies that are actually observed are given by the Bohr formula

$$\nu_{nm} = \frac{E_n - E_m}{\hbar}. \quad (7)$$

The two indices for ω emphasize its origin in a transition between two states: the mth state and the nth state. Equation (7) expressed in terms of ω rather than ν implies $\omega_{nm} = -\omega_{mn}$ and $\omega_{nn} = 0$.

A method of finding out the corresponding $x(t)$ from the Fourier coefficients x_k was worked out by Heisenberg using Bohr's correspondence principle. After some algebraic manipulations, he discovered the product rule of his two-index symbols to be of the following type:

$$(qp)_{mn} = \sum_k q_{mk} p_{kn}. \quad (8)$$

This, in general, implies that qp and pq are not equal; in other words, the product of conjugate quantum variables is not commutative.

This lack of commutativity was a source of worry to Heisenberg, and he felt a little unsure of his theory. All the same, he went ahead, and proceeded to obtain[9] the energy spectrum of a harmonic oscillator as

$$E_n = (n + \tfrac{1}{2})\hbar\omega.$$

This replaced Planck's equation (4), while remaining fully in agreement with the data on molecular spectra. *En route*, he established one very important result for the product difference of conjugate variables q and p as

$$(qp - pq)_{mm} = i\hbar\,, \tag{9}$$

the so-called commutator of Heisenberg.

Heisenberg went to discuss his results with Max Born who recalled that the multiplication rule (8) is the same as that for matrices. This meant that physical variables are not algebraic quantities but may be considered as matrices. Born then looked for a research assistant to help him with matrix manipulations and recruited Pascual Jordan who was already in Göttingen. The collaboration among Born, Jordan and Heisenberg resulted in the famous *three-men paper*,[10] which laid down the foundations of quantum mechanics by matrix formulation. That came to be known as Matrix Mechanics.

Heisenberg visited Cambridge in July 1925 and gave a talk entitled, "Term zoology and Zeeman botany". In this talk, he discussed problems in spectroscopy in terms of the prevailing Bohr–Sommerfeld theory and its limitations. He discussed his theoretical scheme later with Fowler, who showed keen interest and asked for a copy of the paper when published. Heisenberg responded by sending the proof sheets of his paper the following month. These were passed on to Dirac for a closer study. The stage was set.

6. Dirac's Breakthrough

Dirac was then more concerned with the application of Hamiltonian mechanics, as well as its extension to action and angle variables for the problems left unsolved in the Bohr–Sommerfeld theory. Heisenberg's work did not at once arouse his interest. But when a couple of weeks later he read the paper again, he was struck with the novelty of the result regarding the non-commutativity of Heisenberg's variables. Pondering deeply over this during a long walk on a Sunday afternoon, he felt that there was something similar in classical mechanics. But having no suitable book at home to consult, he spent a sleepless night thinking about the problem. Next morning he discovered in the library what he was looking for. It was a discussion on Poisson brackets used more than a hundred years earlier by the French polymath Simeon D. Poisson in connection with perturbation problems in celestial mechanics.

The mathematical treatment devised by Heisenberg has been discussed in great detail in the chapter on Heisenberg. Let us, however, go over a few salient points here. Recall that p and q are not simply scalars. The Poisson bracket of two variables u and v is defined in the case of a single-particle dynamics as

$$\{u, v\} = \sum_i \left(\frac{\partial u}{\partial q_i} \cdot \frac{\partial v}{\partial p_i} - \frac{\partial u}{\partial p_i} \cdot \frac{\partial v}{\partial q_i} \right) . \tag{10}$$

Here, q and p are the coordinate and the momentum, but any pair of conjugate variables could be used as well. Some simple identities follow from this definition. The first,

$$\{u, v\} = -\{v, u\} \tag{11}$$

shows that the Poisson bracket is non-commutative. The second is the Jacobian identity,

$$\{u, \{v, w\}\} + \{v, \{w, u\}\} + \{w, \{u, v\}\} = 0 ; \tag{12}$$

and two other useful relations are as follows:

$$\{u, v_1 + v_2\} = \{u, v_1\} + \{u, v_2\}, \tag{13}$$

$$\{u, v_1 v_2\} = \{u, v_1\} v_2 + v_1 \{u, v_2\} . \tag{14}$$

The quantity $uv - vu$ is the commutator of two variables u and v, and it is usually written as $[u, v]$. The commutator also satisfies the relations (11) to (14). As $[u, v]$ and $\{u, v\}$ are proportional to each other, Dirac could set

$$[u, v] \leftrightarrow i\hbar \{u, v\} . \tag{15}$$

The symbol $\leftrightarrow$ implies a one-to-one correspondence between the commutator of quantum variables on the left side and the Poisson bracket on the right-hand side. The constant h is introduced solely from dimensional considerations and the factor 2π to assure agreement with experiment.

Dirac actually obtained the resulting Eq. (15) from a different consideration. Starting with Heisenberg's product rule (8), he looked for the most general form of the operator d/dx which satisfies the usual rules of differentiation

$$\frac{d}{dx}(u + v) = \frac{d}{dx} u + \frac{d}{dx} v ,$$

$$\frac{d}{dx}(uv) = \left(\frac{d}{dx} u \right) v + u \left(\frac{d}{dx} v \right) .$$

After some algebraic manipulations he found

$$\left(\frac{du}{dx}\right)_{nm} = \sum_k (u_{nk}a_{km} - a_{nk}u_{km});$$

hence,

$$\frac{du}{dx} = ua - au\,,$$

where a is some other quantum variable.

Then Dirac looked for a physical quantity in classical theory which corresponds to $uv - vu$. From the Fourier expansions of u and v in terms of the action-angle variables and the use of the correspondence principle, he found that it corresponds to

$$i\hbar\left(\frac{\partial u}{\partial \omega}\frac{\partial v}{\partial J} - \frac{\partial u}{\partial J}\frac{\partial v}{\partial \omega}\right).$$

Since the Poisson bracket is invariant to the use of any pair of canonically conjugate variables, this quantity can also be written in terms of p and q as

$$i\hbar\left(\frac{\partial u}{\partial q}\frac{\partial v}{\partial p} - \frac{\partial u}{\partial p}\frac{\partial v}{\partial q}\right).$$

The expression in the brackets is the Poisson bracket; hence,

$$uv - vu \leftrightarrow i\hbar\{u, v\}\,.$$

He then considered an arbitrary function $F(q, p)$ which does not explicitly depend on time; then,

$$\dot{F} = \frac{\partial F}{\partial q}\dot{q} + \frac{\partial F}{\partial p}\dot{p}\,.$$

Using (1) and (2) now, it follows that

$$\dot{F} = \frac{\partial F}{\partial q}\frac{\partial H}{\partial p} - \frac{\partial F}{\partial p}\frac{\partial H}{\partial q}\,,$$

whence,

$$\dot{F} = \{F, H\}\,. \tag{16}$$

Using q and p successively in place of F in (16), we get

$$\dot{q} = \{q, H\}\,, \tag{17}$$

$$\dot{p} = \{p, H\}\,. \tag{18}$$

From the definition of the Poisson bracket we obtain $\{q, p\} = 1$; hence, the Heisenberg commutator $[q, p] = i\hbar$ follows.

From this quantum rule, we can get the following commutation relations for a system of many degrees of freedom:

$$\begin{aligned}[q_i, p_i] &= i\hbar\,,\\ [q_i, p_j] &= 0, \quad i \neq j\,,\\ [q_i, q_j] &= 0\,,\\ [p_i, p_j] &= 0\,.\end{aligned}$$

These were independently deduced almost at the same time by Pauli, as well as by Weyl, but were not published because Born and Jordan[11] had included them in an earlier paper.

Furthermore, if F in (16) is identified with H, we get

$$\dot{H} = \{H, H\} = 0\,. \tag{19}$$

Thus, H is a constant of motion and is represented by a diagonal matrix.

In place of the Fourier expansion

$$q = \sum_k q_k e^{ik\omega t}\,,$$

Born, Jordan and Heisenberg[10] had set x to be a matrix having elements of the type $x_{mn}\exp(i\omega_{mn}t)$. Using Eq. (16), Dirac obtained

$$i\hbar\dot{q}_{mn} = i\hbar(i\omega_{mn})q_{mn} = (qH - Hq)_{mn}\,,$$

i.e.

$$\begin{aligned}-\hbar\omega_{mn}q_{mn} &= \sum_k (q_{mk}H_{kn} - H_{mk}q_{kn})\\ &= q_{mn}E_n\delta_{nn} - E_m\delta_{mm}q_{mn} = q_{mn}(E_n - E_m)\,,\end{aligned}$$

where $\delta_{mn} = 0$ for $n \neq m$, but $\delta_{mm} = 1$. Thus, $\hbar\omega_{mn} = E_m - E_n$, which is the Bohr formula (7).

Starting from classical mechanics, Dirac was thus able to obtain the Heisenberg commutator (9) as well as the Bohr formula (7). He also derived all the results of Born and Jordan. By showing that the quantum mechanical commutation relations can be obtained from the classical Poisson bracket, Dirac provided a classical basis for quantum theory. In this regard, his work is often compared with that of Newton, who — by his theory of gravitation — provided a theoretical basis for Kepler's laws. Fowler was greatly satisfied with these results, and so he arranged for their quick publication.[7]

With his paper Dirac established himself as a quantum theorist of the first rank. He should be considered extremely lucky to be able to get his paper quickly published, in view of the extreme efficiency displayed those days by the German scientific juggernaut. In contrast, US scientists faced the problem of slow communication with the scientists in Europe. Slater,[12] who had worked earlier in Cambridge as well as in Göttingen and had settled in U.S.A., had worked out all the results of Dirac, particularly, the interpretation of $[p, q]$ in terms of the Poisson bracket before Dirac's paper was published. One can imagine his frustration when he saw Dirac's work in print just when he was on the point of dispatching his paper for publication.

Kramers[12] in Holland, on the other hand had also noticed the identity of behavior between the Poisson bracket and the Heisenberg commutator, but did not realize it to be of any special significance. He only convinced himself that every quantum mechanical problem always had a classical counterpart.

Within a few weeks Dirac published another paper,[13] in which he introduced the concept of q- (non-commutative) and c- (commutative) numbers, and established the algebraic laws governing them. He also drew further conclusions from the Poisson bracket for more general commutators and defined conditions under which a set of variables is canonical. He obtained the rules for operations with q-numbers and applied them to multiply periodic systems in close agreement with the rules of quantum theory.

He applied this method to the problem of hydrogen atom, and obtained

$$\omega = \frac{\pi m_o e^4}{h}\left(\frac{1}{p^2} - \frac{1}{(P+n\hbar)^2}\right) \tag{20}$$

for the frequencies of spectral lines, n being an integer and m_o the electronic mass. Dirac did not fully work out all the steps because Pauli had only a few weeks earlier solved the hydrogen atom problem using matrix mechanics. Dirac noticed that if the action variable P is an integral multiple of $\hbar$, (20) is in agreement with Bohr's result; however, Dirac could not prove $P/\hbar$ to be integral.

Gregor Wentzel, a young physicist at Munich, also worked independently on the hydrogen spectrum. The special feature of his work was the relativistic treatment. Schrödinger[14] commented to Hendrik Kramers that Dirac and Wentzel calculated pages and pages on the hydrogen spectrum but neither of them categorically stated whether the quantization is in terms of integers or half integers.

Dirac continued to study the role of dynamic variables in quantum theory, and as a result submitted yet another paper[15] within two months. There he discussed the question of the independent canonical variables for an atomic system with more than one electron in a central field problem. Dirac worked out this problem in close analogy with the corresponding classical one, obtaining the quantum variables which satisfy the same algebraic relations. In the process, he finally replaced the classical Poisson bracket by the quantum bracket. In this way Dirac developed the theory of angular momentum and studied the various features of the spectral lines when the source is immersed in a magnetic field.

In a paper presented to the Cambridge Philosophical Society in July 1926, Dirac[16] summarized the properties of q-numbers without realizing that their algebra was exactly equivalent to matrix algebra. His approach appeared to him to be distinctly different than either that of Heisenberg or the matrix method of Born and Jordan. The physicists of that time knew little about matrices or the generalized concept of linear operators. Dirac simply followed his own mathematical instincts. He had started with the action angle variable theory but got nowhere and realized that a new route had to be discovered: q-numbers and their algebra were born.

From the sequence of his papers, it is clear that Dirac worked very hard and with great concentration. During this period he wrote a thesis for which the Cambridge University soon awarded him the Ph.D. He gave a lecture on his new theory at the Kapitza Club in Cambridge and also a regular full graduate course along with those on spectroscopy and quantum theory by Fowler.

7. Schrödinger's Wave Mechanics

The mathematics which Dirac was using was, by and large, unfamiliar to most of the physicists, while the matrix mechanics of Born and Heisenberg was too abstract and unphysical. Both were overshadowed by the developments in which Louis de Broglie and Erwin Schrödinger made valuable contributions.

In 1923, de Broglie[17] suggested that Einstein relations, $E = h\nu$ and $p = h/\lambda$, which determined particle properties of light, could be read backwards to give the wave properties, ν and λ, of a material particle such as an electron. Einstein was highly impressed with this idea of which he himself gave an exposition at the Solvay Congress of 1924.

Schrödinger became the professor of theoretical physics at the University of Zurich in 1921. His seminal contributions to the hydrogen atom modeling have been described by Owen de Lange in the previous chapter, so we shall only take a brief look here.

In a seminar, Schrödinger explained de Broglie's ideas and the interpretation of the Bohr's quantum condition on angular momentum: $pr = nh/2\pi$. This equation may be written as $2\pi r = nh/p$, i.e., $2\pi r = n\lambda$, where λ must be the wavelength of the electron. This implies that the electron forms a stationary wave along its orbit comprising of n nodes.

Schrödinger[18] then introduced a wave function $\psi(x)$ to describe the electron wave and wrote a differential equation for it by replacing p in the Hamiltonian with $-i\hbar d/dx$, so that the well-known Schrödinger equation

$$H(x, -i\hbar d/dx)\psi(x) = E\psi(x) \tag{21}$$

was obtained. Its solution is equivalent to a determination of $\psi(x)$ and the eigenvalue E, which is nothing but the energy of a certain state. Thus, quantization is effectively the result of solving an eigenvalue problem. In just three pages, Schrödinger solved the equation for the hydrogen atom, for which he got the same energy values as in the Bohr formula.

Schrödinger's work was immediately and enthusiastically accepted by the physics community. Physicists were familiar with differential equations and methods of obtaining their solutions. Moreover, Schrödinger's calculations could be easily interpreted in terms of classical concepts; so Planck, Einstein and Max von Laue were very happy[19] that Schrödinger had used continuous functions. There was no need to get familiar with the strange matrix mechanics.

Shortly thereafter, Schrödinger[20] calculated the energy spectrum of the harmonic oscillator and obtained results in close agreement with those of Heisenberg. In the beginning, Schrödinger could not see the connection with Heisenberg's method but felt confident that it contained an element of truth. Within a short time he established[21] the equivalence between the two.

Schrödinger showed that the Heisenberg commutator (9) follows from the choice of the representation of p as $-i\hbar d/dx$. Then he showed that for any observable function F of x and p, it is possible to construct the matrices

$$F_{mn} = \int \psi_m^* F(x, -i\hbar d/dx) \psi_n dx \,, \tag{22}$$

where the asterisk denotes the complex conjugate. When F is set equal to x, p or H, this representation is identical with that of Heisenberg.

Schrödinger also postulated[22] the equation of motion as

$$H\psi(r,t) = i\hbar \frac{\partial}{\partial t} \psi(r,t); \quad r \equiv (x,y,z) \,,$$

whose solution corresponding to the nth energy state must be

$$\psi_n(r,t) = \Psi_n(r) e^{-iE_n t/\hbar} \,. \tag{23}$$

The interpretation of the Ψ_n as probability amplitudes was shortly given by Born.

8. Fermi–Dirac Statistics

All this happened while Dirac was busy completing the program that started with linking the classical Poisson bracket with the commutator. After submitting his Ph.D. thesis, Dirac commenced a study of Schrödinger's work and applied that formalism to investigate the statistical distribution of indistinguishable particles into different energy states.

A two-electron system in which one electron is in state m, and the other in n, may be denoted by $\Psi_m(1)\Psi_n(2)$. One can also think of a state when the electrons are interchanged, and this state is denoted by $\Psi_m(2)\Psi_n(1)$. On the basis of indistinguishability, Heisenberg[23] suggested that these are not to be counted as separate states but as one and the same. Dirac[24] then proposed that the two-particle wave function should be described as

$$\Psi_{mn} = A\Psi_m(1)\Psi_n(2) + B\Psi_m(2)\Psi_n(1)\,.$$

When $A = B$, Ψ_{mn} is symmetric with respect to an interchange of 1 and 2; with $A = -B$, it is antisymmetric. But electrons obey the Pauli exclusion principle, so Ψ_{mn} must be antisymmetric.

Dirac applied this idea to a gas of free particles (electrons) occupying a volume V that can be described by the antisymmetric wave function, and obtained the relation between the number of electrons dN_s having energy in the range E_s to $E_s + dE_s$ as

$$dN_s = \frac{AdE_s}{\exp\dfrac{(\alpha + E_s)}{k_B T} + 1}\,, \tag{24}$$

where k_B is the Boltzmann constant, α is independent of E_s, T is the absolute temperature, and

$$A = \frac{2\pi V}{h^3}\,(2m_o^3 E)^{1/2}\,. \tag{25}$$

For particles with symmetric wave functions, the $+1$ in the denominator of (24) is replaced by -1.

The foregoing results were also published by Enrico Fermi[25] in March 1926. A shorter version of Fermi's paper was published two months earlier in an Italian journal. Dirac had read the March paper, but had forgotten about it as it did not appear to have any relevance with what he was then trying to complete. He discovered the result (24) independently, and linked it up with statistics, which aspect was not included in Fermi's work. Equation (24) is now known as the Fermi–Dirac distribution.

Dirac also developed the time-dependent perturbation theory independently of Schrödinger in this paper.[24] He applied it to establish the equality of the Einstein coefficients of absorption and stimulated emission. There was however no explanation of the spontaneous emission when no external influence was present. He noted that "one cannot take spontaneous emission into account without a more elaborate theory", which he produced in less than six months.

In a previous paper Dirac[15] had made an attempt to explain the Compton effect by introducing $-E$ and t as conjugate variables as could be deduced from the relations

$$p = -i\hbar \frac{\partial}{\partial x}, \quad E = i\hbar \frac{\partial}{\partial t}.$$

In this way he hoped to introduce relativity into quantum theory. But with the availability of the Schrödinger formalism, he used it successfully to derive an expression for the frequency and intensity of the Compton-scattered radiation.[26]

On the advice of Fowler, Dirac spent the next academic year partly with Bohr in Copenhagen, and partly in Göttingen with Born. After the probability interpretation of the wave function by Born in June 1926, a pressing need was felt for the conceptual clarification of quantum mechanics. The question was: how to relate the probabilistic interpretation with Born's matrix formulation. The essential step in this direction was taken by Dirac,[27] in the form of a transformation theory that established the one-to-one correspondence between the approaches of Schrödinger and Heisenberg.

In canonical transformations of classical mechanics, the variables p and q are changed into P and Q with the help of a generating function S. If S is chosen to be a function of old coordinates q and new momenta P, the other variables are found to be

$$p = \frac{\partial S}{\partial q}, \qquad Q = \frac{\partial S}{\partial P}.$$

In matrix mechanics the canonical transformations which leave the commutation relations invariant are written as

$$P = SpS^{-1}, \qquad Q = SqS^{-1}$$

S being the transformation matrix and S^{-1} being its inverse.

Dirac generalized the theory of matrix representation by introducing the delta function $\delta(x-a)$ which is zero for all x except at $x = a$, and which satisfies

$$\int_{-\infty}^{\infty} \delta(x-a)dx = 1.$$

Hence a specific number can be thought of as the value $f(a)$ of a function $f(x)$ at $x = a$, with

$$f(a) = \int_{-\infty}^{\infty} f(x)\delta(x-a)dx\,.$$

Dirac was thus able to formulate matrices with continuous indices — instead of discrete indices — and write the transformation equation for a continuous spectra of variables. He did warn that $\delta(x)$ is not a proper function of x but can be regarded as the limit of a certain sequence of functions. A familiar example of this function is the charge density of the electron.

The most useful transformation in matrix mechanics is for the diagonalization of the Hamiltonian in problems involving perturbations. This can be expressed as

$$S^{-1}HS = E\,.$$

With E as a diagonal matrix, this is equivalent to

$$H(q,\ -i\hbar\,\partial/\partial q)S = ES\,.$$

Now, S can be looked upon as a transformation matrix that diagonalizes H as well as an eigenfunction with the specific eigenvalue E. This enabled Dirac to derive the Schrödinger equation from matrix mechanics. By doing so, he provided a proof of the identity of the Schrödinger and the Heisenberg formulations of quantum mechanics.

Thus by the end of 1926, there were three formulations of quantum mechanics, which had been developed almost simultaneously. In these apparently different viewpoints of Heisenberg, Dirac and Schrödinger, there are several common features, as explained by Cornelius Lanczos.[28] In all of them the phase space of classical mechanics is replaced by the so-called Hilbert space of quadratically integrable functions.

The Schrödinger equation gives a direct representation of the mechanical system, with the help of a state function that changes its position in Hilbert space according to a definite law. The eigenstates of the Schrödinger approach constitute a basic orthogonal reference frame in the Hilbert space, in which the state function and its time variation can be analyzed.

The representation in terms of matrices by Born, Heisenberg and Jordan gives another formalism. The matrix elements are defined as the scalar products which can be calculated in terms of orthogonal base vectors in the Hilbert space. Instead of solving the Schrödinger equation, one solves the matrix equations, which are obtained by re-interpreting Hamiltonian equations in terms of infinite-dimensional matrices.

Dirac's method of using non-commutative algebra can be thought of operating, in the infinite-dimensional Hilbert space, directly with vectors in that space without setting up any orthogonal base system.

But there was still another element of fundamental significance in these formal correspondences which was brought out by Eugene Wigner.[29,30] At that time, Wigner was a research assistant at Göttingen. He had acquired a mastery over group theory, which had been developed about 30 years earlier in Göttingen by Sophus Lie. Wigner realized that the revision of classical mechanics by Schrödinger, Dirac and Heisenberg can also be formulated in terms of group theory. Similarity transformation is one of the fundamental operations in group theory, the commutator arising from restricting this transformation to its infinitesimal analog. The definition of the Poisson bracket has a similar group theoretical counterpart. The Jacobian identity (12) is a must for the existence of a group. A deeper understanding of the Schrödinger theory thus became possible with the addition of the group theoretical viewpoint.

9. Quantum Electrodynamics

Dirac's next contribution[31] was outstanding because therein he was able to solve the longstanding unsolved problem of the quantum theory of radiation. This paper was communicated from Copenhagen to the Royal Society of London by Bohr. With its publication, Dirac provided the foundations of quantum electrodynamics.

In his earlier attempts he had neither been able to explain spontaneous emission nor account for the reaction of the atom, namely the disappearance of the photon. He was able to overcome these deficiencies in this paper. Instead of taking recourse to the correspondence principle, Dirac gave a complete explanation in terms of the quantized vector potential.

In electrodynamics, forces are propagated with a finite speed (*viz*, the speed of light) and not instantaneously as in particle mechanics. Hence, to be consistent with the relativity principle, it would not be appropriate to use the Hamiltonian formalism. However, Dirac found that it possible to use it along with an approximate non-relativistic theory. In his theory, he could consider time as a c-number and permit the variation of mass with velocity.

He considered the atom and the radiation to be enclosed in a cavity so as to have a system with a limited number of degrees of freedom. He then resolved the radiation into its Fourier components. In the absence of interaction, the Hamiltonian of the system can be set down as

$$H = \sum_r E_r + H_o \,,$$

where H_o is the Hamiltonian of the atom. Dirac then assumed that the energy E_r and phase θ_r of the rth component of the radiation field to be conjugate variables, analogous to the momentum and the coordinate, respectively and wrote the canonical equations of motion as

$$\dot{E}_r = -\frac{\partial H}{\partial \theta_r} = 0 \,, \qquad \dot{\theta}_r = \frac{\partial H}{\partial E_r} = 1 \,.$$

Then using (16), he then obtained $\dot{\theta}_r = \{\theta_r, E_r\}$ and also wrote the commutator relation $\theta_r E_r - E_r \theta_r = i\hbar$ by treating θ_r and E_r as q numbers. Taking ν_r as the frequency of the rth component, Dirac went on to show that the ratio $E_r/2\pi\nu_r$ assumes only discrete values differing by $\hbar$, so that

$$E_r = (2\pi\nu_r)n\hbar = nh\nu \,.$$

This procedure endowed particle properties to the radiation.

The Hamiltonian of the interaction of an atom with an assembly of particles moving with the velocity of light and obeying the Bose–Einstein statistics was known then. Dirac proved that the same Hamiltonian describes the atom-radiation interaction he was studying. In this way he showed that there is a perfect harmony between the wave and particle description of the interaction. The method of passing from a system of a single particle and radiation to an assembly of particles is now known as the second quantization.

Dirac pointed out that there is a difference between a light wave and the de Broglie wave associated with a light quantum. The former is real, whereas the latter involves the presence of $i = \sqrt{-1}$. The number of light quanta per unit volume is the energy of the wave divided by $h\nu$. But, a de Broglie wave of amplitude A (multiplied by an exponential imaginary factor) must be interpreted as representing A^2 light quanta per unit volume for all frequencies.

On treating the interaction as a perturbation to the original Hamiltonian, its effect on the atomic transition could be calculated by using the standard quantum mechanical perturbation theory. This gave the finishing touch to the radiation theory that began with Planck's hypothesis and his black body radiation formula. There had been several attempts to derive this formula from first principles. The most satisfactory attempt was by Einstein, who used the concept of the atom continuously emitting and absorbing radiation and thus remaining in equilibrium with it. This derivation of the Planck formula contained the transition probability of spontaneous emission as well as the transition probabilities of induced emission and absorption. These were now derived by Dirac in his radiation theory.

Dirac spent the next few months at Göttingen where the bias was entirely mathematical. There he attended some lecture courses on topics with which he was unacquainted with e.g. Hermann Weyl's course on group theory. During this period he made close acquaintance with John R. Oppenheimer whom he had earlier met in Cambridge.

Oppenheimer and Dirac were staying in the same boarding house in Göttingen. They became very friendly, and used to go for long walks. Oppenheimer was versatile, interested in art, poetry, music as well as physics. Dirac once asked him, "How can you do physics and poetry at the same time? The aim of science is to make difficult things understandable in a simpler way; the aim of poetry is to state simple things in an incomprehensible way. The two are incompatible."

From Göttingen, Dirac published two papers. The first one was on the quantum theory of dispersion.[32] He used the formalism developed in the previous paper and produced a result in agreement with Kramers and Heisenberg. The other paper[33] was on the quantum-mechanical scattering processes. Here Dirac formulated the Schrödinger wave equation in momentum space. He applied this method to absorption processes as well.

Dirac also was invited to attend the fifth Solvay Congress in Brussels in October 1927 which was to discuss the physics of electrons and photons. Also invited were all the stalwarts of old and new quantum theories: Planck, Einstein, Bohr, Ehrenfest, Debye, Born, Heisenberg, Kramers, de Broglie, Pauli and Schrödinger. At the meeting Dirac presented a discussion on the action-angle variables for a link between classical and quantum theories. He also gave a discourse on the essential difference between classical and quantum descriptions of a physical process.

On his return to England, Dirac was elected a fellow of St. John's College. This settled the problem of how to continue with his scientific activity.

10. Quantum Theory of the Electron

Dirac's work on the quantum theory of the electron[34] proved to be his most outstanding contribution. It became a landmark in the history of physics. Most of his earlier results had also been produced by others, in some instances a little before Dirac. So he was anxious to produce something unique.

Dirac had a great fascination for the theory of relativity even before he came to Cambridge. After he saw Heisenberg's paper on quantum mechanics, he tried to make it Lorentz-invariant, but did not succeed. Instead he found, as mentioned earlier, a basis for the Heisenberg commutator from the laws of classical mechanics, thereby placing the quantum theory on a sound mathematical foundation. Finally in 1928, Dirac succeeded in formulating a relativistic wave equation.

Oscar Klein[35] and Walter Gordon[36] independently produced a relativistic wave equation, using the relativistic energy momentum relation and quantum mechanical operators for E and ψ.

$$\left(\nabla^2 - \frac{1}{c^2}\frac{\partial^2}{\partial t^2} - \frac{m_o^2 c^2}{\hbar^2}\right)\psi = 0\,, \tag{26}$$

but Gordon used it to study the Compton effect. This equation is now known as the Klein–Gordon equation. Since it is a differential equation of the second order in time, Dirac noted that it cannot give $\psi(t)$ from the knowledge of $\psi(0)$. So he felt that one must begin with the Schrödinger equation,

$$(H - i\hbar\partial/\partial t)\psi = 0\,, \tag{27}$$

and make it consistent with relativity by using the relativistic energy-momentum relation and using quantum operators for E and p.

George Uhlenbeck and Samuel Goudsmit had proposed in 1925 the electron to possess a spin angular momentum. Since then it was widely believed that spin and relativity were intimately related but it had not been possible to incorporate spin in the Klein–Gordon equation. The nature of the relations between spin, relativity and quantum mechanics was still obscure.

Pauli visited Bohr's institute in January 1927, when both Klein and Dirac were there. Pauli produced a quantum-mechanical spin theory in May 1927 based on the idea of representing spin by three 2×2 matrices — the so-called Pauli spin matrices (σ_1, σ_2, σ_3) — which were related to the electronic spin S via the relation

$$S_r = \frac{1}{2}\hbar\sigma_r, \quad 1 \leq r \leq 3\,.$$

With this, the Pauli wave equation became

$$i\hbar\frac{\partial\psi}{\partial t} = H(x, -i\hbar\partial/\partial x, \sigma)\psi\,. \tag{28}$$

Dirac had also thought about representing spin by the σ matrices, but he did not pursue the idea further. Darwin had obtained an equivalent spin theory using the language of wave mechanics. But neither attempt could integrate relativity with quantum mechanics.

Relativity effects had been considered by some authors as first-order corrections to the non-relativistic Hamiltonian. Dirac did not find these attempts satisfactory. According to him, a correct theory must be relativistically invariant right from the beginning. So he toyed with the idea of producing a relativistic wave equation for a spinless particle first and incorporating spin later, thereby making it applicable to an electron.

During the 1927 Solvay Congress, Dirac mentioned his anxiety to Bohr, who answered that Klein had already solved the problem.

For want of time, Dirac could not then discuss why it was unsatisfactory. Soon after his return to Cambridge, he obtained his now famous equation using the two basic principles that:

(a) the spacetime properties of the equation should transform according to the theory of relativity, and
(b) the quantum properties should transform according to laws of transformation of quantum mechanics.

The Klein–Gordon equation violated the second requirement. Moreover, it did not rule out negative probabilities. For a free electron, Dirac therefore started from (27) and using relativistic energy momentum relation wrote

$$i\hbar\,\frac{\partial\psi(\underline{r},t)}{\partial t} = \pm c\sqrt{(p_1^2+p_2^2+p_3^2+m_o^2c^2}\;\psi(\underline{r},t)\,, \tag{29}$$

where m_o is the rest-mass of the electron, $p_1 = -i\hbar\frac{\partial}{\partial x}$, $p_2 = -i\hbar\frac{\partial}{\partial y}$, $p_3 = -i\hbar\frac{\partial}{\partial z}$, and $\underline{r} = (x,y,z)$.

Dirac's next problem was to somehow linearize the square root on the right side of (29). He took a clue from Pauli's work where he had set

$$\sqrt{p_1^2+p_2^2+p_3^2} = \sigma\cdot p\,,$$

and obtained the wave equation for a two-component ψ as

$$i\hbar\,\frac{\partial\psi}{\partial t} = c\sigma\cdot p\psi\,.$$

This equation is Lorentz-invariant, contains spin 1/2 and is of the first order with respect to time. But it does not contain a mass term so it can apply to a massless particle only. An electron has mass, so Dirac assumed

$$\begin{aligned}\sqrt{p_1^2+p_2^2+p_3^2+m_o^2c^2} &= \alpha_1p_1+\alpha_2p_2+\alpha_3p_3+\alpha_4m_oc\\ &= \alpha\cdot p+\alpha_4m_oc\,,\end{aligned}$$

which transforms (29) to

$$\left(\frac{i\hbar}{c}\frac{\partial}{\partial t} - \alpha\cdot p - \alpha_4m_oc\right)\psi = 0\,, \tag{30}$$

with the + sign on the right-hand side of (29) chosen. Here the α_k (k = 1, 2, 3, 4) are independent of p and $\underline{r}$. Since the square of the operator on the left side of (30) must reproduce the Klein–Gordon equation, the following conditions must hold:

$$\alpha_1^2 = \alpha_2^2 = \alpha_3^2 = \alpha_4^2 = 1\,,$$
$$\alpha_r\alpha_s + \alpha_s\alpha_r = 0,\ 1 \le r \le 4,\ 1 \le s \le 4\,.$$

If $(\alpha_1,\ \alpha_2,\ \alpha_3)$ are taken to be identical with $(\sigma_1,\ \sigma_2,\ \sigma_3)$, what could be the representation of α_4? Dirac felt that there was no reason to stick to 2×2 matrices as Pauli had done, and higher order matrices can be looked for. Dirac then showed that α_k can be represented only by square matrices of even order. He chose to work with 4×4 matrices and put

$$\alpha_k = \begin{bmatrix} 0 & \sigma_k \\ \sigma_k & 0 \end{bmatrix}, \quad 1 \le k \le 3\,,$$
$$\alpha_4 = \begin{bmatrix} \mathbb{1} & 0 \\ 0 & -\mathbb{1} \end{bmatrix},$$

where σ_k are the 2×2 Pauli matrices and $\mathbb{1}$ is the 2×2 unit matrix. Thus all of the symbols in (30) are fully understood. The derivation of (30) is an example of what Wigner has described as the unreasonable effectiveness of mathematics in natural sciences.

Equation (30) can also be cast in the symmetrical form

$$\left(\gamma_1 \frac{\partial}{\partial x_1} + \gamma_2 \frac{\partial}{\partial x_2} + \gamma_3 \frac{\partial}{\partial x_3} + \gamma_4 \frac{\partial}{\partial x_4} + m_o c\right)\psi = 0\,, \qquad (32)$$

where $(x_1,\ x_2,\ x_3)$ are the three spatial coordinates (x, y, z) and $x_4 = ict$. The gamma matrices are defined as

$$\gamma_k = -i\alpha_4\alpha_k, \quad 1 \le k \le 3\,,$$
$$\gamma_4 = \alpha_4\,.$$

Dirac submitted this work to the Royal Society of London in December 1927. In the same paper, he established the invariance of his equation under Lorentz transformation. He also obtained the wave equation for the electron when subjected to an arbitrary electromagnetic field and showed that the electron gets an extra energy

$$\Delta E = \frac{e\hbar}{m_o c}\,B\,,$$

where B is the magnetic flux density. This implies that the electron has a magnetic moment $e\hbar/2m_oc$, the gyromagnetic ratio of 2, and an intrinsic angular momentum $\hbar/2$ which is the same as assumed in Goudsmit's spin hypothesis. This was one of the greatest triumphs of the Dirac equation (32).

From the wave equation of the electron subject to a central force, Dirac showed that L_z, the z component of the orbital angular momentum (z being an arbitrary direction), is no longer a constant of motion. On the other hand it is J_z given by

$$J_z = L_z + S_z$$

which is a constant of motion. (Here S_z is the Z component of the spin angular momentum). He then showed[37] that his theory when applied to the spectra of one electron systems led to the same results as the Pauli–Darwin theory[38] which had explained the doublet nature of their spectra. But Dirac did not actually derive the energy levels of the hydrogen atom which seemed to be a great omission. He was perhaps at the time apprehensive that the higher order corrections may not come out right, which might lead to the failure of what looked like a beautiful theory. He preferred to consolidate his results and publish them quickly with priority. His apprehension proved to be right, as Charles Darwin[39] and Walter Gordon[40] independently obtained in a short time the energy levels of the hydrogen atom, correctly describing their fine structure. Naturally this became a very strong argument in favor of Dirac's theory.

In Bohr's theory the energy of the hydrogen atom was found to depend only on the principal quantum number n. Although Arnold Sommerfeld modified Bohr's theory to include elliptical orbits and introduced the azimuthal quantum number k, the energy was still dependent only on n. In 1916, he used the relativity dynamics to obtain the following formula for the energy of the electron (inclusive of the rest energy),

$$W_{n,k} = \frac{m_o c^2}{\sqrt{1 + \left(\frac{\alpha}{n - k - \sqrt{R^2 - \alpha^2}}\right)^2}} \tag{33}$$

where $\alpha = e^2/\hbar c$ is the fine structure constant, a well-known dimensionless quantity with a value $1/137$. The energy thus depends on both n and k, the latter taking any integral value from 1 to n. To a first approximation the expression for the energy level turns to be

$$\begin{aligned} E_{n,k} &= W_{n,k} - m_o c^2 \\ &= E_n \left[1 + \frac{\alpha^2}{n^2} \left(\frac{n}{k} - \frac{3}{4} \right) \right] \end{aligned} \tag{34}$$

Thus the energy level corresponding to $n = 3$ which is single in Bohr's theory now becomes a close triplet. Similarly, the level corresponding to $n = 2$ becomes a close doublet. A transition between $n = 3$ to $n = 2$ which produces the H_α line at $\lambda = 6563$ Å will then be replaced by a multiplet of closely spaced multiplet of lines, giving rise to the so-called fine structure. The actual transitions are governed by the selection rules $\Delta k = 1$ or 0.

As early as 1887, Michelson detected the presence of a close doublet with a separation of 0.329 cm^{-1} in the structure of the H_α line. Subsequently a weaker component due to the transition $\Delta k = 0$ was detected between the two separated from the right-hand member (of higher frequency) by 0.109 cm^{-1}.

In quantum mechanics, the corrections to the energy level are calculated as the perturbation effect in the non-relativistic Hamiltonian due to (i) used of relativistic energy momentum relation, and (ii) spin orbit interaction. These two corrections are found to contain the same constant factor, as the coefficient suggesting that these effects are closely related. It has already been mentioned before that Dirac succeeded in proving that the spin effect arises from the relativistic formulation of the quantum mechanics. The total correction was found to be

$$\Delta E_{n,j} = E_n \cdot \frac{\alpha^2}{n^2} \left(\frac{n}{j + \frac{1}{2}} - \frac{3}{4} \right) \tag{35}$$

which leads to the same energy levels as given by the Sommerfeld formula, in terms of j rather than k. The energy levels are generally described in the spectroscopic notation where the total angular momentum quantum number j is written as a suffix to the letter S, P, or D, which indicate respectively $p = 0, 1$ or 2 etc. The principal quantum number n is stated as a coefficient of the letter. Equation (35) shows that $2S_{\frac{1}{2}}$ and $2P_{\frac{1}{2}}$ have the same energy (are degenerate). In a similar way $3P_{\frac{3}{2}}$ and $3D_{\frac{3}{2}}$ are also degenerate.

Then comes Dirac with his relativistic wave equation of the electron. With the inclusion of the Coulombic potential energy, the equation becomes,

$$\left(i\hbar\frac{\partial}{\partial t}+\frac{e^2}{r}-c\alpha\cdot p-\alpha_4 m_o c^2\right)\psi(r,t)=0 \qquad (36)$$

As mentioned before, it was solved by Darwin and by Gordon giving the same results as (35). The energy levels thus obtained are shown in the diagram below. The spectroscopic notation for each state is also given.

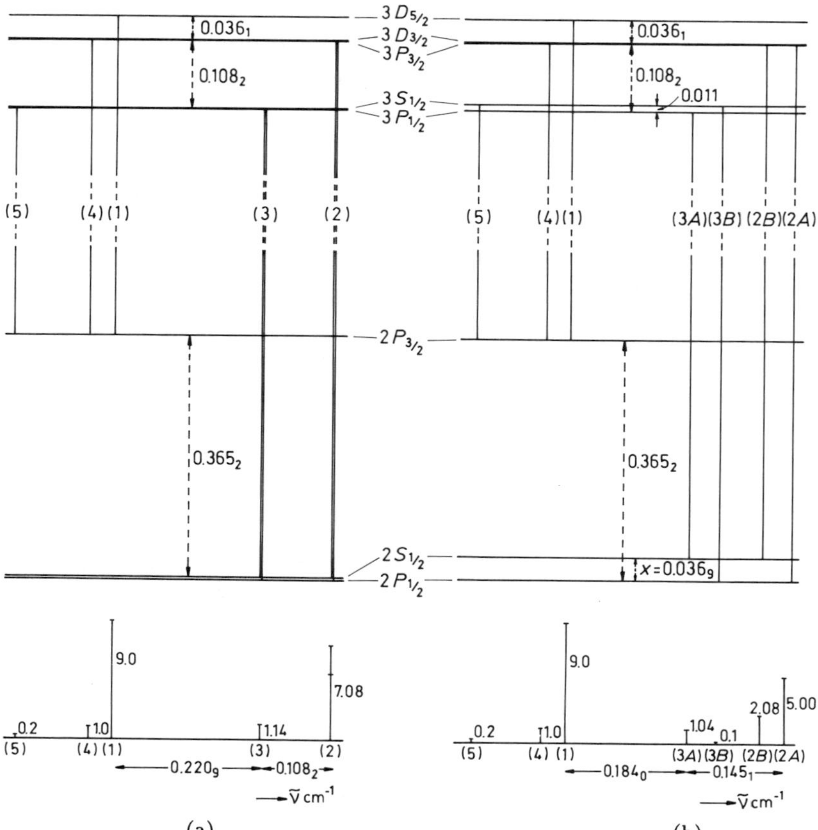

Fig. 1. Energy level diagram of H atom. (a) Dirac theory. (b) Shows shifts in S terms.

It will be noticed that the complete degeneracy for a given n in Bohr's theory is completely lifted by the relativity effects. But the levels $1S_{\frac{1}{2}}$ $2P_{\frac{3}{2}}$ and $3D_{\frac{3}{2}}$ are non-degenerate. However, there still remains a degeneracy between the pair $2S_{\frac{1}{2}}$ and $2P_{\frac{1}{2}}$ as well as between the pairs $3S_{\frac{1}{2}}$ and $3P_{\frac{1}{2}}$ or $3P_{\frac{3}{2}}$ and $3D_{\frac{3}{2}}$.

Several spectroscopic observations indicated that the $2S$ level was slightly higher than the $2P$ level. Willis Lamb[41] along with Robert Retherford, actually observed a radiofrequency resonance at 1057.8 megaHertz between these two levels, which is now known as the Lamb Shift. Lamb was awarded the Nobel prize for the year 1955 for this achievement. Heinrich Kuhn and George Series[42] verified this result optically, and found the energy difference to be 0.03526 cm^{-1}. Hans Bethe[43] explained this as the effect of interaction of the electron with the vacuum fluctuations of the electromagnetic radiation field.

11. The Dirac Equation

The Dirac equation, as (30) soon came to be called, became the cornerstone of modern physics. Its undisputed status emanates from its theoretical power. Although it did not predict any new results and led to an unintelligible conclusion about negative energy, it had a great merit of simplicity and of conceptual clarity and methodological directness. There were however, two difficulties it had to initially face.

First, it was expected to be applicable to a proton also. But the proton's actual magnetic moment is different from the value predictable from the Dirac equation. This defect was corrected much later in particle physics with the introduction of quarks.

Second, in addition to the positive energy solutions, there are an equal number of negative energy solutions, which initially appeared to be without any physical meaning. The appearance of negative energy solutions can be understood from the following considerations.

Since α_s are 4×4 matrices, the wavefunction ψ must be a column matrix with 4 elements. In other words, we have here a four component wavefunction. A free particle can then be represented by a plane wave

$$\psi_j = v_j \exp i(k \cdot r - \omega t)$$

with $\leq 1 \leq j \leq 4$, where the energy momentum eigenvalues are $\hbar\omega$ and $\hbar k$ respectively. These may be denoted by E and p, the latter with components $p_1 p_2 p_3$.

The Dirac equation is thus a set of 4 simultaneous linear differential equations in all the four components. For mutual consistency, we must have

$$E^2 = p^2c^2 + m_o^2c^4 \, .$$

which is the energy-momentum relation of relativistic mechanics. It is now obtained in quantum mechanics as well. It implies that E has two possible values E_+ and E_- given by

$$\begin{aligned} E_+ &= +\sqrt{p^2c^2 + m_o^2c^4} \\ E_- &= -\sqrt{p^2c^2 + m_o^2c^4} \, . \end{aligned}$$

When the particle is at rest, the two energy values are $+m_oc^2$ and $-m_oc^2$ for a particle in motion the energy spectrum has two branches. One extends from $+m_oc^2$ all the way upto ∞. The other goes from $-m_oc^2$ to $-\infty$. There is thus an energy gap of $2m_oc^2$ between the two branches.

In general there are four sets of solutions of Dirac equation for a free particle. In each case two components can be chosen arbitrarily, and the other two are interrelated. In the extreme non-relativistic limit, one of them becomes small enough to be taken as zero, and thus there remains only one significant component. In this way the components get decoupled. In the case of the energy solutions one corresponds to spin-up condition as seen from the eigenvalue consideration. Similar is the case with the negative energy solution. But at high energy the components get mixed up, and there is always a superposition of the spin states.

In spite of the initial difficulties, Dirac gave the following explanation for the negative energy solution. Since the electrons obey the Pauli exclusion principle, Dirac assumed that all negative energy states are filled up. He then showed[38] that transitions from the positive to the negative energy states are highly suppressed. But unoccupied negative energy states, called holes, can be observed. When a positive energy state electron drops into a hole and stays there, the electron and the hole simultaneously disappear with the emission of radiation. Eventually, Dirac found that the negative energy states correspond to positively charged particles which he assumed to be protons, as they were the only positively charged particles known at that time.

A disparity remained between the observed protonic mass and the calculated mass of the positive energy state electron. Dirac calculated[39] the rate of annihilation of the positively charged electron and the hole, by assuming their masses to be equal. But the calculated rate was too high. He was then really worried about the correct interpretation of the hole.

The theory of Dirac was strongly objected to by Bohr, who composed a short story with the title "How to catch an elephant alive?" In that story he suggested the following technique: At the watering spot where elephants come to drink and wash, the hunter should set up a poster explaining Dirac's idea in a single sentence. When the elephant, proverbially a wise animal, reads the poster, it becomes spell-bound for several minutes. The hunter, hiding in the bush nearby comes out and ties the legs of the elephant with heavy ropes. The elephant is then transported to the zoo in Copenhagen.

Pauli made some calculations and showed that if the proton in the hydrogen atom is a Dirac hole, the electron would jump into it within a fraction of a second. The hydrogen atom (and other atoms as well) would be annihilated and a burst of high energy radiation would be released. Pauli then proposed what is known as the second Pauli principle, according to which any theory suggested by a physicist would be immediately applicable to his body. Thus Dirac would be converted into gamma rays before he could tell anybody about his idea.

Finally, in 1931 Dirac declared that the hole must be a new particle,[40] with a positive charge and a mass equal to that of the electron. In this way the concept of anti-particle was finally formulated. In the same paper, he proposed the concept of a magnetic monopole which he showed to follow from the idea of quantization of electrical charge.

The development had a happy ending when shortly thereafter Carl D. Anderson discovered the positive electron, and named it the positron. Patrick M. S. Blackett soon found the examples of e^+ e^- pair creation in cosmic rays. The antiproton and anti-neutron were discovered by Emilio Segre in 1956.

Dirac spent the next two years in preparing his now famous book, *The Principles of Quantum Mechanics*, first published in

1930. It has undergone several editions and has been translated into many languages. He included in it a discussion of many-electron systems and quantum statistical mechanics. The book was of use not only to students taking a course in quantum mechanics, but was studied by experienced physicists as well. It is as highly regarded as Newton's *Principia*.

Dirac's subsequent academic progress was very rapid. He was elected a fellow of the Royal Society of London in 1930. Two years later he was appointed the Lucasian professor of mathematics at Cambridge University, a position once held by Newton. Dirac remained there till his retirement in 1969. The crowning honor was the award of the Nobel Prize in 1933, which he shared with Schrödinger.

Modesty was one of Dirac's greatest virtues. About Heisenberg he had said, "From the initial ideas of Heisenberg, one could easily make a fairly rapid progress. I was just a research student and was lucky to have been born at the right time. It was easy for any second rate physicist to make first rate research." In talking about his work on the relativistic wave equation, he stated that, "It was a direct consequence of Einstein's special relativity. He was the principal hero."

After retirement from Cambridge, Dirac moved to the University of Florida where he continued his activities with untiring zeal and enthusiasm. He died after a brief illness on October 20, 1984.

Acknowledgements

The author expresses his sincere thanks to Dr. A. D. Gangal of the University of Poona, Pune (India) and Dr. K. Arya of the San Jose State University, USA for helpful discussions. Thanks are also due to Dr. A. K. Chandra of the Quantum Chemistry group, National Chemical Laboratory (N.C.L.) Pune and Mr. V. G. Deodhar of N. C. L. Library, Pune, for help in many ways.

References

1. Helge S. Kragh, *Dirac: A Scientific Biography*, Cambridge University Press, Cambridge, 1990, p. 2.
2. Helge S. Kragh, *Dirac: A Scientific Biography*, Cambridge University Press, Cambridge, 1990, p. 9.

3. Paul A. M. Dirac, "Dissociation under a temperature gradient", *Proc. Cambridge Philos. Soc.* **22** (1924) 132–137.
4. Paul A. M. Dirac, "Note on the Doppler principle and Bohr's frequency condition", *Proc. Cambridge Philos. Soc.* **22** (1924) 432–433.
5. Paul A. M. Dirac, "The adiabatic invariants of the quantum integrals", *Proc. R. Soc. London*, Series **A107** (1925) 725–734.
6. Paul A. M. Dirac, "The adiabatic hypothesis for magnetic fields", *Proc. Cambridge Philos. Soc.* **23** (1925) 69–72.
7. Paul A. M. Dirac, "The fundamental equations of quantum mechanics", *Proc. R. Soc. London*, Series **A109** (1925) 642–653.
8. Hans A. Bethe *et al.* (eds.), *From A Life of Physics*, World Scientific, Singapore, 1989, p. 33.
9. Werner Heisenberg, "Über quantentheoretischer Umdeutung und mechanischer Beziehungen", *Z. Phys.* **38** (1925) 879–893.
10. Max Born, Werner Heisenberg and Pascual Jordan, "Zur Quantenmechanik, II", *Z. Phys.* **35** (1925) 557–615.
11. Max Born and Pascual Jordan, "Zur Quantenmechanik", *Z. Phys.* **34** (1925) 858–888.
12. Helge S. Kragh, *Dirac: A Scientific Biography*, Cambridge University Press, Cambridge, 1990, p. 21.
13. Paul A. M. Dirac, "Quantum mechanics and a preliminary investigation of the hydrogen atom", *Proc. R. Soc. London*, Series **A110** (1926) 561–579.
14. Helge S. Kragh, *Dirac: A Scientific Biography*, Cambridge University Press, Cambridge, 1990, p. 24.
15. Paul A. M. Dirac, "Relativity quantum mechanics with an application to Compton scattering", *Proc. R. Soc. London*, Series **A111** (1926) 405–423.
16. Paul A. M. Dirac, "On quantum algebra", *Proc. Cambridge Philos. Soc.* **23** (1926) 412–418.
17. Louis de Broglie, "A tentative theory of light quanta", *Philos. Mag.* **47** (1924) 446–458.
18. Erwin Schrödinger, *Collected Papers on Wave Mechanics*, Chelsea Publishing Company, New York, 1982, pp. 1-12. [See pages 357–371 in the second part of the book for the facsimile reproduction of the German original.]
19. Abdus Salam and Eugene P. Wigner, eds., *Aspects of Quantum Theory*, Cambridge University Press, Cambridge, 1972, p. 35.
20. Erwin Schrödinger, *Collected Papers on Wave Mechanics*, Chelsea Publishing Company, New York, 1982, pp. 13–40.
21. Erwin Schrödinger, *Collected Papers on Wave Mechanics*, Chelsea Publishing Company, New York, 1982, pp. 62–101.
22. Erwin Schrödinger, *Collected Papers on Wave Mechanics*, Chelsea Publishing Company, New York, 1982, pp. 102–123.
23. Werner Heisenberg, "Mechkörperproblem und Rezonanz in der Quantenmechanik", *Z. Phys.* **38** (1925) 411–426.

24. Paul A. M. Dirac, "On the theory of quantum mechanics", *Proc. R. Soc. London*, Series **A112** (1926) 661–677.
25. Enrico Fermi, "Zur Quantelung des idealen einatomigen Gases", *Z. Phys.* **36** (1926) 902–912 .
26. Paul A. M. Dirac, "The Compton effect in wave mechanics", *Proc. Cambridge Philos. Soc.* **23** (1926) 500–507.
27. Paul A. M. Dirac, "The physical interpretation of the quantum dynamics", *Proc. R. Soc. London*, Series **A113** (1927) 621–641.
28. Cornelius Lanczos, "Über eine feldmassige Darstellung der neuen Quantenmechanik", *Z. Phys.* **35** (1926) 812–836.
29. Eugene P. Wigner, "Über nicht kombinierende Terme in der neueren Quantentheories", *Z. Phys.* **40** (1926) 492–500.
30. Eugene P. Wigner, "Einige Folgerungen aus der Schrödingerschen Theorie für die Termstrukturen", *Z. Phys.* **43** (1927) 624–652.
31. Paul A. M. Dirac, "The quantum theory of emission and absorption of radiation", *Proc. R. Soc. London*, Series **A114** (1927) 243–265.
32. Paul A. M. Dirac, "The quantum theory of dispersion", *Proc. R. Soc. London*, Series **A114** (1927) 710–728.
33. Paul A. M. Dirac, "Über die Quantenmechanik der Stoßvorgänge", *Z. Phys.* **44** (1927) 585–595.
34. Paul A. M. Dirac, "The quantum theory of the electron", *Proc. R. Soc. London*, Series **A117** (1928) 610–624. [See pages 373–387 in the second part of the book for the facsimile reproduction.]
35. Oskar Klein, "Quantentheorie und fünfdimensionale Relativitätstheorie", *Z. Phys.* **37** (1926) 895–906.
36. Walter Gordon, "Der Comptoneffekt nach der Schrödingerschen Theorie", *Z. Phys.* **40** (1926) 117–133.
37. Paul A. M. Dirac, "The quantum theory of the electron, Part II", *Proc. R. Soc. London*, Series **A118** (1928) 351–361. [See pages 389–399 in the second part of the book for the facsimile reproduction.]
38. Charles G. Darwin, "The electron as a vector wave", *Proc. R. Soc. London*, Series **A116** (1927) 227.
39. Charles G. Darwin, "The wave equation of the electron", *Proc. R. Soc. London*, Series **A118** (1928) 654.
40. Walter Gordon, "Energieniveaus die wasserstoffatoms nach derDiracschen Quantentheoies des electrons", *Z. Phys.* **48** (1928) 11.
41. Willis E. Lamb, Jr. and Richard C. Retherford, "Fine structure of the H atom by a microwave method", *Phys. Rev.* **72** (1947) 24.
42. Heinrich Kuhn and George W. Series, "Fine structure of H line", *Nature* **162** (1948) 373.
43. Hans A. Bethe, "Electromagnetic shift of energy levels", *Phys. Rev.* **72** (1947) 334.

Vilas Yashwant Rajopadhye had a long teaching career at different colleges affiliated to the Universities of Bombay and Pune. Later he moved to the Indian Institute of Technology, Kharagpur, from where he has now retired. He then worked for a while as a Visiting Professor at the University of Nagpur. Early during his career, he submitted a doctoral thesis to the University of Bombay on the spectroscopy of some organic and inorganic flames and analyzed the combustion processes which produce and cause the excitation of the molecular species present therein. As a visiting research scholar at the University of Bristol later he studied the charge spectrum of the heavy primary cosmic rays and their nuclear interactions. This was followed by a study of meson production by 6 BeV protons, the highest energy available with machines at the time.

Prof. Rajopadhye is the author of two textbooks: one on introductory college physics, and the other on experimental physics for advanced undergraduate students. His present interest is in biographical studies of physicists and the historical development of physics during this century. He has been a member of the Indian Physics Association and the Indian Association of Physics Teachers.

Chapter 10. Hans Sallhofer Speaks

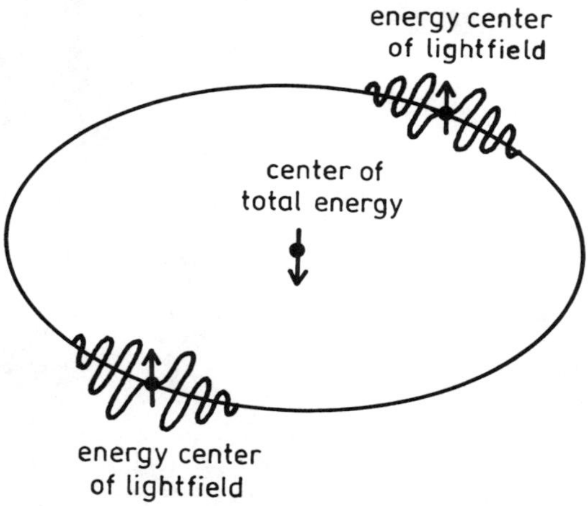

Akhlesh Lakhtakia

H. H. Sallhofer

Chapter 10

Hans Sallhofer Speaks

Akhlesh Lakhtakia

*Department of Engineering Science and Mechanics,
The Pennsylvania State University,
227 Hammond Building, University Park,
PA 16802-1401, USA*

Abstract

In an interview with this book's editor, Hans Sallhofer points out the formal mathematical identity of electrodynamics with wave mechanics. This Maxwell–Dirac isomorphism (MDI) assists him in formulating a model of the hydrogen atom in terms of $\mathbf{E}||\mathbf{B}$ fields. Sallhofer criticizes the lack of predictive content in the Copenhagen interpretation of quantum physics, and comments on certain societal issues affecting science.

1. Introduction

The story of hydrogen models did not end with Dirac. The hydrogen atom is the simplest of all atoms. Hence, every model of matter must be able to explain the structure of the hydrogen atom in order to be considered reasonable. But discussions of many newer models are beyond the scope of this book. Instead, let us turn to a contemporary modeler of hydrogen, a man named Hans Sallhofer, for a frank exchange of ideas. Dr. Sallhofer is a man with strong opinions, which means his views may appear controversial to some but are definitely worthwhile to ponder because he has made an in-depth study of 20th century models of hydrogen.

Born and raised by middle-class parents in an Austrian provincial town, Sallhofer passed through secondary school in the usual way, got his high school diploma at Linz, and finally obtained his doctorate in pure mathematics and theoretical physics (electrodynamics) from the University of Innsbruck in the fifties. His supervisor, Arthur Mach was a former assistant of Erwin Schrödinger's.

After several years of postgraduate studies with Mach (Innsbruck), J. Foster (McGill University, Montreal) and Schrödinger (University of Vienna), Sallhofer entered professional life as an industrial physicist. In 1978, he discovered, after many years of preparatory studies, the concept of Maxwell–Dirac isomorphism (MDI)[1]: He was able to show that electrodynamics and relativistic wave mechanics have identical structures, as exemplified by the following two equations:

$$\left[\gamma \cdot \nabla + i\frac{\omega}{c}\begin{pmatrix} \varepsilon \mathbb{1} & \mathbb{O} \\ \mathbb{O} & \mu \mathbb{1} \end{pmatrix}\right]\psi^{\mathrm{el}} = 0$$

(electrodynamics),

$$\left[\gamma \cdot \nabla + i\frac{\omega}{c}\begin{pmatrix} \left(1 - \dfrac{\Phi - m_0 c^2}{\hbar\omega}\right)\mathbb{1} & \mathbb{O} \\ \mathbb{O} & \left(1 - \dfrac{\Phi + m_0 c^2}{\hbar\omega}\right)\mathbb{1} \end{pmatrix}\right]\psi^{D} = 0$$

(wave mechanics).

The structural correspondence of the two theories turns out to be unequivocal, at least in formal respects, so Sallhofer speaks of an identity of electrodynamics with basic quantum physics.

2. Sallhofer's MDI-Hydrogen Model

The Maxwell–Dirac isomorphism establishes an electrodynamic hydrogen model in correspondence to the wave mechanical Dirac hydrogen model, as shown by Sallhofer recently.[2] Sallhofer's MDI-hydrogen model has six components — three electric and three magnetic components. Compared with Paul Dirac's model, the MDI-hydrogen model has the modest advantage that it expresses the property of half-integer spin not only in the frequency conditions but also in its six components. There are no point charges in Sallhofer's model — especially, it does not have an orbiting electron. Instead, the Keplerian structure necessary for the interpretation of the line spectra is generated by two interacting electromagnetic fields, whose *energy centers* orbit on a Kepler trajectory interacting in antipodal arrangement. The property of half-integer spin of this wave system turns out to be the necessary and sufficient condition for the antipodicity of the two energy centers. The MDI-hydrogen model contains a particle-like, centrally transverse standing wave field, both of whose constituent fields are everywhere and always perpendicular to the radius vector from the particle center. As the electric and magnetic fields are oriented parallel to each other, there is no Poynting radiation; and the spin wave turns out to be a pure phase wave orbiting in half-integer way. At the center the field strengths become integrably infinite. Because of their orthogonality, two MDI-hydrogen atoms may concentrically overlap or not, depending on whether they are in different or in the same quantum states, which is an expression of the Pauli principle.

The implications of the MDI-hydrogen model turn out to be very ambitious. Most prominently, this model requires that the Copenhagen interpretation of quantum physics be abandoned. Sallhofer is uncompromising on this issue because, as he points out, the two-slit experiment proceeds empirically without difficulties for electrodynamically described protons in his model, whereas the Copenhagen quantum physics has to invoke hobgoblins.

There is no one unique way of doing science. This is splendidly exemplified by the differing circumstances that prevailed with Heisenberg's model and Sallhofer's model. While Heisenberg had brought about a theory — quantum mechanics — and Wolfgang Pauli described the hydrogen of quantum mechanics, Sallhofer did not create a new physical theory. Instead, he was able to model the hydrogen atom with the help of his MDI. Thomson introduced Ptolemaic, and Rutherford Copernican structures into the first hydrogen models; Bohr then brought about quantization, Sommerfeld the fine structure, and Goudsmit spin; Schrödinger conceived of the hydrogen atom in terms of waves, and Dirac made that formulation relativistic. But Sallhofer did not add any new element in this vein, his model following merely — and necessarily — from MDI. Sallhofer believes that, as electrodynamics has the final say in the microscopic domain, his model is metaphysically as well as pragmatically consistent with modern electromagnetism. Not surprisingly therefore, he opines that today's quantum physics is unnecessarily hobbled by the Copenhagen interpretation.

3. Sallhofer Speaks

In May 1994 I asked Sallhofer for an interview on the occasion of the international symposium *CHIRAL 94* in Périgueux, France. The Dordogne valley is delightful country, famous for the Lascaux caves as well as goose liver patè. The locale was an appropriate setting for an extended interview. I shared with him my plans for this book and his advice on that matter was invaluable. Sallhofer discussed his MDI-hydrogen model as well as the probabilistic nature of the Copenhagen interpretation and its impact on society. Later, I requested him to edit his answers to my questions in order to eliminate errors that may have crept in.

*Q*1. ***Lakhtakia***: You elected to study physics as an undergraduate student in Austria during the early 1940s. What did you find so attractive about physics?

*A*1. ***Sallhofer***: Above all, I wanted to devote myself to a science that was able to give clear answers to clear questions. First of all, there was humanistic mathematics in the foreground. It appeared to me as a bright transparent crystal palace, where there was only pure truth; where one was free of opinion, interpretation and approximation; where a law, once disclosed, would be valid forever and had been valid all along. Extensive explorations in its well-lit halls gave me the deep experience of absolute truth.

An insight that much distressed me accompanied my thoughts all along: There was so much of all that beauty that one seemingly could not keep track of it at all. I found the total mathematical knowledge overwhelming and, yes, oppressive and discouraging. I realized that I could not stay there. With tears I broke away from *mathematics in itself*, and began to consider mathematics in correlation with a physical phenomenon. After a long search, I came to the conclusion that electrodynamics was the physical science conforming most to mathematics. I settled on electrodynamics.

*Q*2. ***Lakhtakia***: During the late 1940s, you pursued a doctoral degree in physics at Innsbruck. You must have come across great scientists such as Werner Heisenberg and Max Planck. Can you recount some memorable occasions in your life as a graduate student?

*A*2. ***Sallhofer***: I can only remember my encounters with Schrödinger. Heisenberg I met just a few times, Planck I never saw.

Several times I tried to interest Schrödinger in an optical wave function. I tried hard to motivate him in that direction, but he always reacted rather reserved. Although he even accepted my main argument of that time — that Hamilton's analogy turns the classical light equation immediately into his time-dependent equation — it did not turn him on. He did once remark that it would be the jumping-cracker of the century if I were right, but de Broglie had shown long ago that there could not exist an optical wave function. Today I am convinced that Schrödinger would have reacted completely different if I had been able then to shove MDI under his nose.

*Q*3. ***Lakhtakia***: You made the hydrogen atom one of the foci of your research efforts. What attracted you to the hydrogen atom?
*A*3. ***Sallhofer***: As a student I had been deeply impressed by the efforts of the ancient Greeks to gain insight into the essence of matter. Their progressive advance into the structures of the perceptible phenomena, as it is described in old fragments, I saw as my most moving youth adventure. The era of the Greek natural philosophers, the two hundred years between 600 and 400 BC, became my firm favorite in science history, my heroes being Anaximander, Heraclitus, Parmenides and Democritus. I learnt with astonishment and sadness that natural philosophy, after flourishing shortly with the Greek, fell almost into oblivion for two millennia in Europe. And, that only in the sixteenth century life came back again into natural philosophy. I was extremely happy that in the Europe of the Enlightenment, questions on the essence of matter were asked anew, and that this complex of questions reduced eventually to the question: What is a hydrogen atom? That question stayed with me.

*Q*4. ***Lakhtakia***: Force-free magnetic fields, used by Chandrasekhar and others to model black holes, are Beltrami fields.[3] So are electromagnetic fields with $\mathbf{E}||\mathbf{B}$, essential to Fresnel's explanation of circular birefringence. You have also used these fields to explain the non-radiating states of the hydrogen atom. Did your post-doctoral experience on cyclotrons at McGill University during the early 1950s prepare the ground for this model of yours?
*A*4. ***Sallhofer***: The Beltrami character of the MDI-hydrogen model is an imperative result of the solution of the electrodynamic hydrogen boundary value problem. The Beltrami-like hydrogen field is

radiation-free everywhere and at all times for all stationary states, as it should be, because $\mathbf{E} \times \mathbf{B} = 0$ therein.

To your second question: The director of the McGill cyclotron laboratory at that time, John Steward Foster, had advised me to work on a problem in connection with *magic numbers*. In exploring the optimal way of solution, I came across facts that reinforced my conviction that there must exist a general electrodynamic wave function, and therefore also an electrodynamic model of hydrogen.

*Q*5. ***Lakhtakia***: What do you think underlies the Maxwell–Dirac isomorphism?

*A*5. ***Sallhofer***: In the second quarter of the 18th century, Hamilton[4–7] showed that geometric optics and point mechanics are formally identical theories, i.e., he could prove that a light ray in a refraction behaves in the same way as a mass point in a potential. This optics-mechanics isomorphism has been called *Hamilton's analogy* since then, and it found its immediate expression in Hamilton's connecting equation of light refraction:

$$N = c\sqrt{2m\frac{U - \Phi}{U}}\,.$$

Hamilton's contemporaries did not know what to do with the analogy, and most of them therefore regarded it as a mathematical coincidence. Thus Hamilton's analogy soon disappeared from the physics literature and sank into oblivion.

One hundred years later, de Broglie[8] claimed that to any particle there corresponds a wave. *Wave* and *particle* and *correspondence*, that sounds like *optics* and *mechanics* and *analogy*. Schrödinger — also a philosopher and historian of science — knew of Hamilton's analogy and deduced from it, in an intricate way, the Schrödinger equation, without doubt the most important and most cited equation of physics of this century. The theory behind the Schrödinger equation was called *wave mechanics*, and its solutions entered the index of concepts of quantum physics as *Schrödinger functions* or *wave functions*. The waves of the wave functions were called *matter waves*.

What a particle (e.g., an orbiting electron) is, one then thought one knew; but matter waves? What waves are they? The controversy about them — the so-called *interpretation controversy* — developed in the 1930s and remains undecided even now. At the end of the 1950s, I tried to point out to Schrödinger that, according to Hamilton's analogy, there should correspond an optical model to Bohr's mechanical hydrogen model, from which one could infer that the matter waves in question should be light waves. And, that indeed one immediately would get the time-dependent Schrödinger equation if one inserted, under the assumption of harmonic solutions, the aforementioned Hamilton light refraction analogy into the classical light equation:

$$\nabla^2\Psi - \frac{N^2}{c^2}\frac{\partial^2\Psi}{\partial t^2} = 0\,.$$

But Schrödinger had been fixed so much on Rutherford's orbiting electron that he could not think of light, but had to believe in charge: He intoned a theory according to which in the hydrogen atom, a negative charge is situated around a positive nucleus as a charge cloud, resulting from the wave function, that behaves like an orbiting electron. This theory was stillborn, and all his further efforts towards an interpretation retaining the orbiting electron were in vain. In his disapproval of the optical interpretation, he liked to quote de Broglie who thought he had proved the non-existence of optical wave functions.[9]

Eventually, others offered less contradictory interpretations. Max Born, Niels Bohr, Werner Heisenberg and Pascual Jordan were able to agree on the *Copenhagen interpretation* in Copenhagen, and it became internationally accepted. Note, though, that the Copenhagen quantum physics had been strongly rejected by Max Planck, Hendrik Lorentz, Max von Laue, Einstein, and even by Schrödinger.

According to Hamilton's analogy, there corresponds to any system of Newtonian point mechanics a light system of classical optics. Dirac formulated relativistically the Hamilton-analogous counterpart of Bohr's point-mechanical hydrogen model. He thus had to get a relativistic formulation of classical optics. Relativistically formulated classical optics, however, is electrodynamics. Dirac's relativistic formulation therefore must be isomorphic to electrodynamics, as I showed some years ago.[10,11] In conclusion, I would therefore answer your question in this manner: MDI is rooted in Hamilton's analogy.

Q6. ***Lakhtakia***: What may one conclude from the Maxwell–Dirac isomorphism?

A6. ***Sallhofer***: The Maxwell–Dirac isomorphism does not establish a new physical theory. That it should exist had to be expected from Hamilton's analogy. When it was finally found, it became clear to me that the central expression of quantum physics — the wave function — must describe electrodynamic fields. MDI insists therefore on full disclosure, as befits electrodynamic phenomena. In contrast, the Copenhagen quantum physics offers a semi-causal wave-particle metaphor with a wave function that lacks clarity, and that squanders the larger part of the wave function's potential of expression by its probabilistic interpretation.

The MDI-hydrogen model[2] shows that electrodynamics has the say also in the microscopic domain. And there it speaks with an incomparably more concrete and instructive language than does the Copenhagen quantum physics. Thus, for example, it brings about half-integer hydrogen spin not only in the frequency conditions but also in the electrodynamic field components, whereas the Copenhagen people cannot express the half-integer property in the spinor components. The most important advantage of the electrodynamic picture is its six-component reality as compared to the four-component probability of the Copenhagen interpretation. One may proceed with the assumption that electrodynamics will be able to describe also all other elementary particles in six-component clarity.

Let me digress. In October 1993, the US Congress voted down with compelling majority the Superconducting Super Collider (SSC), the world's largest accelerator then under construction at Waxahachie, Texas. That action was the largest revocation of a previously approved research investment in the history of the United States and the world. The construction of the SSC would have cost an estimated US$15 billion, maybe even US$25 billion. Approximately US$2 billion had been used up already at the time of suspension of the project. Of course, the suspension had been preceded by extensive legislative hearings: many noted experts in high-energy physics had testified. The scientists could not dispel the impression that their meager results had been in no tolerable relation to the enormous costs for a long time already; that the reduction of the inhabitants of the particle zoo to just a few has failed altogether; that, on the contrary, the range of the supposed basic units (i.e., the so-called Standard Model) again has started to widen. Therefore, I do not consider the spectacular revocation of funds as a usual measure to reduce expenditures. Revoking the SSC project may have rung the death knell of the Copenhagen interpretation. The Copenhagen people have tried for decades to compensate for the lack of determinism by monetary brute force in experimental research. That their *cost what it may* approach for any small gain in knowledge would be caught by the politicians one day had to be expected.

For consolation in the post-Waxahachie era, essays of advice have appeared, and US physicists have been recommended to, among other things, line up at the European accelerators, for instance, in Geneva. That may work out well because the funders of the intricately financed CERN have not woken up yet and smelled the coffee. I cannot help thinking that MDI quantum physics — because of the much stronger predictive power of its wave function — is in a position to significantly reduce the costs of research, as it would not employ accelerators for reconnaissance but, at most, for controlling and as rearguard.

The wave-particle problem does not burden my approach. MDI shows that the electrodynamic wave function exists; or, on the one-component Schrödinger level, that the Schrödinger function is optical. Ehrenfest[12] has shown that the energy center of a Schrödinger function obeys Newtonian mechanics. Conclusion: a particle, in general, consists of electromagnetic waves whose energy center manifests itself as a mass point.

The conclusions from MDI, asked for by you at the beginning, I would summarize as follows:

(i) MDI offers a completely clear, six-component electrodynamic wave function;
(ii) Copenhagen quantum physics seriously compromises the information content of the wave function because of its approximate interpretation, but MDI is not incapacitated thus; and
(iii) MDI shows that not only light but matter also is electromagnetic, and is therefore not burdened by the wave-particle problem.

With regard to the hydrogen atom, MDI leads to absolutely concrete concepts by solving the electrodynamic hydrogen boundary value problem. The evolution of hydrogen models in the book you are editing will very likely show

(i) that the development of the line of models proceeded continuously in the sense that every previous model is contained in its successor as a simple approximation;
(ii) that the Kepler structure is an invariant of this evolution; and
(iii) that in all models in this line, except for mine, the two masses of the Kepler structure may be assumed as arbitrarily large — and either different or the same — without compromising compatibility with the observed spectrum of hydrogen.

In my model, there are two equal Kepler masses, which in fact expresses itself in the symmetric ambiguity of the electrodynamic solution's half-integer spin factor. Therefore, not the Kepler structure with different masses (proton/electron) of the models preceding it but the mass-equality of the Standard Model (u-quark/u-quark) belongs to the MDI-hydrogen model. That is, the two u-quarks are related to the two light fields forming the Kepler structure, whereas the d-quark corresponds to a centrally symmetric integral function of the complete solution.

*Q*7. ***Lakhtakia***: You think the Copenhagen interpretation is doomed to failure. What do you find so fateful about its beginnings?
*A*7. ***Sallhofer***: The Copenhagen interpretation is the work of three theoreticians — Born, Bohr and Heisenberg; maybe, in addition, Jordan. That Bohr showed a propensity for supernatural concepts, for ghost fields, in a difficult situation is not astonishing. He never believed in nature being comprehensible. He is known to having taken the view that the opposite of a deep truth may again be a deep truth.[13] Heisenberg considered the Copenhagen interpretation an "emergency delivery".[14]

Max Born, as you know, made his credo for the probability interpretation dependent on the outcome of a theoretical experiment. He had a Schrödinger wave field run against an obstacle — an obstructing grating. When he found that part of the wave overcame the obstacle, that is, it could be observed on the other side of the grating, he concluded that the Schrödinger function could not represent a matter-carrying wave. Why? Since it behaves like light, yes, like light! But as it described in principle an indivisible particle, it could only represent a position probability.

Born thus saw that the Schrödinger function behaves like a light function. Yet he never wondered whether the Schrödinger equation might possibly describe light. It supposedly was his prejudice towards particle pictures, his bias towards particles in his basic concepts, that forced him away from light fields towards probabilistic fields.

That reminds me of Schrödinger's bias towards particles. As I mentioned earlier, I myself had pointed out several times to him that Hamilton's analogy represented an indication that the Schrödinger function is a light function. But Schrödinger would not hear of light, in accordance with de Broglie. I chuckle whenever I think that Schrödinger did not ever know what *his* function should stand for. He rejected the Copenhagen interpretation as well as the light interpretation.

Anyhow, Born's probability wave, that found its cheap completion in the Copenhagen interpretation, marked a very decisive turn in the history of quantum physics. It produced two consequences of wide import.

First the good news: During the great helplessness regarding the interpretation of the wave function at the end of the 1920s, Born's probability interpretation made the Schrödinger equation intelligible. It also turned out that the probability interpretation provided the best prerequisites for a simple and quick extension of the theory into a perfectly working technology. The atomic pile and bomb are remarkable instances of the effective applicability of probabilistic nuclear technology to large scale technical domains.

Today, however, this good news isn't such good news any more. Many conservationists and humanists are vocal that the world would be in a better position if Schrödinger and Born, Hahn and Fermi had never been born. Then we would have experienced neither a Three Mile Island nor a Chernobyl, and would not be subject to the tremendous potential hazards that spread out over the civilized world like a shroud, prey on our minds, and let our souls sink into the twilight of omnipresent fear. Then the Second World War would have been decided on the battlefields and not by a demonstrative mass assassination that took the lives of hundreds of thousands of Japanese women and children and, above all, left behind nameless misery and inconceivable prospects for the future. Let scientists not dismiss these thinkers as emotional and immature, for they have a great social influence on science.

Now the bad news: Light that could have been a conceivable interpretation was ruled out when Born castrated the wave function. From then on a tired gelding pulls the carriage of knowledge, in front of which a fiery stallion had been harnessed before Copenhagen. Born at that time gave a completely *ad hoc* interpretation: The wave function specifies the position probability of the particle. This interpretation implies a reassurance: should it turn out some day that the wave fields carry matter (or energy), a larger energy concentration — for instance — could well be related to a higher position probability. Should it come about that the energy center of a light field is a photon, one would not be too far off. In a wave function interpreted in such a way, one may arbitrarily mutually interchange finite spherical harmonics, or replace them by other finite goniometric functions; raise or lower the exponents in the Sommerfeld polynomials, or interchange the coefficients of the powers; and so forth, without anyone interfering appreciably with the quality of the probabilistic statement. The clear numerical predictions of the beautiful analyticity of the wave function sank into the spatial and temporal insignificance of the probabilistic utilization of coordinates; that is, the precise data of the analytic factors of the wave functions were reduced in their significance to an abstraction with almost vanishing predictive content.

And exactly because of that, to come back to your question, I find the Copenhagen interpretation had a fatal beginning. For one was faced with ignoring the immediate analytic top information and had to look out for other, deeply embedded, poorly defined and unprofitable reference points. Especially, symmetries and the search for their associated conservation laws came in vogue. Today, the potential of symmetries has long been exhausted. Supposed symmetries have turned out to be flops after decades of established status, and the experimental look into the kaleidoscope of structures of the elementary particles has proved to be the most expensive enterprise of scientific research by far, providing increasingly less results. The whole situation of today's high energy physics suggests that the theoreticians especially are in the epicyclical phase of Ptolemaic theory.

*Q*8. ***Lakhtakia***: Why did you not stay at a university after your post-doctoral stints? Why did you decide to go for industrial research?
*A*8. ***Sallhofer***: My decision in this respect was made in the time after the Second World War. In post-war Europe of that time, it could not be overlooked that all available funds were assigned by the governments of the states involved in the War to rebuilding the damaged cities and industries. The budgets of universities and research institutes were reduced to bare essentials. All funds from the generous Marshall Plan went into rebuilding the economy — industry and industrial research. Thus, I came to the conclusion that it would be more sensible to work in industry.

*Q*9. ***Lakhtakia***: Do you consider the cost factor a serious problem of today's scientific research?
*A*9. ***Sallhofer***: Since quantum physicists are considering equipment costing ten billion dollars and more, the question of expenditures naturally comes to the fore. I do not want to assume that experimentalists have become insensitive to the exorbitant costs, the immense heaps of money they draw out of the pockets of taxpayers. But the initial investment and the maintenance costs of the latest accelerators have now reached a magnitude that has raised the issue of the benefit to the wider society. Then the issue of successful experimentation becomes political, that is, it becomes the responsibility of politicians. Since a politician has to assess the benefit of research not only from the point of view of scientists but with the taxpaying voters in mind, asking for continually higher physics budgets may easily trigger cost reduction instead, even when benefit can be shown. The reaction of the politicians will turn out worse, of course, if those *demanding* monies are standing there empty-handed.

*Q*10. ***Lakhtakia***: Let me clarify: Do you think that the expenditures of particle physics research produce merely minor benefits nowadays, may be none?
*A*10. ***Sallhofer***: One thing is for certain: The existing benefit is in an unacceptable proportion to the expenditures.

Viewed within a longer time frame, the basic successes of the atomic sciences have all been achieved in this century; and the major insights came in the first third. Total research expenditures during this third: nil. The immense successes immediately expanded physics teaching facilities at the universities. Although the major insights had been gained already in my opinion, an ever larger number of physicists pondered additional possibilities to classify the great ideas of the first third of this century — and maybe convert them into tangible profit, following the example of the chemical sciences, for instance.

During the 1940s the first accelerators appeared on the scene. They had in them the tendency to grow. A parameter to estimate their construction and operating costs: the ring diameter. Accelerators started with diameters of craftsman's size and expanded to over hundreds of meters to kilometers in dimensions. The progressively rising budgetary demands of high-energy physicists in general found open hands; Einstein's famous letter to the US President and its consequences raised the reputation of physics far above that of any other science.

Then came the fifties. Confrontation escalated between Capital and Labor, between the East and the West, and the specter of a Third World War was raised. He who has reasons for fear needs good armorers. Politicians became generous, the diameters of accelerators grew with the East–West tension, in the West as well as in the East.

In the sixties, theoreticians rigged up the Standard Model out of the glut of accelerator data of the fifties and sixties. The reductionists were riding high: just 12 basic entities. That happy state, unfortunately, did not last for long. Soon there were 60 basic entities, and at the Standard Model's horizon more candidates lurked about.

During the eighties, the US President and Man of Capital, Reagan staged the Strategic Defense Initiative (SDI): a defense with first-strike capability, a kiss of death eventually to friend Gorbachev. The Labor Paradise, having armed itself to death, threw in the towel and sank into a democratic anarchy. So, the West has no partner in the tension any more. What do you think will happen when the East–West tension decreases even further?

*Q*11. ***Lakhtakia***: Do you think that the waning East–West tension has a direct influence on the allocation of funds to big physics in the USA?
*A*11. ***Sallhofer***: I have not yet completely answered your previous question regarding benefit. The profit of the immense expenditures for the nutcracker physics of the last thirty years is crystallized in the Standard Model. And here the question splits into two:

(i) What is the purely epistemological benefit of the Standard Model?
(ii) Does the Standard Model have any prospects of technological extensions that promise economic profit in the foreseeable future?

The whole epistemological benefit is essentially realized in the moderate categorization success of the Standard Model. Attempts at reduction are met at first, but then the reduced palette of basic entities explodes from 12 to 60. Further enrichment cannot be ruled out. Even the most tangible and omnipresent member of the palette of twelve — the electron — remains a stranger to the Standard Modelers. They do not know its electrodynamic equation.

In the history of accelerators, a prominent role is played by the strong force. The Standard Modelers keep saying: Present-day accelerators cannot yet bring it about, but the next generation will. It is justifiably doubtful whether the strong force will ever be clarified by fragmentation.

The price of the Standard Model, measured as the sum of the investment and the maintenance costs of all accelerators, is by far the largest questionable expenditure in the history of sciences up to now, in view of the transitory character of the model. Main cause of the excess: The multiplicity of particles because of the preference given to the Copenhagen interpretation as against a more deterministic interpretation. I rate the epistemological benefit of the Standard Model as moderate, at best.

The situation looks worse with respect to economic expectations. Up to now there are no indications whatsoever that the Standard Model will create any technology.

Now to your last question, whether danger of war and money for big physics are correlated. If one bears in mind that Einstein's letter to President Franklin Roosevelt and the ensuing application of nuclear physics theories abruptly changed the course of Second World War, then one may understand that politicians have absorbed in their minds atomic physics as a decisive instrument of power. The representatives of the people, in addition, know that nuclear energy represents the one and only potential that may guarantee the survival of our civilization after the exhaustion of the fossil sources of energy. So, big physics for years could bask in the assurance that the politicians would willingly grant their ever-increasing budgetary wishes.

Now, however, with the waning of the East–West tension, the armorers are not anymore those people against whom one has to behave well unconditionally. And so, in the late eighties and early nineties, the climate became markedly frostier for physicists demanding an uninterruptible and unquestioning flow of money. A milestone in this cooling is represented by the US Congressional cancellation of funds for the SSC project.

*Q*12. ***Lakhtakia***: As you have already mentioned, this cancellation was preceded by extensive testimony of high-energy physicists. According to your version, it would have been appropriate to put steam behind the members of the US Congress by pointing out security aspects of weapons technology. But completely opposite statements by competent high-energy people became known. For instance, the founding director of Fermilab, Robert Wilson testified that the SSC project would not enhance US security.
*A*12. ***Sallhofer***: In response to questioning from Senator John Pastore, Wilson actually replied that the SSC project has "nothing to do directly with defending our country except to make it worth defending."[15] Today such answers, of course, have to provoke outbursts inevitably. Such answers are arrogant. Quantum physicists are not a super-intelligent race whom the lowly masses must toil for. And I strongly concur with Weinberg, of electroweak force's fame, when he wrote that "it is evident that we as physicists have not been successful in conveying to Congress and the public the significance of the scientific goals of the Super Collider."[16]

*Q*13. ***Lakhtakia***: What would you have said in Wilson's place?
*A*13. ***Sallhofer***: I do not know. Maybe I would have mentioned Zhirinovsky, the right-wing Russian party leader, who has good chances now (1994) to come to power this or the next year, and

who, according to his own statements, wants to head a confrontation course again. If the politicians would then have allowed continuation of the digging in Waxahachie, one could have renamed the Super Collider *Super Zhirinovsky.*

But, no kidding, Americans understandably are not afraid of Russia anymore. East Germany, during its forced union with socialism between 1945 and 1990, fell behind so much in many aspects of the quality of life in comparison to the West that West Germany has to supply now more than US$1 trillion within ten years in order to bring East Germany up to the standards of the West. This fact was also the ultimate reason for the negative attitude of the European Union when Gorbachev tried to bring the East Bloc into a *common European house.* If East Germany with a population of 16 million costs more than US$1,000 billion, then the East Bloc with its population of, say, 300 million would cost Western Europe US$20 to 30 trillion, correspondingly. That means, the East neglected its infrastructure and light industry to an unimaginable amount of US$25 trillion during its socialistic period in order to keep the defense balance with the West. If Zhirinovsky would now take over again, this economic disparity would aggravate further.

However, the nuclear potential of Russia still remains a factor. Although about 3800 Russian intercontinental missiles — that were aimed mainly at the USA and Western Europe — have indeed been decommissioned in the last few years, they could, of course, be made operational again very quickly. The prospect that Zhirinovsky could have massive destructive power in his hands is alarming.

*Q*14. ***Lakhtakia***: Where, in your opinion, lies the cause for the big difference between the economic efficiencies of the East and the West?

*A*14. ***Sallhofer***: Any employee, whether in the East or in the West, wants to increase the financial compensation for work done. In the West, she/he can achieve that by increasing personal productivity, and the GNP is thereby raised too. In the East s/he could bring about a raise in wages only by reducing personal productivity in view of the uniform wages throughout the land, which would reduce the GNP. So, a more efficient *elbower society* established itself in the West, in the East a less efficient *shirker society.* The shirker has at his disposal less infrastructure and a lower quality of life, but also no heart attacks and no fear of losing employment.[a] I suppose, to Marx and Lenin — both of whom had not been able to experience the socialistic experiment in the Soviet Union — this difference in efficiencies must have been clear. That is why, I think, they insisted strongly on world revolution first, then socialism: a uniformly socialistic world would not have the possibility of a comparison.

I think one should also not act as if Chinese Socialism would automatically take the same road as the Soviet one. The Chinese are much more aware of the elbower-shirker tension. They got themselves their own measure for elbower efficiency with their *special economic zones*, in order to gauge against it the deficiency of their shirker society, to find a way that keeps the deficiency small. Should the Chinese socialists be successful, one will have to count on the confrontation between Capital and Labor being possibly reactivated.

Lakhtakia: Thank you for giving me your precious time. I enjoyed the candor with which your replied to my questions.

References

1. Hans Sallhofer, "Elementare Herleitung der Dirac-Gleichung", *Z. Naturforsch.* **A33** (1990) 1378–1379.
2. Hans Sallhofer, "Hydrogen in electrodynamics. VI. The general solution", *Z. Naturforsch.* **A45** (1990) 1361–1366.
3. Akhlesh Lakhtakia, "Viktor Trkal, Beltrami fields, and Trkalian flows", *Czechoslovak J. Phys.* **44** (1994) 89–96.
4. William R. Hamilton, "Theory of systems of rays", *Trans. R. Irish Acad.* **15** (1828) 69–174.
5. William R. Hamilton, "Supplement to an essay on the theory of systems of rays", *Trans. R. Irish Acad.* **16** (1830) 1–67.
6. William R. Hamilton, "Second supplement to an essay on the theory of systems of rays", *Trans. R. Irish Acad.* **16** (1830) 93–126.
7. William R. Hamilton, "Third supplement to an essay on the theory of systems of rays", *Trans. R. Irish Acad.* **17** (1837) 1–144.
8. Louis de Broglie, "Sur le parallélisme entre la dynamique du point matriel et l'optique géométrique", *J. Phys. Radium* **1** (1926) 1–6.
9. Louis de Broglie, "Uné nouvelle conception del la lumière", *Actualités Scientifiques et Industrielles (Paris)* **181** (1934) 1–48.

[a]An old maxim of Soviet workers is: They pretend to give us some money, we pretend to do some work.

10. Hans Sallhofer, "Elementary derivation of the Dirac equation. X", *Z. Naturforsch.* **A41** (1986) 468–470.
11. Hans Sallhofer, "Maxwell–Dirac–Isomorphism. XI", *Z. Naturforsch.* **A41** (1986) 1087–1088.
12. Paul Ehrenfest, "Bemerkung über die angenäeherte Gülitigkeit der klassichen Mechanik innerhlab der Quantenmechanik", *Z. Phys.* **45** (1927) 455–457.
13. Werner Heisenberg, *Der Teil und Das Ganze*, Deutscher Taschenbuch Verlag, Münich, 1991, p. 124.
14. Werner Heisenberg, *Physics and Philosophy*, George Allen & Unwin, London, 1971, p. 43.
15. Leon Lederman with Dick Teresi, *The God Particle*, Bantam Doubleday Dell, New York, 1993, p. 199.
16. Steven Weinberg, *Der Traum von der Einheit des Uiniversums*, Bertelsmann, Munich, 1993, p. 283.

Akhlesh Lakhtakia is an Associate Professor of Engineering Science and Mechanics at the Pennsylvania State University. He has published more than 260 journal articles; has contributed chapters to six research books; has edited, co-edited, authored or co-authored six books; has reviewed for 40 journals; and is the Editor-in-Chief of the international journal *Speculations in Science and Technology.* He served as an international lecturer for the International Commission for Optics and the Optical Society of America, was twice a Visiting Professor of Physics at Universidad de Buenos Aires, is a Fellow of the Optical Society of America, and is a past-Chair of the IEEE EMC Technical Committee on Nonsinusoidal Fields.

PART II: FACSIMILE REPRINTS OF TEN ORIGINAL PAPERS

Reprint 1

THE

LONDON, EDINBURGH, AND DUBLIN

PHILOSOPHICAL MAGAZINE

AND

JOURNAL OF SCIENCE.

[SIXTH SERIES.]

MARCH 1904.

XXIV. *On the Structure of the Atom: an Investigation of the Stability and Periods of Oscillation of a number of Corpuscles arranged at equal intervals around the Circumference of a Circle; with Application of the results to the Theory of Atomic Structure.* By J. J. THOMSON, *F.R.S., Cavendish Professor of Experimental Physics, Cambridge* *.

THE view that the atoms of the elements consist of a number of negatively electrified corpuscles enclosed in a sphere of uniform positive electrification, suggests, among other interesting mathematical problems, the one discussed in this paper, that of the motion of a ring of n negatively electrified particles placed inside a uniformly electrified sphere. Suppose when in equilibrium the n corpuscles are arranged at equal angular intervals round the circumference of a circle of radius a, each corpuscle carrying a charge e of negative electricity. Let the charge of positive electricity contained within the sphere be ve, then if b is the radius of this sphere, the radial attraction on a corpuscle due to the positive electrification is equal to ve^2a/b^3; if the corpuscles are at rest this attraction must be balanced by the repulsion exerted by the other corpuscles. Now the repulsion along OA, O being the centre of the sphere, exerted on a corpuscle at A by one at B, is equal to $\frac{e^2}{\mathrm{AB}^2}\cos \mathrm{OAB}$, and, if OA=OB, this is equal to $\frac{e^2}{4\mathrm{OA}^2 \sin\frac{1}{2}\mathrm{AOB}}$: hence, if we have n corpuscles arranged at equal angular intervals $2\pi/n$ round the circumference of a circle, the radial repulsion on one corpuscle

* Communicated by the Author.

Reprinted from Philosophical Magazine and Journal of Science, Series 6, Vol. 7, No. 39, pp. 237–265 (1904).

due to the other $(n-1)$ is equal to

$$\frac{e^2}{4a^2}\left(\operatorname{cosec}\frac{\pi}{n}+\operatorname{cosec}\frac{2\pi}{n}+\operatorname{cosec}\frac{3\pi}{n}+\dots\quad+\operatorname{cosec}\frac{(n-1)\pi}{n}\right).$$

If the corpuscles are at rest this must be equal to the radial attraction. Hence, if

$$S_n=\operatorname{cosec}\frac{\pi}{n}+\operatorname{cosec}\frac{2\pi}{n}+\dots\quad\operatorname{cosec}\frac{(n-1)\pi}{n},$$

$$\frac{\nu e^2 a}{b^3}=\frac{e^2}{4a^2}S_n,$$

or

$$\frac{a^3}{b^3}=\frac{S_n}{4\nu}. \quad . \quad . \quad . \quad . \quad . \quad . \quad . \quad . \quad . \quad . \quad (1)$$

The following are the values of S_n from $n=2$ to $n=6$.

$S_2=1$, $S_3=2{\cdot}3094$, $S_4=3{\cdot}8284$, $S_5=5{\cdot}5056$, $S_6=7{\cdot}3094$.

In the important case when $\nu=n$, *i. e.* when the positive charge on the sphere is equal to the sum of all the negative charges in the ring of corpuscles, we get by (1) the following values for a/b :—

n.	$\frac{a}{b}$.
2	·5
3	·5773
4	·6208
5	·6505
6	·6726

If the ring of corpuscles, instead of being at rest, is rotating with an angular velocity ω, the condition for steady motion is

$$\frac{\nu e^2 a}{b^3}=ma\omega^2+\frac{e^2}{4a^2}S_n,$$

or

$$\frac{\nu a^3}{b^3}=\frac{m}{e^2}\omega^2+\frac{S_n}{4};$$

here m is the mass of a corpuscle.

We shall now proceed to find the forces acting on a corpuscle when the corpuscles are slightly displaced from their positions of equilibrium. Let the position of the corpuscles be fixed by the polar coordinates r and θ in the plane of the undisturbed orbit, and by the displacement z at right angles to this plane; let r_s, θ_s, z_s be the coordinates of the sth corpuscle; then, since the corpuscles are but slightly displaced from their positions of equilibrium, $r_s=a+\rho_s$ where ρ_s is small compared with a, z_s is also small compared with a, and $\theta_s-\theta_{s-1}=\frac{2\pi}{n}+\phi_s-\phi_{s-1}$, where n is the number of corpuscles and the ϕ's are small quantities.

The radial repulsion exerted by the sth corpuscle on the pth is equal to

$$-e^2\frac{d}{dr_p}\frac{1}{(r_p^2+r_s^2-2r_pr_s\cos(\theta_s-\theta_p)+(z_p-z_s)^2)^{\frac{1}{2}}};$$

expanding this, retaining only the first powers of ρ, ϕ, and z, we find that if R_{ps} is this repulsion

$$R_{ps}=\frac{e^2}{4a^2\sin\psi}\left\{1-\frac{\rho_p}{a}\left(\frac{3}{2}-\frac{1}{2\sin^2\psi}\right)\right.$$
$$\left.-\frac{\rho_s}{a}\left(\frac{1}{2}+\frac{1}{2\sin^2\psi}\right)-\tfrac{1}{2}(\phi_s-\phi_p)\cot\psi\right\},$$

where $\psi=(p-s)\frac{\pi}{n}$.

The tangential force Θ_{ps} tending to increase θ_p is equal to

$$-\frac{e^2}{r_pd\theta_p}\frac{1}{\{r_p^2+r_s^2-2r_pr_s\cos(\theta_s-\theta_p)+(z_p-z_s)^2\}^{\frac{1}{2}}};$$

expanding this and retaining only the first powers of the small quantities, we get

$$\Theta_{ps}=-\frac{e^2}{4a^2}\frac{\cos\psi}{\sin^2\psi}\left\{1-\frac{3}{2}\frac{\rho_p}{a}-\frac{1}{2}\frac{\rho_s}{a}-(\phi_s-\phi_p)(\cot\psi+\tfrac{1}{2}\tan\psi)\right\}.$$

Z_{ps}, the force at right angles to the undisturbed plane of the orbit, is easily seen to be given by the equation

$$Z_{ps}=\frac{e^2}{8a^3\sin^3\psi}(z_p-z_s).$$

The total radial force R_p exerted on the pth corpuscle by all the other corpuscles, is equal to

$$\frac{e^2}{4a^2}S-\rho_pA'-\Sigma\rho_{p+s}A_{p.p+s}-a\Sigma\phi_{p+s}B_{p.p+s},$$

where
$$S=\frac{1}{\sin\frac{\pi}{n}}+\frac{1}{\sin\frac{2\pi}{n}}+\cdots\frac{1}{\sin\frac{(n-1)\pi}{n}};$$

$$A'=\frac{e^2}{4a^3}\left(\frac{3}{2}\left(\frac{1}{\sin\frac{\pi}{n}}+\frac{1}{\sin\frac{2\pi}{n}}+\cdots\frac{1}{\sin\frac{(n-1)\pi}{n}}\right)\right.$$
$$\left.-\frac{1}{2}\left(\frac{1}{\sin^3\frac{\pi}{n}}+\frac{1}{\sin^3\frac{2\pi}{n}}+\cdots\frac{1}{\sin^3\frac{(n-1)\pi}{n}}\right)\right);$$

240 Prof. J. J. Thomson *on the*

$$A_{p.p+s}=\frac{e^2}{8a^3}\left(\frac{1}{\sin\frac{s\pi}{n}}+\frac{1}{\sin^3\frac{s\pi}{n}}\right);$$

$$B_{p.p+s}=\frac{e^3}{8a^3}\frac{\cos\frac{s\pi}{n}}{\sin^2\frac{s\pi}{n}}.$$

The coefficient of ϕ_p in the expression for R_p vanishes, since

$$\Sigma\frac{\cos\frac{s\pi}{n}}{\sin^2\frac{s\pi}{n}}=0.$$

As $A_{p.p+s}$, $B_{p.p+s}$ do not involve p, it is more convenient to use the symbols A_s and B_s for these quantities, and to write

$$R_p=\frac{e^2}{4a^2}S-\rho_p A'-\Sigma\rho_{p+s}A_s-a\Sigma\phi_{p+s}B_s.$$

The tangential force Θ_p acting on the p'th particle may similarly be written

$$\Theta_p=\Sigma\rho_{p+s}B_s-a\phi_p C+a\Sigma\phi_{p+s}C_s,$$

where

$$C=\frac{e^2}{4a^3}\left(\frac{\cos\frac{\pi}{n}}{\sin^2\frac{\pi}{n}}\left(\cot\frac{\pi}{n}+\frac{1}{2}\tan\frac{\pi}{n}\right)+\frac{\cos\frac{2\pi}{n}}{\sin^2\frac{2\pi}{n}}\left(\cot\frac{2\pi}{n}+\frac{1}{2}\tan\frac{2\pi}{n}\right)+\ldots\right),$$

$$C_s=\frac{e^2}{4a^3}\frac{\cos\frac{s\pi}{n}}{\sin^2\frac{s\pi}{n}}\left(\cot\frac{s\pi}{n}+\frac{1}{2}\tan\frac{s\pi}{n}\right);$$

while Z_p, the force at right angles to the plane of the orbit, is given by the equation

$$Z_p=z_p D-\Sigma z_{p+s}D_s,$$

where

$$D=\frac{e^2}{8a^3}\left(\frac{1}{\sin^3\frac{\pi}{n}}+\frac{1}{\sin^3\frac{2\pi}{n}}+\cdots\frac{1}{\sin^3\frac{(n-1)\pi}{n}}\right),$$

and

$$D_s=\frac{e^2}{8a^3}\frac{1}{\sin^3\frac{s\pi}{n}}.$$

The equations of motion of the pth corpuscle are

$$m\left(\frac{d^2r_p}{dt^2}-r_p\left(\frac{d\theta_p}{dt}\right)^2\right)=-\frac{\nu e^2 r_p}{b^3}+R_p\,; \quad . \quad . \quad (\alpha)$$

$$m\left(r\frac{d^2\theta_p}{dt^2}+2\frac{dr}{dt}\frac{d\theta_p}{dt}\right)=\Theta_p\,; \quad . \quad . \quad . \quad . \quad . \quad (\beta)$$

$$m\frac{d^2z_p}{dt^2}=-\frac{\nu e^2}{b^3}z_p+Z_p. \quad . \quad . \quad . \quad . \quad . \quad . \quad . \quad (\gamma)$$

Retaining only the first powers of small quantities, we get from these equations, if ω is the value of $\frac{d\theta}{dt}$ when the motion is steady,

$$\frac{\nu e^2 a}{b^3}=ma\omega^2+\frac{e^2}{4a^2}S,$$

$$m\frac{d^2\rho_p}{dt^2}-2ma\omega\frac{d\theta_p}{dt}=\rho_p\left(m\omega^2-\frac{\nu e^2}{b^3}\right)+R_p-\frac{e^2}{4a^2}S.$$

If ρ_p and θ_p vary as $e^{\iota qt}$, this equation may be written

$$(A-mq^2)\rho_p+A_1\rho_{p+1}+A_2\rho_{p+2}+\dots$$
$$-2ma\omega\iota q\phi_p+aB_1\phi_{p+1}+aB_2\phi_{p+2}+\dots=0,$$

where

$$A=\frac{e^2}{4a^3}S+A'=\frac{e^2}{8a^3}\left\{5\left(\frac{1}{\sin\frac{\pi}{n}}+\frac{1}{\sin\frac{2\pi}{n}}+\cdots\right)-\left(\frac{1}{\sin^3\frac{\pi}{n}}+\frac{1}{\sin^3\frac{2\pi}{n}}+\cdots\right)\right\}$$

Writing 1, 2, 3 for p we get

$$\left.\begin{aligned}
&(A-mq^2)\rho_1+A_1\rho_2+A_2\rho_3\dots+A_{n-1}\rho_n-2ma\omega\iota q\phi_1+aB_1\phi_2+aB_2\phi_3+\dots=0\\
&(A-mq^2)\rho_2+A_1\rho_3+A_2\rho_4+\dots\dots\dots \quad -2ma\omega\iota q\phi_2+aB_1\phi_3+aB_2\phi_4+\dots=0\\
&\quad . \quad . \quad . \quad . \quad\quad\quad\quad . \quad . \quad . \quad . \quad . \quad . \quad .\\
&(A-mq^2)\rho_n+A_1\rho_1+A_2\rho_2+\dots\dots\dots \quad -2ma\omega\iota q\phi_n+aB_1\phi_1+aB_2\phi_2+\dots=0
\end{aligned}\right\}(A)$$

By equation β we have

$$2m\omega\iota q\frac{\rho_p}{a}-B_1\frac{\rho_{p+1}}{a}-B_2\frac{\rho_{p+2}}{a}+\dots(C-mq^2)\phi_p-C_1\phi_{p+1}-C_2\phi_{p+2}-\dots=0.$$

Writing 1, 2, 3 in succession for p we get

$$\left.\begin{aligned}
&2\iota m\omega q\frac{\rho_1}{a}-B_1\frac{\rho_2}{a}-B_2\frac{\rho_3}{a}\dots \quad +(C-mq^2)\phi_1-C_1\phi_2-C_2\phi_3-\dots=0\\
&2\iota m\omega q\frac{\rho_2}{a}-B_1\frac{\rho_3}{a}-B_2\frac{\rho_4}{a}\dots \quad +(C-mq^2)\phi_2-C_2\phi_3-C_2\phi_4-\dots=0\\
&\quad . \quad . \quad . \quad . \quad\quad\quad . \quad . \quad . \quad . \quad . \quad .\\
&2\iota m\omega q\frac{\rho_n}{a}-B_1\frac{\rho_1}{a}-B_2\frac{\rho_2}{a}\dots \quad +(C-mq^2)\phi_n-C_1\phi_1-C_2\phi_2-\dots=0
\end{aligned}\right\}(B)$$

To solve equations A and B we notice that if ω be any root of the equation $x^n=1$, *i. e.* if ω be one of the nth roots of unity, equations A will be satisfied by

$$\rho_2=\omega\rho_1,\ \rho_3=\omega\rho_2,\ \rho_4=\omega\rho_3 \dots \quad \phi_2=\omega\phi_1,\ \phi_3=\omega\phi_2,\ \phi_4=\omega\phi_3 \dots$$

provided

$$\rho_1(A-mq^2+\omega A_1+\omega^2 A_2+\dots\omega^{n-1}A_{n-1}) + \phi_1 a(-2\iota m\omega q+\omega B_1+\omega^2 B_2+\omega^{n-1}B_{n-1})=0\,; \quad (1)$$

while equations B will be satisfied by the same values provided

$$\rho_1(2\iota m\omega q-\omega B_1-\omega^2 B_2-\omega^{n-1}B_{n-1}) + \phi_1 a(C-mq^2-\omega C_1-\omega^2 C_2-\omega^{n-1}C_{n-1})=0. \quad (2)$$

Hence, if both sets of equations are satisfied by these values, we have, eliminating ρ_1 and ϕ_1 from (1) and (2),

$$((A-mq^2)+\omega A_1+\omega^2 A_2+\dots\omega^{n-1}A_{n-1})(C-mq^2-\omega C_1-\omega^2 C_2-\omega^{n-1}C_{n-1}) = -(-2\iota m\omega q+\omega B_1+\omega^2 B_2+\dots\omega^{n-1}B_n)^2, \quad (1)$$

a biquadratic equation to determine q the frequency of the oscillations of the system. Now ω is of the form

$$\cos\frac{2k\pi}{n}+\iota\sin\frac{2k\pi}{n},$$

where k is an integer between 0 and $n-1$. Substituting this value for ω, we find

$$\omega A_1+\omega^2 A_2+\omega^{n-1}A_{n-1}=\frac{e^2}{8a^3}\left\{\cos\frac{2k\pi}{n}\left(\frac{1}{\sin\frac{\pi}{n}}+\frac{1}{\sin^3\frac{\pi}{n}}\right)\right.$$

$$\left.+\cos\frac{4k\pi}{n}\left(\frac{1}{\sin\frac{2\pi}{n}}+\frac{1}{\sin^3\frac{2\pi}{n}}\right)+\cos\frac{6k\pi}{n}\left(\frac{1}{\sin\frac{3\pi}{n}}+\frac{1}{\sin^3\frac{3\pi}{n}}\right)+\dots\right\}$$

We shall denote this by L_k; it will be noticed that L_k contains no imaginary terms. We find also that

$$\omega C_1+\omega^2 C_2+\omega^3 C_3+\omega^{n-1}C_{n-1}=\frac{e^2}{4a^3}\left(\cos\frac{2k\pi}{n}\frac{\cos\frac{\pi}{n}}{\sin^2\frac{\pi}{n}}\left(\cot\frac{\pi}{n}+\frac{1}{2}\tan\frac{\pi}{n}\right)\right.$$

$$+\cos\frac{4k\pi}{n}\frac{\cos\frac{2\pi}{n}}{\sin^2\frac{2\pi}{n}}\left(\cot\frac{2\pi}{n}+\frac{1}{2}\tan\frac{2\pi}{n}\right)$$

$$\left.+\dots\right).$$

We shall denote this by N_k.

Again,

$$\omega B_1+\omega^2B_2+\omega^{n-1}B_{n-1}=\frac{\iota\,.\,e^2}{8a^3}\left(\sin\frac{2k\pi}{n}\frac{\cos\frac{\pi}{n}}{\sin^2\frac{\pi}{n}}+\sin\frac{4k\pi}{n}\frac{\cos\frac{2\pi}{n}}{\sin^2\frac{2\pi}{n}}\right.$$

$$\left.+\sin\frac{6k\pi}{n}\frac{\cos\frac{3\pi}{n}}{\sin^2\frac{3\pi}{n}}+\ldots\right)$$

$$=\iota M_k,\text{ say.}$$

Substituting these values, equation (1) becomes

$$((A-mq^2)+L_k)(C-mq^2-N_k)=(M_k-2m\omega q)^2. \quad . \quad (2)$$

From the value of C given on p. 240 we see that C is the value of N_k when $k=0$, and so may be denoted by N_0, and that $A=\frac{3}{4}\frac{e^2}{a^3}S-L_0$; hence equation (2) may be written

$$\left(\frac{3}{4}\frac{e^2}{a^3}S+L_k-L_0-mq^2\right)(N_0-N_k-mq^2)=(M_k-2m\omega q)^2. \quad (3)$$

k in this equation may have any value from 0 to $(n-1)$; but we see that if we write $n-k$ for k, the values of q given by the two equations differ only in sign, and so give the same frequencies; thus all the values of q can be got by putting $k=0,1,\ldots\frac{n-1}{2}$, if n be odd, or $k=0,1,\frac{n}{2}$ if n be even; thus if n be odd there are $\frac{n+1}{2}$ equations of the type (3). When $k=0$, $M_k=0$, and (3) reduces to a quadratic equation; so that the number of roots of these $\frac{n+1}{2}$ equations is $4\times\frac{n+1}{2}-2=2n$; if n be even there are $\frac{n}{2}+1$ equations; but as $M_k=0$ when $k=0$ and $k=\frac{n}{2}$, two of these reduce to quadratics; so that the number of roots of these equations is $4\left(\frac{n}{2}+1\right)-4=2n$. Thus in each case the number of roots is equal to $2n$, the number of degrees of freedom of the corpuscles in the plane of their undisturbed orbit.

Let us now consider the motion at right angles to this plane. By equation γ we have

$$m\frac{d^2z_p}{dt^2}=-\frac{\nu e^2}{b^3}z_p+Dz_p-\Sigma D_s z_{p+s};$$

or if z_p is proportional to $\epsilon^{\iota qt}$,

$$z_p\left(\frac{\nu e^2}{b^3}-\mathrm{D}-mq^2\right)+\Sigma \mathrm{D}_s z_{p+s}=0\ ;$$

thus

$$\left.\begin{aligned}
&z_1\left(\frac{\nu e^2}{b^3}-\mathrm{D}-mq^2\right)+z_2\mathrm{D}_1+z_3\mathrm{D}_2+\ldots=0,\\
&z_2\left(\frac{\nu e^2}{b^3}-\mathrm{D}-mq^2\right)+z_3\mathrm{D}_1+z_4\mathrm{D}_2+\ldots=0,\\
&\cdot\quad\cdot\quad\cdot\quad\cdot\quad\cdot\quad\cdot\\
&z_n\left(\frac{\nu e^2}{b^3}-\mathrm{D}-mq^2\right)+z_1\mathrm{D}_1+z_2\mathrm{D}_2+\ldots=0.
\end{aligned}\right\}\quad . \quad . \quad (\mathrm{C})$$

We see that ω again being one of the nth roots of unity, the solution of equations C is

$$z_2=\omega z_1, \qquad z_3=\omega z_2, \qquad z_4=\omega z_3 \ldots$$

$$\frac{\nu e^2}{b^3}-\mathrm{D}-mq^2+\omega \mathrm{D}_1+\omega^2\mathrm{D}_2+\omega^{n-1}\mathrm{D}_{n-1}=0. \quad . \quad . \quad (4)$$

Putting $\omega=\cos\frac{2k\pi}{n}+\iota\sin\frac{2k\pi}{n}$, we find, substituting the values for D given above, that

$$\omega\mathrm{D}_1+\omega^2\mathrm{D}_2+\omega^{n-1}\mathrm{D}_{n-1}=\frac{e^2}{8a^3}\left(\frac{\cos\frac{2k\pi}{n}}{\sin^3\frac{\pi}{n}}+\frac{\cos\frac{4k\pi}{n}}{\sin^3\frac{2\pi}{n}}+\frac{\cos\frac{6k\pi}{n}}{\sin^3\frac{3\pi}{n}}+\ldots\right)$$

Denoting this by P_k and noticing that $\mathrm{D}=\mathrm{P}_0$, we find that equation (4) becomes

$$\frac{\nu e}{b^3}+\mathrm{P}_k-\mathrm{P}_0-mq^2=0.$$

Putting in succession $k=0, 1, \ldots n-1$, we get n values of q giving the n frequencies corresponding to the displacements at right angles to the plane of the undisturbed orbit.

We shall now proceed to calculate the frequencies for systems containing various numbers of corpuscles. The four quantities L_k, M_k, N_k, P_k which occur in the frequency equation may be expressed in terms of three quantities S_k, T_k, U_k, where

$$\mathrm{S}_k=\Sigma_1^{n-1}\cos\frac{2ks\pi}{n}\frac{1}{\sin\frac{s\pi}{n}},$$

$$T_k=\Sigma_1^{n-1}\cos\frac{2ks\pi}{n}\,\frac{1}{\sin^3\frac{s\pi}{n}},$$

$$U_k=\Sigma_1^{n-1}\sin\frac{2ks\pi}{n}\,\frac{\cos\frac{s\pi}{n}}{\sin^2\frac{s\pi}{n}};$$

for we have

$$L_k=(S_k+T_k)\frac{e^2}{8a^3},\quad N_k=(2T_k-S_k)\frac{e^2}{8a^3},$$

$$M_k=U_k\frac{e^2}{8a^3},\qquad P_k=T_k\frac{e^2}{8a^3}.$$

Case of two corpuscles.

When $n=2$ we have

$$L_0=\frac{2e^2}{8a^3},\quad M_0=0,\quad N_0=\frac{e^2}{8a^3},\quad P_0=\frac{e^2}{8a^3},$$

$$L_1=-\frac{2e^2}{8a^3},\quad M_1=0,\quad N_1=-\frac{e^2}{8a^3},\quad P_1=-\frac{e^2}{8a^3}.$$

Hence for vibrations in the plane of the orbit we have, when $k=0$,

$$\left(\frac{3}{4}\frac{e^2}{a^3}-mq^2\right)(-mq^2)=4m^2\omega^2q^2;$$

the roots of this equation are

$$q=0,\qquad q=\sqrt{\frac{3}{4}\frac{e^2}{ma^3}+4\omega^2}=\sqrt{\frac{3ve^2}{mb^3}+\omega^2}.$$

When $k=1$, the frequency equation is

$$\left(\frac{1}{4}\frac{e^2}{a^3}-mq^2\right)^2=4m^2\omega^2q^2;$$

the roots of this equation are

$$q=\quad\omega\pm\sqrt{\frac{1}{4}\frac{e^2}{ma^3}+\omega^2}=\omega\pm\sqrt{\frac{ve^2}{mb^3}}$$

and

$$q=-\omega\pm\sqrt{\frac{1}{4}\frac{e^2}{ma^3}+\omega^2}=-\omega\pm\sqrt{\frac{ve^2}{mb^3}},$$

the second set of values only differing in sign from the first.

For the vibrations perpendicular to the plane of the orbit, we have for $k=0$,

$$q=\sqrt{\frac{\nu e^2}{mb^3}};$$

for $k=1$,

$$q=\sqrt{\frac{\nu e^2}{mb^3}-\frac{e^2}{4ma^3}}=\omega.$$

Thus the six frequencies corresponding to the six degrees of freedom of the two corpuscles are

$$0,\ \omega,\ \sqrt{\frac{\nu e}{mb^3}},\ \sqrt{\frac{\nu e^2}{mb^3}}-\omega,\ \sqrt{\frac{\nu e^2}{mb^3}}+\omega,\ \sqrt{\frac{3\nu e^2}{mb^3}+\omega^2}.$$

When the corpuscles are not rotating round the circle, two of these roots are zero, three equal to $\sqrt{\frac{\nu e^2}{mb^3}}$, and the sixth equal to $\sqrt{\frac{3\nu e^2}{mb^3}}$. Thus the effect of rotation on the triple frequency $\sqrt{\nu e^2/mb^3}$ is to separate the roots, one remaining unaltered, one increasing, and the other diminishing.

Case of three corpuscles.
When $n=3$.

$$S_0=\frac{4}{\sqrt{3}},\quad T_0=\frac{16}{3\sqrt{3}},\quad U_0=0,\quad L_0=\frac{28}{3\sqrt{3}}\frac{e^2}{8a^3},$$

$$N_0=\frac{20}{3\sqrt{3}}\frac{e^2}{8a^3},\quad M_0=0,\quad P_0=\frac{16}{3\sqrt{3}}\frac{e^2}{8a^3},$$

$$S_1=-\frac{2}{\sqrt{3}},\quad T_1=-\frac{8}{3\sqrt{3}},\quad U_1=\frac{2}{\sqrt{3}},\quad L_1=-\frac{14}{3\sqrt{3}}\frac{e^2}{8a^3},$$

$$N_1=-\frac{10}{3\sqrt{3}}\frac{e^2}{8a^3},\quad M_1=\frac{2}{\sqrt{3}}\frac{e^2}{8a^3},\quad P_1=-\frac{8}{3\sqrt{3}}\frac{e^2}{8a^3},$$

$$S_2=S_1,\quad T_2=T_1,\quad U_2=-U_1,\quad L_2=L_1,\quad N_2=N_1,$$
$$M_2=-M_1,\quad P_2=P_1.$$

For the vibrations in the plane of the orbit, when $k=0$, the frequency equation is

$$\left(\sqrt{3}\frac{e^2}{a^3}-mq^2\right)(-mq^2)=4m^2\omega^2q^2;$$

the solution of this is

$$q=0 \text{ and } q=\left\{\sqrt{3}\frac{e^2}{ma^3}+4\omega^2\right\}^{\frac{1}{2}}=\left\{\frac{3\nu e^2}{mb^3}+\omega^2\right\}^{\frac{1}{2}}.$$

When $k=1$, the frequency equation is

$$\left(\frac{5}{4\sqrt{3}}\frac{e^2}{a^3}-mq^2\right)^2=\left(\frac{2}{\sqrt{3}}\frac{e^2}{8a^3}-2m\omega q\right)^2.$$

The solution of this equation is

$$q=\quad \omega\pm\sqrt{\frac{1}{\sqrt{3}}\frac{e^2}{ma^3}+\omega^2}=\omega\pm\sqrt{\frac{\nu e^2}{mb^3}},$$

$$q=-\omega\pm\sqrt{\frac{\sqrt{3}}{2}\frac{e^2}{ma^3}+\omega^2}=-\omega\pm\sqrt{\frac{3}{2}\frac{\nu e^2}{mb^3}-\tfrac{1}{2}\omega^2}.$$

When $k=2$ the frequencies are the same as when $k=1$; we have thus six frequencies corresponding to the six degrees of freedom of the three corpuscles in the plane of their undisturbed orbit.

For the vibration at right angles to the plane of this orbit, when $k=0$ the frequency equation is

$$\frac{\nu e^2}{b^3}-mq^2=0,$$

or

$$q=\sqrt{\frac{\nu e^2}{mb^3}}.$$

When $k=1$, the frequency equation is

$$\frac{\nu e^2}{b^3}-\frac{e^2}{\sqrt{3}a^3}-mq^2=0,$$

or

$$q=\pm\omega.$$

In the case of three corpuscles, as in that of two, we see that when there is no rotation three of the periods are equal; these are separated when the corpuscles are in rotation.

Case of four corpuscles.

When $n=4$,

$$S_0=1+2\sqrt{2},\quad T_0=4\sqrt{2}+1,\quad U_0=0,\quad L_0=(6\sqrt{2}+2)\frac{e^2}{8a^3},$$

$$N_0=(6\sqrt{2}+1)\frac{e^2}{8a^3},\quad M_0=0,\quad P_0=(4\sqrt{2}+1)\frac{e^2}{8a^3},$$

$$S_1=-1,\qquad T_1=-1,\qquad U_1=2\sqrt{2},\qquad L_1=-2\frac{e^2}{8a^3},$$

$$N_1=-\frac{e^2}{8a^3},\quad M_1=2\sqrt{2}\frac{e^2}{8a^3},\quad P_1=-\frac{e^2}{8a^3},$$

$$S_2=-2\sqrt{2}+1,\quad T_2=-4\sqrt{2}+1,\quad U_2=0.\quad L_2=(-6\sqrt{2}+2)\frac{e^2}{8a^3},$$

$$N_2=(-6\sqrt{2}+1)\frac{e^2}{8a^3},\quad M_2=0,\quad P_2=(-4\sqrt{2}+1)\frac{e^2}{8a^3}.$$

When $k=0$, the frequency equation is

$$\left(\frac{3}{4}\frac{e^2}{a^3}(1+2\sqrt{2})-mq^2\right)(-mq^2)=4m^2\omega^2q^2;$$

the solution of which is

$$q=0,\quad q=\sqrt{\frac{3}{4}\frac{e^2}{a^3}(1+2\sqrt{2})+4\omega^2}=\sqrt{\frac{3\nu e^2}{mb^3}+\omega^2}.$$

When $k=1$, the frequency equation is

$$\left((6\sqrt{2}+2)\frac{e^2}{8a^3}-mq^2\right)^2=\left(2\sqrt{2}\frac{e^2}{8a^3}-2m\omega q\right)^2;$$

the solution of this is

$$q=\quad\omega\pm\sqrt{\frac{2\sqrt{2}+1}{4}\frac{e^2}{ma^3}+\omega^2}=\quad\omega\pm\sqrt{\frac{\nu e^2}{mb^3}},$$

$$q=-\omega\pm\sqrt{\frac{4\sqrt{2}+1}{4}\frac{e^2}{ma^3}+\omega^2}=-\omega\pm\sqrt{\frac{4\sqrt{2}+1}{2\sqrt{2}+1}\frac{\nu e^2}{mb^2}-\frac{2\sqrt{2}\omega^2}{2\sqrt{2}+1}}.$$

When $k=2$, the frequency equation is

$$\left(\frac{3}{4}\frac{e^2}{a^3}-mq^2\right)\left(\frac{3}{\sqrt{2}}\frac{e^2}{a^3}-mq^2\right)=4m^2\omega^2q^2.$$

Regarding this as a quadratic in q^2, we see that the roots are positive, so that the values of q are real and the arrangement is stable. The roots of the equation are

$$q^2=\frac{3}{8\sqrt{2}}(4+\sqrt{2})\frac{e^2}{ma^3}+2\omega^2$$

$$\pm\sqrt{\frac{9(4-\sqrt{2})^2}{128}\frac{e^4}{m^2a^6}+\frac{3(\sqrt{2}+4)}{2\sqrt{2}}\frac{e^2}{ma^3}\omega^2+4\omega^4}.$$

Let us now consider the motion at right angles to the plane of the orbit. When $k=0$, the frequency equation is

$$\frac{\nu e^2}{b^3}-mq^2=0,$$

or

$$q=\sqrt{\frac{\nu e^2}{mb^3}}.$$

When $k=1$, the frequency equation is

$$\frac{\nu e^2}{b^3}-(4\sqrt{2}+2)\frac{e^2}{8a^3}-mq^2=0,$$

or

$$q=\pm\omega.$$

When $k=2$, the frequency equation is

$$\frac{\nu e^2}{b^3}-\frac{8\sqrt{2}e^2}{8a^3}-mq^2=0,$$

or

$$q^2=\frac{8\sqrt{2}}{4\sqrt{2}+2}\omega^2-\frac{(4\sqrt{2}-2)}{4\sqrt{2}+2}\frac{\nu e^2}{mb^3}.$$

Thus, unless

$$\omega^2>\frac{4\sqrt{2}-2}{8\sqrt{2}}\frac{\nu e^2}{mb^3}>\cdot 325\frac{\nu e^2}{mb^3}, \quad . \quad . \quad . \quad (1)$$

q^2 is negative, and the equilibrum is unstable, the four corpuscles then arranging themselves at the corner of a regular tetrahedron. When, however, ω is large enough to satisfy condition (1), four corpuscles will be in equilibrium when in steady motion in one plane at the corners of a square.

Case of five corpuscles.
When $n=5$, we have

$$S_0=5\cdot 5056, \quad T_0=12\cdot 1732, \quad U_0=0, \quad L_0=\frac{e^2}{8a^3}(17\cdot 6788),$$

$$N_0=\frac{e}{8a^3}(18\cdot 8408), \quad M_0=0, \quad P_0=\frac{e^2}{8a^3}(12\cdot 1732),$$

$$S_1=-\cdot 65, \quad T_1=1\cdot 1609, \quad U_1=4\cdot 856, \quad L_1=\frac{e^2}{8a^3}(\cdot 511),$$

$$N_1=\frac{e^2}{8a^3}(2\cdot 9716), \quad M_1=\frac{e^2}{8a^3}4\cdot 856, \quad P_1=\frac{e^2}{8a^3}(1\cdot 1609),$$

$$S_2=-2\cdot 103, \quad T_1=-7\cdot 249, \quad U_2=2\cdot 103, \quad L_2=-\frac{e^2}{8a^3}(9\cdot 352),$$

$$N_2=-\frac{e^2}{8a^3}12\cdot 4, \quad M_2=\frac{e^2}{8a^3}2\cdot 103, \quad P_2=-\frac{e^2}{8a^3}7\cdot 249.$$

The frequency equation when $k=0$ is

$$\left(\frac{3}{4}\;5\cdot 5056\frac{e^2}{a^3}-mq^2\right)(-mq^2)=4m^2\omega^2q^2,$$

the solution of which is

$$q=0, \quad q=\sqrt{\frac{3\nu e}{b^3}+\omega^2}.$$

When $k=1$, the frequency equation is

$$\left(15\cdot 87\frac{e^2}{8a^3}-mq^2\right)^2=\left(\frac{e^2}{8a^3}4\cdot 856-2m\omega q\right)^2,$$

250 Prof. J. J. Thomson *on the*

or

$$q=\omega\pm\sqrt{10\cdot918\frac{e^2}{8a^3}+\omega^2}=\omega\pm\sqrt{\frac{\nu e^2}{mb^3}},$$

$$q=-\omega\pm\sqrt{20\cdot726\frac{e^2}{8a^3}+\omega^2}.$$

When $k=2$,

$$\left(6\frac{e^2}{8a^3}-mq^2\right)\left(31\cdot24\frac{e^2}{8a^3}-mq^2\right)=\left(2\cdot103\frac{e^2}{8a^3}-2m\omega q\right)^2.$$

By applying the usual methods we find that all the roots of this equation are real, so that the steady motion of the five particles is stable for displacements in the plane of the orbit.

Let us now consider displacements at right angles to the plane of the orbit. When $k=0$ the frequency equation is

$$\frac{\nu e^2}{b^3}-mq^2=0,$$

the solution of which is

$$q=\sqrt{\frac{\nu e^2}{mb^3}}.$$

When $k=1$, the frequency equation is

$$m\omega^2-mq^2=0,$$

hence

$$q=\omega.$$

When $k=2$, the frequency equation is

$$\frac{\nu e^2}{b^3}-19\cdot42\frac{e^2}{8a^3}-mq^2=0,$$

or

$$\frac{19\cdot42}{11}m\omega^2+\frac{\nu e^2}{b^3}-\frac{19\cdot42}{11}\frac{\nu e^2}{b^3}-mq^2=0,$$

$$\frac{19\cdot42}{11}m\omega^2-\frac{8\cdot42}{11}\frac{\nu e^2}{b^3}-mq^2=0.$$

Hence, in order that the equilibrium may be stable,

$$\omega^2 \text{ must be} > \frac{8\cdot42}{19\cdot42}\frac{\nu e^2}{b^3}>\cdot433\frac{\nu e^2}{mb^3}.$$

Thus the five corpuscles are unstable when in one plane unless the angular velocity exceeds a certain value; the arrangement is stable, however, when the angular velocity is large.

Case of six corpuscles.

When $n=6$,

$$S_0=5+\frac{4}{\sqrt{3}},\quad T_0=17+\frac{16}{3\sqrt{3}},\quad U_0=0,\quad L_0=\left(22+\frac{28}{3\sqrt{3}}\right)\frac{e^2}{8a^3},$$

$$N_0=\left(29+\frac{20}{3\sqrt{3}}\right)\frac{e^2}{8a^3},\quad M_0=0,\quad P_0=\left(17+\frac{16}{3\sqrt{3}}\right)\frac{e^2}{8a^3},$$

$$S_1=1-\frac{2}{\sqrt{3}},\quad T_1=7-\frac{8}{3\sqrt{3}},\quad U_1=6+\frac{2}{\sqrt{3}},\quad L_1=\left(8-\frac{14}{3\sqrt{3}}\right)\frac{e^2}{8a^3},$$

$$N_1=\left(13-\frac{10}{3\sqrt{3}}\right)\frac{e^2}{8a^3},\quad M_1=\left(7-\frac{8}{3\sqrt{3}}\right)\frac{e^2}{8a^3},\quad P_1=\left(7-\frac{8}{3\sqrt{3}}\right)\frac{e^2}{8a^3},$$

$$S_2=-1-\frac{2}{\sqrt{3}},\quad T_2=-7-\frac{8}{3\sqrt{3}},\quad U_2=6-\frac{2}{\sqrt{3}},\quad L_2=\left(-8-\frac{14}{3\sqrt{3}}\right)\frac{e^2}{8a^3},$$

$$N_2=\left(-13-\frac{10}{3\sqrt{3}}\right)\frac{e^2}{8a^3},\quad M_2=\left(6-\frac{2}{\sqrt{3}}\right)\frac{e^2}{8a^3},\quad P_2=\left(-7-\frac{8}{3\sqrt{3}}\right)\frac{e^2}{8a^3},$$

$$S_3=-5+\frac{4}{\sqrt{3}},\quad T_3=-17+\frac{16}{3\sqrt{3}},\quad U_3=0,\quad L_3=\left(-22+\frac{28}{3\sqrt{3}}\right)\frac{e^2}{8a^3},$$

$$N_3=\left(-29+\frac{20}{3\sqrt{3}}\right)\frac{e^2}{8a^3},\quad M_3=0,\quad P_3=\left(-17+\frac{16}{3\sqrt{3}}\right)\frac{e^2}{8a^3}.$$

It is not necessary to write down all the frequency equations because, as we shall show, the arrangement of six corpuscles is unstable. For when $k=3$ the frequency equation is

$$\left(\frac{6}{8}\left(5+\frac{4}{\sqrt{3}}\right)\frac{e^2}{8a^3}-44\,\frac{e^2}{8a^3}-mq^2\right)\left(58\,\frac{e^2}{8a^3}-mq^2\right)=4m^2\omega^2q^2,$$

or

$$\left(-\frac{(14-8\sqrt{3})}{8}\frac{e^2}{a^3}-mq^2\right)\left(58\,\frac{e^2}{8a^3}-mq^2\right)=4m^2\omega^2q^2. \quad (1)$$

As $14-8\sqrt{3}$ is positive, we see that one of the roots of this equation for q^2 is negative, so that q is imaginary; this shows that the steady motion of 6 corpuscles in a ring is unstable, however rapid the rotation. We can, however, make the motion stable by putting a corpuscle at the centre; if we have a negative charge equal to that of p corpuscles at the centre of the ring the radial force it exerts on the sth corpuscle is $\frac{pe^2}{(a+\rho)^2}$, or $\frac{pe^2}{a^2}-\frac{2pe^2\rho}{a^3}$. Introducing this term into the expression for the radial force we find the frequency equation becomes

$$\left(\frac{3}{4}\frac{e^2}{a^3}S_0+\frac{3pe^2}{a^3}+L_k-L_0-mq^2\right)(N_0-N_k-mq^2)=(M_k-2m\omega q)^2.$$

Using this frequency equation, and supposing that $p=1$, *i. e.* that there is only one corpuscle at the centre of the hexagon, we get instead of (1),

$$\left(\frac{10+8\sqrt{3}}{8}\frac{e^2}{a^3}-mq^2\right)\left(\frac{58}{8}\frac{e^2}{a^3}-mq^2\right)=4m^2\omega^2q^2. \quad . \quad (2)$$

The roots of this equation in q^2 are both positive, so that q is real and the equilibrium is stable.

Let us now investigate the conditions for stability for displacements at right angles to the plane of the orbit.

For the motion at right angles to the plane of the ring, the frequency equation when $k=3$ is

$$\frac{\nu e^2}{b^3}-\frac{pe^2}{a^3}-\frac{34}{8}\frac{e^2}{a^3}-mq^2=0.$$

For this to represent the displacement of a stable system q^2 must be positive, so that if $p=1$

$$\frac{\nu e^2}{b^3}-\frac{e^2}{a^3}-\frac{34}{8}\frac{e^2}{a^3}$$

must be positive; we have, however,

$$\frac{\nu e^2}{b^3}=m\omega^2+\frac{e^2}{a^3}+\frac{e^2}{4a^3}\left(5+\frac{4}{\sqrt{3}}\right);$$

so that for

$$\frac{\nu e^2}{b^3}-\frac{e^2}{a^3}-\frac{34}{8}\frac{e^2}{a^3}$$

to be positive

$$m\omega^2 \text{ must be greater than } \frac{12-4/\sqrt{3}}{21}\frac{\nu e^2}{b^3},\ i.\,e.\ \cdot 46\,\frac{\nu e^2}{b^3}.$$

Let us now consider the stability of the corpuscle at the centre of the ring: if it is displaced through a distance z at right angles to the ring, the equation of motion of the corpuscles is

$$m\frac{d^2z}{dt^2}=-\frac{\nu e^2}{b^3}z+\frac{6e^2}{a^3}z.$$

Thus if the motion is stable

$$\frac{\nu e^2}{b^3}>\frac{6e^2}{a^3},$$

or

$$m\omega^2 > \frac{15-4/\sqrt{3}}{24}\frac{\nu e^2}{b^3},\ i.\ e.\ \cdot 53\,\frac{\nu e^2}{b^3}.$$

This value of ω^2 is greater than that required to make the equilibrium of the ring stable for displacements at right angles to its plane; if the central corpuscle, instead of being in the plane of the ring, was one side of the centre of the sphere of positive electrification while the ring was on the other side, the rotation required to make the equilibrium of the detached corpuscle stable would be less than when it was in the plane of the ring; for equilibrium the distance of the detached corpuscle from the centre of the sphere must be six times the distance of the plane of the ring from that point.

Conditions for the stability of rings containing more than six corpuscles.

I find that a single corpuscle in the centre is sufficient to make rings of 7 and 8 corpuscles stable; in the latter case, however, one of the values of q^2 though positive is exceedingly small. When the number of corpuscles exceeds 8 the number of central corpuscles required to ensure stability increases very rapidly with the number of corpuscles in the ring.

The frequency equation is

$$\left(\frac{3}{4}\frac{e^2 S_0}{a^3}+\frac{3pe^2}{a^3}-(L_0-L_k)-mq^2\right)\left(N_0-N_k-mq^2\right)=(M_k-2m\omega q)^2.$$

Now N_0-N_k is always positive and M is small compared with L and N; hence this equation will have real roots if

$$\frac{3}{4}\frac{e^2 S_0}{a^3}+\frac{3pe^2}{a^3}-(L_0-L_k)$$

is positive. The greatest value of L_0-L_k is got by putting $k=n/2$ when n is even, and $=(n-1)/2$ when n is odd: hence the condition that the values of q should be real, *i. e.* that the equilibrium of the ring should be stable, is

$$\frac{3pe^2}{a^3}>(L_0-L_{\frac{n}{2}})-\frac{3}{4}\frac{e^2 S_0}{a^3}\quad\text{when } n \text{ is even,}$$

and

$$\frac{3pe^2}{a^3}>(L_0-L_{\frac{n-1}{2}})-\frac{3}{4}\frac{e^2 S_0}{a^3}\quad\text{when } n \text{ is odd.}$$

From this equation we can calculate the least value of p which will make a ring of n corpuscles stable. The values of

p for a series of values of n are given in the following table:—

n ...	5	6	7	8	9	10	15	20	30	40
p ...	0	1	1	1	2	3	15	39	101	232

For large values of n the values of p are proportional to n^3. When p is greater than one, the internal corpuscles necessary to produce equilibrium cannot all be at the centre of the sphere, they will separate until their repulsions are balanced by the attraction of the positive electricity in the sphere. Thus when there are two internal corpuscles, as when $n=9$, these two will separate and will form a pair with the line joining them parallel to the plane of the ring. If we assume, as is approximately the case, that the pair of equal corpuscles exerts at external points the same force as a double charge placed at a point midway between them, the preceding theory will apply, and the system consisting of the ring of 9 and the pair of corpuscles will be in stable equilibrium. When $n=10$, the internal corpuscles must be three in number; these three will arrange themselves at the corners of an equilateral triangle, and the system of 13 corpuscles will consist of a ring of 10 and a triangle of 3, the planes of the ring and triangle being parallel but not coincident; the corpuscles are all supposed to be in rapid rotation round the diameter of the sphere drawn at right angles to the planes of the ring. For a ring of 12 corpuscles we require 7 inside, but 7 corpuscles, as we have seen, cannot form a single ring, but will arrange themselves as a ring of 6 with one at the centre. Thus the system of 19 corpuscles will consist of an outer ring of 12, an inner ring of 6 in a plane parallel to the outer ring, and one corpuscle along the axis of rotation.

In this way we see that when we have a large number of corpuscles in rapid rotation they will arrange themselves as follows:—The corpuscles form a series of rings, the corpuscles in one ring being approximately in a plane at right angles to the axis of rotation, the number of particles in the rings diminishing as the radius of the ring diminishes. If the corpuscles can move at right angles to the plane of their orbit, the rings will be in different planes adjusting themselves so that the repulsion between the rings is balanced by the attraction exerted by the positive electrification of the sphere in which they are placed. We have thus in the first place a sphere of uniform positive electrification, and inside this sphere a number of corpuscles arranged in a series of parallel rings, the number of corpuscles in a ring varying from ring to ring: each corpuscle is travelling at a high speed round the circumference of the ring in which it is situated, and the rings are so arranged that those which contain a large number of corpuscles are near the surface of the sphere, while those in which there are a smaller number of corpuscles are more in the inside.

If the corpuscles, like the poles of the little magnets in Mayer's experiments with the floating magnets, are constrained to move in one plane, they would, even if not in rotation, be in equilibrium when arranged in the series of rings just described. The rotation is required to make the arrangement stable when the corpuscles can move at right angles to the plane of the ring.

Application of the preceding Results to the Theory of the Structure of the Atom.

We suppose that the atom consists of a number of corpuscles moving about in a sphere of uniform positive electrification: the problems we have to solve are (1) what would be the structure of such an atom, *i. e.* how would the corpuscles arrange themselves in the sphere; and (2) what properties would this structure confer upon the atom. The solution of (1) when the corpuscles are constrained to move in one plane is indicated by the results we have just obtained—the corpuscles will arrange themselves in a series of concentric rings. This arrangement is necessitated by the fact that a large number of corpuscles cannot be in stable equilibrium when arranged as a single ring, while this ring can be made stable by placing inside it an appropriate number of corpuscles. When the corpuscles are not constrained to one plane, but can move about in all directions, they will arrange themselves in a series of concentric shells; for we can easily see that, as in the case of the ring, a number of corpuscles distributed over the surface of a shell will not be in stable equilibrium if the number of corpuscles is large, unless there are other corpuscles inside the shell, while the equilibrium can be made stable by introducing within the shell an appropriate number of other corpuscles.

The analytical and geometrical difficulties of the problem of the distribution of the corpuscles when they are arranged in shells are much greater than when they are arranged in rings, and I have not as yet succeeded in getting a general solution. We can see, however, that the same kind of properties will be associated with the shells as with the rings; and as our solution of the latter case enables us to give definite results, I shall confine myself to this case, and endeavour to show that the properties conferred on the atom by this ring structure are analogous in many respects to those possessed by the atoms of the chemical elements, and that in particular the properties of the atom will depend upon its atomic weight in a way very analogous to that expressed by the periodic law.

Let us suppose, then, that we have N corpuscles each carrying a charge e of negative electricity, placed in a sphere of positive electrification, the whole charge in the sphere being equal to Ne; let us find the distribution of the corpuscles when they are arranged in what we may consider to be the simplest way, *i.e.* when the number of rings is a minimum, so that in each ring there are as nearly as possible as many corpuscles as it is possible for the corpuscles inside to hold in equilibrium. Let us suppose that the number of internal corpuscles required to make the equilibrium of a ring of n corpuscles stable is $f(n)$. The value of $f(n)$ for a series of values of n is given in the table on page 254; in that table $f(n)$ is denoted by p. The number of corpuscles in the outer ring n_1 will then be determined by the condition that $\mathrm{N}-n_1$, the number of corpuscles inside, must be just sufficient to keep the ring of n_1 corpuscles in equilibrium, *i.e.*, n_1 will be determined by the equation

$$\mathrm{N}-n_1=f(n_1). \quad . \quad . \quad . \quad . \quad . \quad . \quad (1)$$

If the value of n_1 got from this equation is not an integer we must take the integral part of the value.

To get n_2, the number of corpuscles in the second ring, we notice that there must be $\mathrm{N}-n_1-n_2$ corpuscles inside; hence n_2 is given by the equation

$$\mathrm{N}-n_1-n_2=f(n_2).$$

Similarly, n_3, n_4, ..., the number of corpuscles in the 3rd, 4th, &c. rings reckoned from the outside, are given by

$$\mathrm{N}-n_1-n_2-n_3=f(n_3),$$
$$\mathrm{N}-n_1-n_2-n_3-n_4=f(n_4).$$

These equations can be solved very rapidly by a graphical method. Draw the graph whose abscissa $=f(n)$ and whose ordinate is n. The values of $f(n)$ for a series of values of n are given on page 254; from these values the curve fig. 1 has been constructed.

To find how a number of corpuscles equal to N will arrange themselves, measure off on the axis of abscissæ a distance from O equal to N. Let OP be this distance, through P draw PQ inclined at an angle of 135° to the horizontal axis, cutting the curve in Q, draw the ordinate QM; then the integral part of QM will be the value of n_1, the number of

Fig. 1.

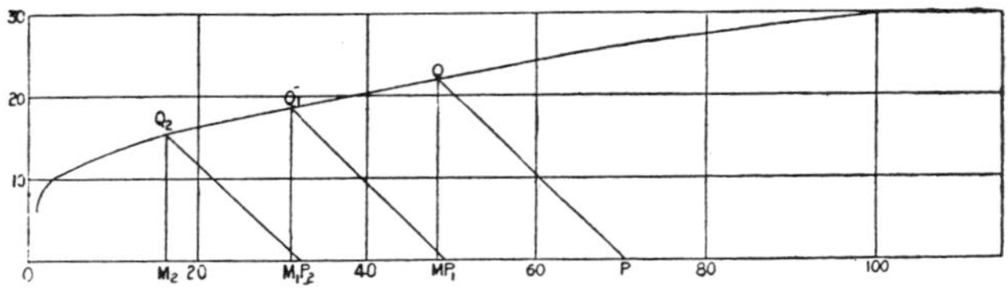

corpuscles in the first ring reckoned from the outside. For evidently

$$OM = f(QM),$$

and $OM = ON - NM$, and since PQ is inclined at 45° to the axis, $NM = OM$; hence

$$ON - QM = f(QM).$$

Comparing this with equation (1) we see that the integral part of QM is the value of n_1.

To get the value of n_2, the number of corpuscles in the second ring, we mark off the abscissa $OP_1 = N - n_1$ (if QM is an integer P_1 will coincide with M), then from P_1 draw P_1Q_1 parallel to PQ cutting the curve in Q_1; the integral part of Q_1M_1 will be the value of n_2. To get n_3 mark off the abscissa $OP_2 = N - n_1 - n_2$, and draw P_2Q_2 parallel to PQ; the integral part of Q_2M_2 will be the value of n_3. In this way we can in a very short time find the configuration.

The following table, which gives the way in which various numbers of corpuscles group themselves, has been calculated in this way; the numbers range downwards from 60 at intervals of 5.

Number of corpuscles	60.	55.	50.	45.	40.	35.
Number in successive rings...	20 16 13 8 3	19 16 12 7 1	18 15 11 5 1	17 14 10 4	16 13 8 3	16 12 6 1
Number of corpuscles	30.	25.	20.	15.	10.	5.
Number in successive rings...	15 10 5	13 9 3	12 7 1	10 5	8 2	5

We give also the entire series of arrangement of corpuscles for which the outer ring consists of 20 corpuscles.

Number of corpuscles	59.	60.	61.	62.	63.	64.	65.	66.	67.
Number in successive rings...	20	20	20	20	20	20	20	20	20
	16	16	16	17	17	17	17	17	17
	13	13	13	13	13	13	14	14	15
	8	8	9	9	10	10	10	10	10
	2	3	3	3	3	4	4	5	5

59 is the smallest number of corpuscles which can have an outer ring of 20, while when the number of corpuscles is greater than 67 the outer ring will contain more than 20 corpuscles.

Let us now consider the connexion between these results and the properties possessed by the atoms of the chemical elements. We suppose that the mass of an atom is the sum of the masses of the corpuscles it contains, so that the atomic weight of an element is measured by the number of corpuscles in its atom. An inspection of the results just given will show that systems built up of rings of corpuscles in the way we have described, will possess properties analogous to some of those possessed by the atom. In the first place, we see that the various arrangements of the corpuscles can be classified in families, the grouping of the corpuscles in the various members of the family having certain features in common. Thus, for example, we see that the group of 60 corpuscles consists of the same rings of corpuscles as the group of 40 with an additional ring of 20 corpuscles round it, while the group of 40 consists of the same series of rings as the group of 24 with an additional ring outside, while 24 is the group 11 with an additional ring. To continue the series for larger numbers of corpuscles, take the curve $x=f(y)$ when $f(n)$ is the number of corpuscles that must be placed

Fig. 2.

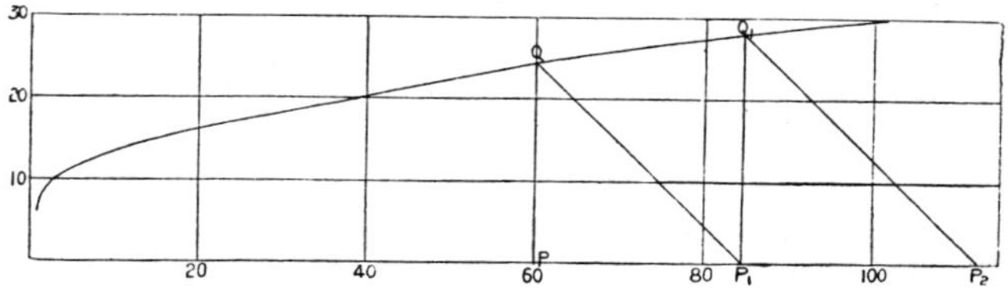

inside a ring of n corpuscles to make it stable. Let Q be the point on this curve corresponding to 60 corpuscles, *i. e.* OP=60, from Q draw QP_1 inclined at an angle of 135° to the axis of x; then the number of corpuscles represented by OP_1 will be arranged like the 60 corpuscles with an addition ring of Q_1P_1 corpuscles (fig. 2). To find the next member of the family, draw Q_1P_2 parallel to QP_1 cutting the axis of x in P_2, then OP_2 will represent the number of corpuscles in the next member of the family; and by continuing the process we can find the successive members. Thus we see that we can divide the various groups of atoms into series such that each member of the series is derived from the preceding member (*i. e.* the member next below it in atomic weight) by adding to it another ring of corpuscles. We should expect the atoms formed by a series of corpuscles of this kind to have many points of resemblance. Take, for example, the vibrations of the corpuscles; these may be divided into two sets:—(1) Those arising from the rotation of the corpuscles around their orbits: if all the corpuscles in one atom have the same angular velocity, the frequency of the vibrations produced by the rotation of the ring of corpuscles is proportional to the number of corpuscles in the ring; and thus in the spectrum of each element in the series there would be a series of frequencies bearing the same ratio to each other, the ratio of the frequencies being the ratios of the numbers in the various rings.

The second system of vibrations are those arising from the displacement of the ring from its circular figure. If now the distance of a corpuscle in the outer ring from a corpuscle in the collection of rings inside it is great compared with the distance of the second corpuscle from its nearest neighbour on its own ring, the effect of the outer ring of corpuscles on the inner set of rings will only "disturb" the vibrations of the latter without fundamentally altering the character of their vibrations. Thus for these vibrations, as well as for those due to the rotations, the sequence of frequencies would present much the same features for the various elements in the series; there would be in the spectrum corresponding groups of associated lines. We regard a series of atoms formed in this way, *i. e.* when the atom of the pth member is formed from that of the $(p-1)$th by the addition of a single ring of corpuscles, as belonging to elements in the same group in the arrangement of the elements according to the periodic law; *i. e.*, they form a series which, if arranged according to Mendeléef's table, would all be in the same vertical column.

The gradual change in the properties of the elements which takes place as we travel along one of the horizontal rows in Mendeléef's arrangement of the elements, is also illustrated by the properties possessed by these groups of corpuscles. Thus consider the series of arrangements of the corpuscles given on p. 258, in all of which the outer ring contains 20 corpuscles An outer row of 20 corpuscles first occurs with 59 corpuscles; in this case the number of corpuscles inside is only just sufficient to make the outer ring stable; this ring will therefore be on the verge of instability, and when the corpuscles in this ring are displaced the forces of restitution urging them back to their original position will be small. Thus when this ring is subjected to disturbances from an external source, one or more corpuscles may easily be detached from it; such an atom therefore will easily lose a negatively electrified corpuscle, and thus acquire a charge of positive electricity; such an atom would behave like the atom of a strongly electropositive element. When we pass from 59 to 60 corpuscles the outer ring is more stable, because there is an additional corpuscle inside it; the corresponding atom will thus not be so electropositive as that containing only 59 corpuscles. The addition of each successive corpuscle will make it more difficult to detach corpuscles from the outer ring, and will therefore make the atom less electropositive. When the stability of the outer ring gets very great, it may be possible for one or more corpuscles to be on the surface of the atom without breaking up the ring; in this case the atom could receive a charge of negative electricity, and would behave like the atom of an electronegative element. The increase in the stability of the ring, and consequently in the electronegative character of the atom, would go on increasing until we had as many as 67 corpuscles, when the stability of the outer ring would be at a maximum. A great change in the properties of the atom would occur with 68 corpuscles, for now the number of corpuscles in the outer ring increases to 21; these 21 corpuscles are, however, only just stable, and would, like the outer ring of 20 in the arrangement of the 59 corpuscles, readily lose a corpuscle and so make the atom strongly electropositive.

The properties of the groups of 59 and 67 corpuscles, which are respectively at the beginning and end of the series which has an outer ring of 20 corpuscles, deserve especial consideration. The arrangement of corpuscles in the group of 59, although very near the verge of instability, and therefore very liable to lose a corpuscle and thereby acquire a positive charge, would not be able to retain this charge. For when it had lost a corpuscle, the 58 corpuscles left would arrange themselves in the grouping corresponding to 58 corpuscles which is the last to have an outer ring of 19 corpuscles; this ring is therefore exceedingly stable so that no further corpuscles would escape from it, while the positive charge on the system due to the escape of the 59th corpuscle would attract the surrounding corpuscles. Thus this arrangement could not remain permanently charged; for as soon as one corpuscle had escaped it would be replaced by another. An atom constituted in this way would be neither electropositive nor electronegative, but one incapable of receiving permanently a charge of electricity.

The group containing 60 corpuscles would be the most electropositive of the series; but this could only lose one corpuscle; *i. e.* acquire a charge of one unit of positive electricity; for if it lost two we should have 58 corpuscles—as when the group of 59 had lost one corpuscle—and in this case the system would be even more likely than the other to attract external corpuscles, for it would have a charge of two units of positive electricity instead of one. Thus the system containing 60 corpuscles would get charged with one, but only one, unit of positive electricity: it would therefore act like the atom of a monovalent electropositive element.

The group containing 61 corpuscles would not part with its corpuscles so readily as the group of 60, but on the other hand it could afford to lose two, as it is not until it has lost three that its corpuscles are reduced to 58, when, as we have seen, it begins to acquire fresh corpuscles. Thus this system might get charged with two units of positive electricity, and would act like the atom of a divalent electropositive element. Similarly the group of 62, though less liable even than the 61 to lose its corpuscles, could, on the other hand, lose 3 without beginning to recover its corpuscles; it could thus acquire a charge of 3 units of positive electricity, and would act like the atom of a trivalent electropositive element.

Let us now go to the groups at the other end of the series and consider the properties of the last of the series, the group of 67 corpuscles. The outer ring would be very stable, but if the system acquired another corpuscle, the 68 corpuscles would arrange themselves with a ring of 21 corpuscles on the outside; as 68 is the smallest number of corpuscles with an outer ring of 21, the ring is very nearly unstable and easily loses a corpuscle. Thus the group of 67 corpuscles, as soon as it acquires a negative charge, would lose it again, and the system, like the group of 59, would be incapable of being permanently charged with electricity—it would act like the atom of an element of no valency.

The group of 66 would be the most electronegative of the series, but this would only be able to retain a charge of one unit of negative electricity; for if it acquired 2 units there would be 68 corpuscles, an arrangement which, as we have seen, rapidly loses its corpuscles. This group of 66 would therefore act like the atom of a monovalent electronegative element.

The group of 65 would be less liable than that of 66 to acquire negative corpuscles, but, on the other hand, it would under suitable circumstances be able to retain 2 corpuscles, and thus be charged with 2 units of negative electricity, and would act like the atom of a divalent electronegative element.

Similarly, the group of 64 would act like the atom of a trivalent electronegative element, and so on.

Thus, if we consider the series of arrangements of corpuscles having on the outside a ring containing a constant number of corpuscles, we have at the beginning and end systems which behave like the atoms of an element whose atoms are incapable of retaining a charge of either positive or negative electricity; then (proceeding in the order of increasing number of corpuscles) we have first a system which behaves like the atom of a monovalent electropositive element, next one which behaves like the atom of a divalent electropositive element, while at the other end of the series we have a system which behaves like an atom with no valency, immediately preceding this, one which behaves like the atom of a monovalent electronegative element, while this again is preceded by one behaving like the atom of a divalent electronegative element.

This sequence of properties is very like that observed in the case of the atoms of the elements.

Thus we have the series of elements:

He	Li	Be	B	C	N	O	F	Ne.
Ne	Na	Mg	Al	Si	P	S	Cl	Arg.

The first and last element in each of these series has no valency, the second is a monovalent electropositive element, the last but one is a monovalent electronegative element, the third is a divalent electropositive element, the last but two a divalent electronegative element, and so on.

When atoms like the electronegative ones, in which the corpuscles are very stable, are mixed with atoms like the electropositive ones, in which the corpuscles are not nearly so firmly held, the forces to which the corpuscles are subject by the action of the atoms upon each other may result in the detachment of corpuscles from the electropositive atoms and their transference to the electronegative. The electronegative atoms will thus get a charge of negative electricity, the electropositive atoms one of positive, the oppositely charged atoms will attract each other, and a chemical compound of the electropositive and electronegative atoms will be formed.

Just as an uncharged conducting sphere will by electrostatic induction attract a corpuscle in its neighbourhood, so a corpuscle outside an atom will be attracted, even though the atom has not become positively charged by losing a corpuscle. When the outside corpuscle is dragged into the atom there will be a diminution in the potential energy, the amount of this diminution depending on the number of corpuscles in the atom. If now we have an atom A such that loss of potential energy due to the fall into the atom of a corpuscle from outside is greater than the work required to drag a corpuscle from an atom B of a different kind, then an intimate mixture of A and B atoms will result in the A atoms dragging corpuscles from the B atoms, thus the A atoms will get negatively, the B atoms positively electrified, and the oppositely electrified atoms will combine, forming a compound such as A_-B_+; in such a case as this chemical combination might be expected whenever the atoms were brought into contact. Even when the loss of potential energy when a corpuscle falls into A is less than the work required to drag a corpuscle right away from B, the existence of a suitable physical environment may lead to chemical combination between A and B. For when a corpuscle is dragged out of and away from an atom a considerable portion of the work is spent on the corpuscle after it has left the atom, while of the work gained when a corpuscle falls into an atom, the proportion done outside to that done inside the atom is smaller than the proportion for the corresponding quantities when the corpuscle is dragged out of an atom. Thus, though the work required to move a corpuscle from B to an infinite distance may be greater than that gained when a corpuscle moves from an infinite distance into A, yet the work gained when a corpuscle went from the surface of A into its interior might be greater than the work required to move a corpuscle from the interior to the surface of B. In this case anything which diminished the forces on the corpuscle when they got outside the atom, as, for example, the presence of a medium of great specific inductive capacity such as water, or contact with a metal such as platinum black, would greatly increase the chance of chemical combination.

The Existence of Secondary Groups of Corpuscles within the Atom.

The expression given on p. 238 for the radius of a ring of corpuscles shows that it depends on $\nu e/b^3$, where νe is the amount of positive electrification within a sphere of radius b: thus ve/b^3 is equal to $\frac{4\pi}{3}\rho$, where ρ is the density of the positive electrification in the sphere: thus, if the density of the electrification be kept constant, the radius of the ring will be independent of the size of the sphere. Now let us take a large sphere and place within it a ring of such a size that the ring would be in stable equilibrium if its centre were at the centre of the sphere. To fix our ideas, let us take the case of three corpuscles at the corners of an equilateral triangle, and place this triangle so that its centre O′ is no longer at the centre of the sphere: we can easily see that the corpuscles will remain at the corners of an equilateral triangle of the same size, and that the triangle will move like a rigid body acted upon by a force proportional to the distance of its centre from O the centre of the sphere. To prove this we notice that the repulsion between the corpuscles is the same as when the centre of the triangle is at O. The attraction of the sphere on a corpuscle P is proportional to OP, and so may be resolved into two forces, one proportional to O′P along PO′ (O′ is the centre of the triangle) and the other proportional to OO′ acting along O′O. Now the corpuscles are by hypothesis in equilibrium under their mutual repulsions, and the attraction to the centre proportional to O′P: thus the relative position of the corpuscles will remain unaltered, and the system of three corpuscles will move as a rigid body under a central force acting on its centre of gravity proportional to the distance of that point from the centre of the sphere.

The three corpuscles will, at a point whose distance from their centre is large compared with a side of the triangle, produce the same effect as if the charges on the three corpuscles were condensed at the centre of the triangle; they will thus at such points act like a unit, and the results we have previously obtained for single corpuscles may be extended to the case when the single corpuscles are replaced by rings of corpuscles which would by themselves be in equilibrium. It should be noted that the atom in which these systems are placed must be large enough to allow these rings of corpuscles—sub-atoms we may call them, to be separated by distances considerably greater than the distance between the corpuscles in one of the rings.

If we regard the atoms of the heavier elements as produced by the coalescence of lighter atoms, it is reasonable to suppose that the corpuscles in the heavier atoms may be arranged in secondary groups or sub-atoms, each of these groups acting as a unit. When the corpuscles are done up in bundles in this way, it is possible to have stability when these bundles are arranged in a ring with a smaller number of corpuscles inside than when the corpuscles in the bundles are arranged at equal intervals round the circumference of the ring. Thus, take the case of a ring of 30 corpuscles; if these were arranged at equal intervals, 101 corpuscles would be required inside the ring to make it stable. If, however, the 30 corpuscles were grouped in ten sets of three each, only $3 \times 3 = 9$ corpuscles in the interior would be required to make the arrangement stable.

Constitution of the Atom of a Radioactive Element.

Our study of the stability of systems of corpuscles has made us acquainted with systems which are stable when the corpuscles are rotating with an angular velocity greater than a certain value, but which become unstable when the velocity falls below this value. Thus, to take an instance, we saw (p. 249) that four corpuscles can be stable in one plane at the corners of a square, if they are rotating with an angular velocity greater than $\cdot 325 ve^2/mb^3$, but become unstable if the velocity falls below this velocity, the corpuscles in this case tending to place themselves at the corners of a tetrahedron. Consider now the properties of an atom containing a system of corpuscles of this kind, suppose the corpuscles were originally moving with velocities far exceeding the critical velocity; in consequence of the radiation from the moving corpuscles, their velocities will slowly—very slowly—diminish; when, after a long interval, the velocity reaches the critical velocity, there will be what is equivalent to an explosion of the corpuscles, the corpuscles will move far away from their original positions, their potential energy will decrease, while their kinetic energy will increase. The kinetic energy gained in this way might be sufficient to carry the system out of the atom, and we should have, as in the case of radium, a part of the atom shot off. In consequence of the very slow dissipation of energy by radiation the life of the atom would be very long. We have taken the case of the four corpuscles as the type of a system which, like a top, requires for its stability a certain amount of rotation. Any system possessing this property would, in consequence of the gradual dissipation of energy by radiation, give to the atom containing it radioactive properties similar to those conferred by the four corpuscles.

Reprint 2 [669]

LXXIX. *The Scattering of α and β Particles by Matter and the Structure of the Atom.* *By* Professor E. RUTHERFORD, *F.R.S., University of Manchester* *.

§ 1. IT is well known that the α and β particles suffer deflexions from their rectilinear paths by encounters with atoms of matter. This scattering is far more marked for the β than for the α particle on account of the much smaller momentum and energy of the former particle. There seems to be no doubt that such swiftly moving particles pass through the atoms in their path, and that the deflexions observed are due to the strong electric field traversed within the atomic system. It has generally been supposed that the scattering of a pencil of α or β rays in passing through a thin plate of matter is the result of a multitude of small scatterings by the atoms of matter traversed. The observations, however, of Geiger and Marsden † on the scattering of α rays indicate that some of the α particles must suffer a deflexion of more than a right angle at a single encounter. They found, for example, that a small fraction of the incident α particles, about 1 in 20,000, were turned through an average angle of 90° in passing through a layer of gold-foil about ·00004 cm. thick, which was equivalent in stopping-power of the α particle to 1·6 millimetres of air. Geiger ‡ showed later that the most probable angle of deflexion for a pencil of α particles traversing a gold-foil of this thickness was about 0°·87. A simple calculation based on the theory of probability shows that the chance of an α particle being deflected through 90° is vanishingly small. In addition, it will be seen later that the distribution of the α particles for various angles of large deflexion does not follow the probability law to be expected if such large deflexions are made up of a large number of small deviations. It seems reasonable to suppose that the deflexion through a large angle is due to a single atomic encounter, for the chance of a second encounter of a kind to produce a large deflexion must in most cases be exceedingly small. A simple calculation shows that the atom must be a seat of an intense electric field in order to produce such a large deflexion at a single encounter.

Recently Sir J. J. Thomson § has put forward a theory to

* Communicated by the Author. A brief account of this paper was communicated to the Manchester Literary and Philosophical Society in February, 1911.

† Proc. Roy. Soc. lxxxii. p. 495 (1909).

‡ Proc. Roy. Soc. lxxxiii. p. 492 (1910).

§ Camb. Lit. & Phil. Soc. xv. pt. 5 (1910).

Reprinted from Philosophical Magazine and Journal of Science, Series 6, Vol. 21, No. 125, pp. 669–688 (1911).

explain the scattering of electrified particles in passing through small thicknesses of matter. The atom is supposed to consist of a number N of negatively charged corpuscles, accompanied by an equal quantity of positive electricity uniformly distributed throughout a sphere. The deflexion of a negatively electrified particle in passing through the atom is ascribed to two causes—(1) the repulsion of the corpuscles distributed through the atom, and (2) the attraction of the positive electricity in the atom. The deflexion of the particle in passing through the atom is supposed to be small, while the average deflexion after a large number m of encounters was taken as $\sqrt{m} \cdot \theta$, where θ is the average deflexion due to a single atom. It was shown that the number N of the electrons within the atom could be deduced from observations of the scattering of electrified particles. The accuracy of this theory of compound scattering was examined experimentally by Crowther * in a later paper. His results apparently confirmed the main conclusions of the theory, and he deduced, on the assumption that the positive electricity was continuous, that the number of electrons in an atom was about three times its atomic weight.

The theory of Sir J. J. Thomson is based on the assumption that the scattering due to a single atomic encounter is small, and the particular structure assumed for the atom does not admit of a very large deflexion of an α particle in traversing a single atom, unless it be supposed that the diameter of the sphere of positive electricity is minute compared with the diameter of the sphere of influence of the atom.

Since the α and β particles traverse the atom, it should be possible from a close study of the nature of the deflexion to form some idea of the constitution of the atom to produce the effects observed. In fact, the scattering of high-speed charged particles by the atoms of matter is one of the most promising methods of attack of this problem. The development of the scintillation method of counting single α particles affords unusual advantages of investigation, and the researches of H. Geiger by this method have already added much to our knowledge of the scattering of α rays by matter.

§ 2. We shall first examine theoretically the single encounters † with an atom of simple structure, which is able to

* Crowther, Proc. Roy. Soc. lxxxiv. p. 226 (1910).

† The deviation of a particle throughout a considerable angle from an encounter with a single atom will in this paper be called "single" scattering. The deviation of a particle resulting from a multitude of small deviations will be termed "compound" scattering.

produce large deflexions of an α particle, and then compare the deductions from the theory with the experimental data available.

Consider an atom which contains a charge $\pm Ne$ at its centre surrounded by a sphere of electrification containing a charge $\mp Ne$ supposed uniformly distributed throughout a sphere of radius R. e is the fundamental unit of charge, which in this paper is taken as $4{\cdot}65 \times 10^{-10}$ E.S. unit. We shall suppose that for distances less than 10^{-12} cm. the central charge and also the charge on the α particle may be supposed to be concentrated at a point. It will be shown that the main deductions from the theory are independent of whether the central charge is supposed to be positive or negative. For convenience, the sign will be assumed to be positive. The question of the stability of the atom proposed need not be considered at this stage, for this will obviously depend upon the minute structure of the atom, and on the motion of the constituent charged parts.

In order to form some idea of the forces required to deflect an α particle through a large angle, consider an atom containing a positive charge Ne at its centre, and surrounded by a distribution of negative electricity Ne uniformly distributed within a sphere of radius R. The electric force X and the potential V at a distance r from the centre of an atom for a point inside the atom, are given by

$$X = Ne\left(\frac{1}{r^2} - \frac{r}{R^3}\right)$$

$$V = Ne\left(\frac{1}{r} - \frac{3}{2R} + \frac{r^2}{2R^3}\right).$$

Suppose an α particle of mass m and velocity u and charge E shot directly towards the centre of the atom. It will be brought to rest at a distance b from the centre given by

$$\tfrac{1}{2}mu^2 = NeE\left(\frac{1}{b} - \frac{3}{2R} + \frac{b^2}{2R^3}\right).$$

It will be seen that b is an important quantity in later calculations. Assuming that the central charge is 100 e, it can be calculated that the value of b for an α particle of velocity $2{\cdot}09 \times 10^9$ cms. per second is about $3{\cdot}4 \times 10^{-12}$ cm. In this calculation b is supposed to be very small compared with R. Since R is supposed to be of the order of the radius of the atom, viz. 10^{-8} cm., it is obvious that the α particle before being turned back penetrates so close to the central charge, that the field due to the uniform distribution of negative electricity may be neglected. In general, a simple calculation shows that for all deflexions greater than a degree, we may without sensible error suppose the deflexion due to the field of the central charge alone. Possible single deviations due to the negative electricity, if distributed in the form of corpuscles, are not taken into account at this stage of the theory. It will be shown later that its effect is in general small compared with that due to the central field.

Consider the passage of a positive electrified particle close to the centre of an atom. Supposing that the velocity of the particle is not appreciably changed by its passage through the atom, the path of the particle under the influence of a repulsive force varying inversely as the square of the distance will be an hyperbola with the centre of the atom S as the external focus. Suppose the particle to enter the atom in the direction PO (fig. 1), and that the direction of motion

Fig. 1.

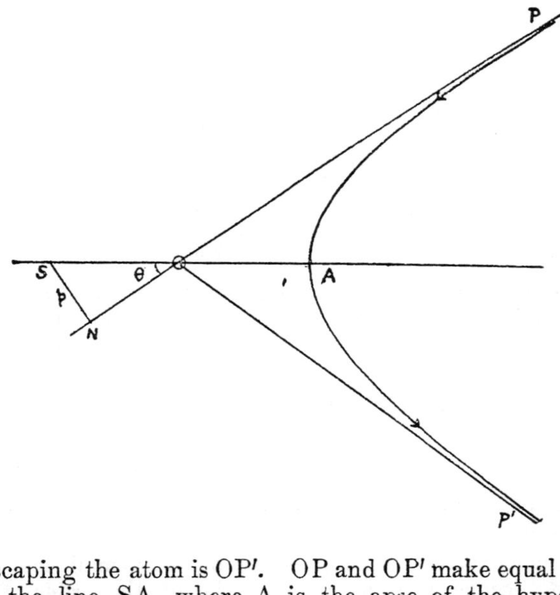

on escaping the atom is OP′. OP and OP′ make equal angles with the line SA, where A is the apse of the hyperbola. p = SN = perpendicular distance from centre on direction of initial motion of particle.

Let angle $POA=\theta$.

Let V=velocity of particle on entering the atom, v its velocity at A, then from consideration of angular momentum

$$pV=SA\,.\,v.$$

From conservation of energy

$$\tfrac{1}{2}mV^2=\tfrac{1}{2}mv^2-\frac{NeE}{SA},$$

$$v^2=V^2\left(1-\frac{b}{SA}\right).$$

Since the eccentricity is $\sec\theta$,

$$SA=SO+OA=p\operatorname{cosec}\theta(1+\cos\theta)$$
$$=p\cot\theta/2,$$
$$p^2=SA(SA-b)=p\cot\theta/2(p\cot\theta/2-b),$$
$$\therefore\quad b=2p\cot\theta.$$

The angle of deviation ϕ of the particle is $\pi-2\theta$ and

$$\cot\phi/2=\frac{2p}{b}{}^{*}\quad\ldots\ldots\quad(1)$$

This gives the angle of deviation of the particle in terms of b, and the perpendicular distance of the direction of projection from the centre of the atom.

For illustration, the angle of deviation ϕ for different values of p/b are shown in the following table :—

p/b	10	5	2	1	·5	·25	·125
ϕ	5°·7	11°·4	28°	53°	90°	127°	152°

§ 3. *Probability of single deflexion through any angle.*

Suppose a pencil of electrified particles to fall normally on a thin screen of matter of thickness t. With the exception of the few particles which are scattered through a large angle, the particles are supposed to pass nearly normally through the plate with only a small change of velocity. Let n=number of atoms in unit volume of material. Then the number of collisions of the particle with the atom of radius R is πR^2nt in the thickness t.

* A simple consideration shows that the deflexion is unaltered if the forces are attractive instead of repulsive.

The probabilty m of entering an atom within a distance p of its centre is given by

$$m = \pi p^2 nt.$$

Chance dm of striking within radii p and $p+dp$ is given by

$$dm = 2\pi pnt \,.\, dp = \frac{\pi}{4} ntb^2 \cot \phi/2 \operatorname{cosec}^2 \phi/2 \, d\phi, \quad (2)$$

since $$\cot \phi/2 = 2p/b.$$

The value of dm gives the *fraction* of the total number of particles which are deviated between the angles ϕ and $\phi + d\phi$.

The fraction ρ of the total number of particles which are deflected through an angle greater than ϕ is given by

$$\rho = \frac{\pi}{4} ntb^2 \cot^2 \phi/2. \quad . \quad . \quad . \quad . \quad . \quad (3)$$

The fraction ρ which is deflected between the angles ϕ_1 and ϕ_2 is given by

$$\rho = \frac{\pi}{4} ntb^2 \left(\cot^2 \frac{\phi_1}{2} - \cot^2 \frac{\phi_2}{2}\right). \quad . \quad . \quad . \quad (4)$$

It is convenient to express the equation (2) in another form for comparison with experiment. In the case of the α rays, the number of scintillations appearing on a *constant* area of a zinc sulphide screen are counted for different angles with the direction of incidence of the particles. Let r = distance from point of incidence of α rays on scattering material, then if Q be the total number of particles falling on the scattering material, the number y of α particles falling on unit area which are deflected through an angle ϕ is given by

$$y = \frac{Q dm}{2\pi r^2 \sin \phi \,.\, d\phi} = \frac{ntb^2 \,.\, Q \,.\, \operatorname{cosec}^4 \phi/2}{16 r^2}. \quad . \quad . \quad (5)$$

Since $b = \dfrac{2NeE}{mu^2}$, we see from this equation that the number of α particles (scintillations) per unit area of zinc sulphide screen at a given distance r from the point of incidence of the rays is proportional to

(1) $\text{cosec}^4 \phi/2$ or $1/\phi^4$ if ϕ be small;
(2) thickness of scattering material t provided this is small;
(3) magnitude of central charge Ne;
(4) and is inversely proportional to $(mu^2)^2$, or to the fourth power of the velocity if m be constant.

In these calculations, it is assumed that the α particles scattered through a large angle suffer only one large deflexion. For this to hold, it is essential that the thickness of the scattering material should be so small that the chance of a second encounter involving another large deflexion is very small. If, for example, the probability of a single deflexion ϕ in passing through a thickness t is 1/1000, the probability of two successive deflexions each of value ϕ is $1/10^6$, and is negligibly small.

The angular distribution of the α particles scattered from a thin metal sheet affords one of the simplest methods of testing the general correctness of this theory of single scattering. This has been done recently for α rays by Dr. Geiger *, who found that the distribution for particles deflected between 30° and 150° from a thin gold-foil was in substantial agreement with the theory. A more detailed account of these and other experiments to test the validity of the theory will be published later.

§ 4. *Alteration of velocity in an atomic encounter.*

It has so far been assumed that an α or β particle does not suffer an appreciable change of velocity as the result of a single atomic encounter resulting in a large deflexion of the particle. The effect of such an encounter in altering the velocity of the particle can be calculated on certain assumptions. It is supposed that only two systems are involved, viz., the swiftly moving particle and the atom which it traverses supposed initially at rest. It is supposed that the principle of conservation of momentum and of energy applies, and that there is no appreciable loss of energy or momentum by radiation.

* Manch. Lit. & Phil. Soc. 1910.

Let m be mass of the particle,

v_1 = velocity of approach,

v_2 = velocity of recession,

M = mass of atom,

V = velocity communicated to atom as result of encounter.

Let OA (fig. 2) represent in magnitude and direction the momentum mv_1 of the entering particle, and OB the momentum of the receding particle which has been turned through an angle AOB $=\phi$. Then BA represents in magnitude and direction the momentum MV of the recoiling atom.

Fig. 2.

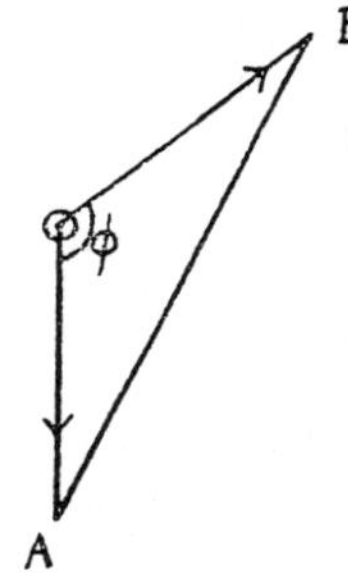

$$(MV)^2=(mv_1)^2+(mv_2)^2-2m^2v_1v_2\cos\phi. \quad (1)$$

By the conservation of energy

$$MV^2=mv_1{}^2-mv_2{}^2. \quad . \quad . \quad (2)$$

Suppose $M/m=K$ and $v_2=\rho v_1$, where ρ is <1.

From (1) and (2),

$$(K+1)\rho^2-2\rho\cos\phi=K-1,$$

or
$$\rho=\frac{\cos\phi}{K+1}+\frac{1}{K+1}\sqrt{K^2-\sin^2\phi}.$$

Consider the case of an α particle of atomic weight 4, deflected through an angle of 90° by an encounter with an atom of gold of atomic weight 197.

Since $K=49$ nearly,

$$\rho=\sqrt{\frac{K-1}{K+1}}=\cdot 979,$$

or the velocity of the particle is reduced only about 2 per cent. by the encounter.

In the case of aluminium $K=27/4$ and for $\phi=90°$ $\rho=\cdot 86$.

It is seen that the reduction of velocity of the α particle becomes marked on this theory for encounters with the lighter atoms. Since the range of an α particle in air or other matter is approximately proportional to the cube of the velocity, it follows that an α particle of range 7 cms. has its range reduced to 4·5 cms. after incurring a single deviation of 90° in traversing an aluminium atom. This is of a magnitude to be easily detected experimentally. Since the value of K is very large for an encounter of a β particle with an atom, the reduction of velocity on this formula is very small.

Some very interesting cases of the theory arise in considering the changes of velocity and the distribution of scattered particles when the α particle encounters a light atom, for example a hydrogen or helium atom. A discussion of these and similar cases is reserved until the question has been examined experimentally.

§ 5. *Comparison of single and compound scattering.*

Before comparing the results of theory with experiment, it is desirable to consider the relative importance of single and compound scattering in determining the distribution of the scattered particles. Since the atom is supposed to consist of a central charge surrounded by a uniform distribution of the opposite sign through a sphere of radius R, the chance of encounters with the atom involving small deflexions is very great compared with the chance of a single large deflexion.

This question of compound scattering has been examined by Sir J. J. Thomson in the paper previously discussed (§ 1). In the notation of this paper, the average deflexion ϕ_1 due to the field of the sphere of positive electricity of radius R and quantity Ne was found by him to be

$$\phi_1 = \frac{\pi}{4} \cdot \frac{NeE}{mu^2} \cdot \frac{1}{R}.$$

The average deflexion ϕ_2 due to the N negative corpuscles supposed distributed uniformly throughout the sphere was found to be

$$\phi_2 = \frac{16}{5} \frac{eE}{mu^2} \cdot \frac{1}{R} \sqrt{\frac{3N}{2}}.$$

The mean deflexion due to both positive and negative electricity was taken as

$$(\phi_1^2 + \phi_2^2)^{1/2}.$$

In a similar way, it is not difficult to calculate the average deflexion due to the atom with a central charge discussed in this paper.

Since the radial electric field X at any distance r from the centre is given by

$$X=Ne\left(\frac{1}{r^2}-\frac{r}{R^3}\right),$$

it is not difficult to show that the deflexion (supposed small) of an electrified particle due to this field is given by

$$\theta=\frac{b}{p}\left(1-\frac{p^2}{R^2}\right)^{3/2},$$

where p is the perpendicular from the centre on the path of the particle and b has the same value as before. It is seen that the value of θ increases with diminution of p and becomes great for small values of ϕ.

Since we have already seen that the deflexions become very large for a particle passing near the centre of the atom, it is obviously not correct to find the average value by assuming θ is small.

Taking R of the order 10^{-8} cm., the value of p for a large deflexion is for α and β particles of the order 10^{-11} cm. Since the chance of an encounter involving a large deflexion is small compared with the chance of small deflexions, a simple consideration shows that the average small deflexion is practically unaltered if the large deflexions are omitted. This is equivalent to integrating over that part of the cross section of the atom where the deflexions are small and neglecting the small central area. It can in this way be simply shown that the average small deflexion is given by

$$\phi_1=\frac{3\pi}{8}\frac{b}{R}.$$

This value of ϕ_1 for the atom with a concentrated central charge is three times the magnitude of the average deflexion for the same value of Ne in the type of atom examined by Sir J. J. Thomson. Combining the deflexions due to the electric field and to the corpuscles, the average deflexion is

$$(\phi_1^2+\phi_2^2)^2 \quad \text{or} \quad \frac{b}{2R}\left(5\cdot 54+\frac{15\cdot 4}{N}\right)^{1/2}.$$

It will be seen later that the value of N is nearly proportional to the atomic weight, and is about 100 for gold. The effect due to scattering of the individual corpuscles expressed by the second term of the equation is consequently small for heavy atoms compared with that due to the distributed electric field.

Neglecting the second term, the average deflexion per atom is $\frac{3\pi b}{8\mathrm{R}}$. We are now in a position to consider the relative effects on the distribution of particles due to single and to compound scattering. Following J. J. Thomson's argument, the average deflexion θ_t after passing through a thickness t of matter is proportional to the square root of the number of encounters and is given by

$$\theta_t = \frac{3\pi b}{8\mathrm{R}} \sqrt{\pi \mathrm{R}^2 . n . t} = \frac{3\pi b}{8} \sqrt{\pi n t},$$

where n as before is equal to the number of atoms per unit volume.

The probability p_1 for compound scattering that the deflexion of the particle is greater than ϕ is equal to $e^{-\phi^2/\theta_t^2}$.

Consequently $$\phi^2 = -\frac{9\pi^3}{64} b^2 nt \log p_1.$$

Next suppose that single scattering alone is operative. We have seen (§ 3) that the probability p_2 of a deflexion greater than ϕ is given by

$$p_2 = \frac{\pi}{4} b^2 . n . t \cot^2 \phi/2.$$

By comparing these two equations

$$p_2 \log p_1 = -\cdot 181 \phi^2 \cot^2 \phi/2,$$

ϕ is sufficiently small that

$$\tan \phi/2 = \phi/2,$$

$$p_2 \log p_1 = -\cdot 72.$$

If we suppose $p_2 = \cdot 5$, then $p_1 = \cdot 24$.

If $p_2 = \cdot 1$, $p_1 = \cdot 0004$.

It is evident from this comparison, that the probability for any given deflexion is always greater for single than for compound scattering. The difference is especially marked when only a small fraction of the particles are scattered through any given angle. It follows from this result that the distribution of particles due to encounters with the atoms is for small thicknesses mainly governed by single scattering. No doubt compound scattering produces some effect in equalizing the distribution of the scattered particles; but its effect becomes relatively smaller, the smaller the fraction of the particles scattered through a given angle.

§ 6. *Comparison of Theory with Experiments.*

On the present theory, the value of the central charge Ne is an important constant, and it is desirable to determine its value for different atoms. This can be most simply done by determining the small fraction of α or β particles of known velocity falling on a thin metal screen, which are scattered between ϕ and $\phi+d\phi$ where ϕ is the angle of deflexion. The influence of compound scattering should be small when this fraction is small.

Experiments in these directions are in progress, but it is desirable at this stage to discuss in the light of the present theory the data already published on scattering of α and β particles.

The following points will be discussed :—

(*a*) The "diffuse reflexion" of α particles, *i. e.* the scattering of α particles through large angles (Geiger and Marsden).

(*b*) The variation of diffuse reflexion with atomic weight of the radiator (Geiger and Marsden).

(*c*) The average scattering of a pencil of α rays transmitted through a thin metal plate (Geiger).

(*d*) The experiments of Crowther on the scattering of β rays of different velocities by various metals.

(*a*) In the paper of Geiger and Marsden (*loc. cit.*) on the diffuse reflexion of α particles falling on various substances it was shown that about 1/8000 of the α particles from radium C falling on a thick plate of platinum are scattered back in the direction of the incidence. This fraction is deduced on the assumption that the α particles are uniformly scattered in all directions, the observations being made for a deflexion of about 90°. The form of experiment is not very suited for accurate calculation, but from the data available it can be shown that the scattering observed is about that to be expected on the theory if the atom of platinum has a central charge of about 100 e.

(*b*) In their experiments on this subject, Geiger and Marsden gave the relative number of α particles diffusely reflected from thick layers of different metals, under similar conditions. The numbers obtained by them are given in the table below, where z represents the relative number of scattered particles, measured by the number of scintillations per minute on a zinc sulphide screen.

Metal.	Atomic weight.	z.	$z/A^{3/2}$.
Lead	207	62	208
Gold	197	67	242
Platinum	195	63	232
Tin	119	34	226
Silver	108	27	241
Copper	64	14·5	225
Iron	56	10·2	250
Aluminium	27	3·4	243
		Average	233

On the theory of single scattering, the fraction of the total number of α particles scattered through any given angle in passing through a thickness t is proportional to $n.A^2t$, assuming that the central charge is proportional to the atomic weight A. In the present case, the thickness of matter from which the scattered α particles are able to emerge and affect the zinc sulphide screen depends on the metal. Since Bragg has shown that the stopping power of an atom for an α particle is proportional to the square root of its atomic weight, the value of nt for different elements is proportional to $1/\sqrt{A}$. In this case t represents the greatest depth from which the scattered α particles emerge. The number z of α particles scattered back from a thick layer is consequently proportional to $A^{3/2}$ or $z/A^{3/2}$ should be a constant.

To compare this deduction with experiment, the relative values of the latter quotient are given in the last column. Considering the difficulty of the experiments, the agreement between theory and experiment is reasonably good *.

The single large scattering of α particles will obviously affect to some extent the shape of the Bragg ionization curve for a pencil of α rays. This effect of large scattering should be marked when the α rays have traversed screens of metals of high atomic weight, but should be small for atoms of light atomic weight.

(*c*) Geiger made a careful determination of the scattering of α particles passing through thin metal foils, by the scintillation method, and deduced the most probable angle

* The effect of change of velocity in an atomic encounter is neglected in this calculation.

through which the α particles are deflected in passing through known thicknesses of different kinds of matter.

A narrow pencil of homogeneous α rays was used as a source. After passing through the scattering foil, the total number of α particles deflected through different angles was directly measured. The angle for which the number of scattered particles was a maximum was taken as the most probable angle. The variation of the most probable angle with thickness of matter was determined, but calculation from these data is somewhat complicated by the variation of velocity of the α particles in their passage through the scattering material. A consideration of the curve of distribution of the α particles given in the paper (*loc. cit.* p. 496) shows that the angle through which half the particles are scattered is about 20 per cent greater than the most probable angle.

We have already seen that compound scattering may become important when about half the particles are scattered through a given angle, and it is difficult to disentangle in such cases the relative effects due to the two kinds of scattering. An approximate estimate can be made in the following way :— From (§ 5) the relation between the probabilities p_1 and p_2 for compound and single scattering respectively is given by

$$p_2 \log p_1 = -\cdot 721.$$

The probability q of the combined effects may as a first approximation be taken as

$$q = (p_1^2 + p_2^2)^{1/2}.$$

If $q = \cdot 5$, it follows that

$$p_1 = \cdot 2 \quad \text{and} \quad p_2 = \cdot 46.$$

We have seen that the probability p_2 of a single deflexion greater than ϕ is given by

$$p_2 = \frac{\pi}{4} n \, . \, t \, . \, b^2 \cot^2 \phi/2.$$

Since in the experiments considered ϕ is comparatively small

$$\frac{\phi \sqrt{p_2}}{\sqrt{\pi n t}} = b = \frac{2\mathrm{N}e\mathrm{E}}{mu^2}.$$

Geiger found that the most probable angle of scattering of the α rays in passing through a thickness of gold equivalent in stopping power to about $\cdot 76$ cm. of air was $1° \, 40'$. The angle ϕ through which half the α particles are turned thus corresponds to $2°$ nearly.

$$t = \cdot 00017 \text{ cm.} ; \; n = 6 \cdot 07 \times 10^{22} ;$$

$$u \text{ (average value)} = 1 \cdot 8 \times 10^9.$$

$$\mathrm{E}/m = 1 \cdot 5 \times 10^{14} \, . \text{ E.S. units} ; \; e = 4 \cdot 65 \times 10^{-10}.$$

Taking the probability of single scattering $= \cdot 46$ and substituting the above values in the formula, the value of N for gold comes out to be 97.

For a thickness of gold equivalent in stopping power to 2·12 cms. of air, Geiger found the most probable angle to be $3^\circ 40'$. In this case $t = \cdot 00047$, $\phi = 4^\circ \cdot 4$, and average $u = 1 \cdot 7 \times 10^9$, and N comes out to be 114.

Geiger showed that the most probable angle of deflexion for an atom was nearly proportional to its atomic weight. It consequently follows that the value of N for different atoms should be nearly proportional to their atomic weights, at any rate for atomic weights between gold and aluminium.

Since the atomic weight of platinum is nearly equal to that of gold, it follows from these considerations that the magnitude of the diffuse reflexion of α particles through more than 90° from gold and the magnitude of the average small angle scattering of a pencil of rays in passing through gold-foil are both explained on the hypothesis of single scattering by supposing the atom of gold has a central charge of about $100\,e$.

(*d*) *Experiments of Crowther on scattering of β rays.*—We shall now consider how far the experimental results of Crowther on scattering of β particles of different velocities by various materials can be explained on the general theory of single scattering. On this theory, the fraction of β particles p turned through an angle greater than ϕ is given by

$$p = \frac{\pi}{4} n \,.\, t \,.\, b^2 \cot^2 \phi/2.$$

In most of Crowther's experiments ϕ is sufficiently small that $\tan \phi/2$ may be put equal to $\phi/2$ without much error. Consequently

$$\phi^2 = 2\pi\, n \,.\, t \,.\, b^2 \quad \text{if } p = 1/2.$$

On the theory of compound scattering, we have already seen that the chance p_1 that the deflexion of the particles is greater than ϕ is given by

$$\phi^2/\log p_1 = -\frac{9\pi^3}{64} n \,.\, t \,.\, b^2.$$

Since in the experiments of Crowther the thickness t of matter was determined for which $p_1 = 1/2$,

$$\phi^2 = \cdot 96\pi\, n\, t\, b^2.$$

For a probability of 1/2, the theories of single and compound scattering are thus identical in general form, but differ by a numerical constant. It is thus clear that the main relations on the theory of compound scattering of Sir J. J. Thomson, which were verified experimentally by Crowther, hold equally well on the theory of single scattering.

For example, if t_m be the thickness for which half the particles are scattered through an angle ϕ, Crowther showed that $\phi/\sqrt{t_m}$ and also $\frac{mu^2}{E}\cdot\sqrt{t_m}$ were constants for a given material when ϕ was fixed. These relations hold also on the theory of single scattering. Notwithstanding this apparent similarity in form, the two theories are fundamentally different. In one case, the effects observed are due to cumulative effects of small deflexions, while in the other the large deflexions are supposed to result from a single encounter. The distribution of scattered particles is entirely different on the two theories when the probability of deflexion greater than ϕ is small.

We have already seen that the distribution of scattered α particles at various angles has been found by Geiger to be in substantial agreement with the theory of single scattering, but cannot be explained on the theory of compound scattering alone. Since there is every reason to believe that the laws of scattering of α and β particles are very similar, the law of distribution of scattered β particles should be the same as for α particles for small thicknesses of matter. Since the value of mu^2/E for the β particles is in most cases much smaller than the corresponding value for the α particles, the chance of large single deflexions for β particles in passing through a given thickness of matter is much greater than for α particles. Since on the theory of single scattering the fraction of the number of particles which are deflected through a given angle is proportional to kt, where t is the thickness supposed small and k a constant, the number of particles which are undeflected through this angle is proportional to $1-kt$. From considerations based on the theory of compound scattering, Sir J. J. Thomson deduced that the probability of deflexion less than ϕ is proportional to $1-e^{-\mu/t}$ where μ is a constant for any given value of ϕ.

The correctness of this latter formula was tested by Crowther by measuring electrically the fraction I/I_0 of the scattered β particles which passed through a circular opening subtending an angle of 36° with the scattering material. If

$$I/I_0=1-e^{-\mu/t},$$

the value of I should decrease very slowly at first with

increase of t. Crowther, using aluminium as scattering material, states that the variation of I/I_0 was in good accord with this theory for small values of t. On the other hand, if single scattering be present, as it undoubtedly is for α rays, the curve showing the relation between I/I_0 and t should be nearly linear in the initial stages. The experiments of Madsen * on scattering of β rays, although not made with quite so small a thickness of aluminium as that used by Crowther, certainly support such a conclusion. Considering the importance of the point at issue, further experiments on this question are desirable.

From the table given by Crowther of the value $\phi/\sqrt{t_m}$ for different elements for β rays of velocity $2{\cdot}68\times10^{10}$ cms. per second, the values of the central charge Ne can be calculated on the theory of single scattering. It is supposed, as in the case of the α rays, that for the given value of $\phi/\sqrt{t_m}$ the fraction of the β particles deflected by single scattering through an angle greater than ϕ is ·46 instead of ·5.

The values of N calculated from Crowther's data are given below.

Element.	Atomic weight.	$\phi/\sqrt{t_m}$.	N.
Aluminium	27	4·25	22
Copper	63·2	10·0	42
Silver	108	15·4	78
Platinum	194	29·0	138

It will be remembered that the values of N for gold deduced from scattering of the α rays were in two calculations 97 and 114. These numbers are somewhat smaller than the values given above for platinum (viz. 138), whose atomic weight is not very different from gold. Taking into account the uncertainties involved in the calculation from the experimental data, the agreement is sufficiently close to indicate that the same general laws of scattering hold for the α and β particles, notwithstanding the wide differences in the relative velocity and mass of these particles.

As in the case of the α rays, the value of N should be most simply determined for any given element by measuring

* Phil. Mag. xviii. p. 909 (1909).

the small fraction of the incident β particles scattered through a large angle. In this way, possible errors due to small scattering will be avoided.

The scattering data for the β rays, as well as for the α rays, indicate that the central charge in an atom is approximately proportional to its atomic weight. This falls in with the experimental deductions of Schmidt *. In his theory of absorption of β rays, he supposed that in traversing a thin sheet of matter, a small fraction α of the particles are stopped, and a small fraction β are reflected or scattered back in the direction of incidence. From comparison of the absorption curves of different elements, he deduced that the value of the constant β for different elements is proportional to nA^2 where n is the number of atoms per unit volume and A the atomic weight of the element. This is exactly the relation to be expected on the theory of single scattering if the central charge on an atom is proportional to its atomic weight.

§ 7. *General Considerations.*

In comparing the theory outlined in this paper with the experimental results, it has been supposed that the atom consists of a central charge supposed concentrated at a point, and that the large single deflexions of the α and β particles are mainly due to their passage through the strong central field. The effect of the equal and opposite compensating charge supposed distributed uniformly throughout a sphere has been neglected. Some of the evidence in support of these assumptions will now be briefly considered. For concreteness, consider the passage of a high speed α particle through an atom having a positive central charge Ne, and surrounded by a compensating charge of N electrons. Remembering that the mass, momentum, and kinetic energy of the α particle are very large compared with the corresponding values for an electron in rapid motion, it does not seem possible from dynamic considerations that an α particle can be deflected through a large angle by a close approach to an electron, even if the latter be in rapid motion and constrained by strong electrical forces. It seems reasonable to suppose that the chance of single deflexions through a large angle due to this cause, if not zero, must be exceedingly small compared with that due to the central charge.

It is of interest to examine how far the experimental evidence throws light on the question of the extent of the

* *Annal. d. Phys.* iv. 23. p. 671 (1907).

distribution of the central charge. Suppose, for example, the central charge to be composed of N unit charges distributed over such a volume that the large single deflexions are mainly due to the constituent charges and not to the external field produced by the distribution. It has been shown (§ 3) that the fraction of the α particles scattered through a large angle is proportional to $(NeE)^2$, where Ne is the central charge concentrated at a point and E the charge on the deflected particle. If, however, this charge is distributed in single units, the fraction of the α particles scattered through a given angle is proportional to Ne^2 instead of N^2e^2. In this calculation, the influence of mass of the constituent particle has been neglected, and account has only been taken of its electric field. Since it has been shown that the value of the central point charge for gold must be about 100, the value of the distributed charge required to produce the same proportion of single deflexions through a large angle should be at least 10,000. Under these conditions the mass of the constituent particle would be small compared with that of the α particle, and the difficulty arises of the production of large single deflexions at all. In addition, with such a large distributed charge, the effect of compound scattering is relatively more important than that of single scattering. For example, the probable small angle of deflexion of a pencil of α particles passing through a thin gold foil would be much greater than that experimentally observed by Geiger (§ *b–c*). The large and small angle scattering could not then be explained by the assumption of a central charge of the same value. Considering the evidence as a whole, it seems simplest to suppose that the atom contains a central charge distributed through a very small volume, and that the large single deflexions are due to the central charge as a whole, and not to its constituents. At the same time, the experimental evidence is not precise enough to negative the possibility that a small fraction of the positive charge may be carried by satellites extending some distance from the centre. Evidence on this point could be obtained by examining whether the same central charge is required to explain the large single deflexions of α and β particles; for the α particle must approach much closer to the centre of the atom than the β particle of average speed to suffer the same large deflexion.

The general data available indicate that the value of this central charge for different atoms is approximately proportional to their atomic weights, at any rate for atoms heavier than aluminium. It will be of great interest to examine experimentally whether such a simple relation holds also for the lighter atoms. In cases where the mass of the deflecting atom (for example, hydrogen, helium, lithium) is not very different from that of the α particle, the general theory of single scattering will require modification, for it is necessary to take into account the movements of the atom itself (see § 4).

It is of interest to note that Nagaoka * has mathematically considered the properties of a "Saturnian" atom which he supposed to consist of a central attracting mass surrounded by rings of rotating electrons. He showed that such a system was stable if the attractive force was large. From the point of view considered in this paper, the chance of large deflexion would practically be unaltered, whether the atom is considered to be a disk or a sphere. It may be remarked that the approximate value found for the central charge of the atom of gold ($100\,e$) is about that to be expected if the atom of gold consisted of 49 atoms of helium, each carrying a charge $2\,e$. This may be only a coincidence, but it is certainly suggestive in view of the expulsion of helium atoms carrying two unit charges from radioactive matter.

The deductions from the theory so far considered are independent of the sign of the central charge, and it has not so far been found possible to obtain definite evidence to determine whether it be positive or negative. It may be possible to settle the question of sign by consideration of the difference of the laws of absorption of the β particle to be expected on the two hypotheses, for the effect of radiation in reducing the velocity of the β particle should be far more marked with a positive than with a negative centre. If the central charge be positive, it is easily seen that a positively charged mass if released from the centre of a heavy atom, would acquire a great velocity in moving through the electric field. It may be possible in this way to account for the high velocity of expulsion of α particles without supposing that they are initially in rapid motion within the atom.

Further consideration of the application of this theory to these and other questions will be reserved for a later paper, when the main deductions of the theory have been tested experimentally. Experiments in this direction are already in progress by Geiger and Marsden.

University of Manchester,
April 1911.

* Nagaoka, Phil. Mag. vii. p. 445 (1904).

Reprint 3

THE

LONDON, EDINBURGH, AND DUBLIN

PHILOSOPHICAL MAGAZINE

AND

JOURNAL OF SCIENCE.

[SIXTH SERIES.]

JULY 1913.

I. *On the Constitution of Atoms and Molecules.*

By N. Bohr, *Dr. phil. Copenhagen* *.

Introduction.

In order to explain the results of experiments on scattering of α rays by matter Prof. Rutherford † has given a theory of the structure of atoms. According to this theory, the atoms consist of a positively charged nucleus surrounded by a system of electrons kept together by attractive forces from the nucleus; the total negative charge of the electrons is equal to the positive charge of the nucleus. Further, the nucleus is assumed to be the seat of the essential part of the mass of the atom, and to have linear dimensions exceedingly small compared with the linear dimensions of the whole atom. The number of electrons in an atom is deduced to be approximately equal to half the atomic weight. Great interest is to be attributed to this atom-model; for, as Rutherford has shown, the assumption of the existence of nuclei, as those in question, seems to be necessary in order to account for the results of the experiments on large angle scattering of the α rays ‡.

In an attempt to explain some of the properties of matter on the basis of this atom-model we meet, however, with difficulties of a serious nature arising from the apparent

* Communicated by Prof. E. Rutherford, F.R.S.

† E. Rutherford, Phil. Mag. xxi. p. 669 (1911).

‡ See also Geiger and Marsden, Phil. Mag. April 1913.

Reprinted from Philosophical Magazine and Journal of Science, Series 6, Vol. 26, No. 151, pp. 1–25 (1913).

instability of the system of electrons: difficulties purposely avoided in atom-models previously considered, for instance, in the one proposed by Sir J. J. Thomson *. According to the theory of the latter the atom consists of a sphere of uniform positive electrification, inside which the electrons move in circular orbits.

The principal difference between the atom-models proposed by Thomson and Rutherford consists in the circumstance that the forces acting on the electrons in the atom-model of Thomson allow of certain configurations and motions of the electrons for which the system is in a stable equilibrium; such configurations, however, apparently do not exist for the second atom-model. The nature of the difference in question will perhaps be most clearly seen by noticing that among the quantities characterizing the first atom a quantity appears—the radius of the positive sphere—of dimensions of a length and of the same order of magnitude as the linear extension of the atom, while such a length does not appear among the quantities characterizing the second atom, viz. the charges and masses of the electrons and the positive nucleus; nor can it be determined solely by help of the latter quantities.

The way of considering a problem of this kind has, however, undergone essential alterations in recent years owing to the development of the theory of the energy radiation, and the direct affirmation of the new assumptions introduced in this theory, found by experiments on very different phenomena such as specific heats, photoelectric effect, Röntgen-rays, &c. The result of the discussion of these questions seems to be a general acknowledgment of the inadequacy of the classical electrodynamics in describing the behaviour of systems of atomic size †. Whatever the alteration in the laws of motion of the electrons may be, it seems necessary to introduce in the laws in question a quantity foreign to the classical electrodynamics, *i. e.* Planck's constant, or as it often is called the elementary quantum of action. By the introduction of this quantity the question of the stable configuration of the electrons in the atoms is essentially changed, as this constant is of such dimensions and magnitude that it, together with the mass and charge of the particles, can determine a length of the order of magnitude required.

This paper is an attempt to show that the application of the above ideas to Rutherford's atom-model affords a basis

* J. J. Thomson, Phil. Mag. vii. p. 237 (1904).

† See f. inst., 'Théorie du rayonnement et les quanta.' Rapports de la réunion à Bruxelles, Nov. 1911. Paris, 1912.

for a theory of the constitution of atoms. It will further be shown that from this theory we are led to a theory of the constitution of molecules.

In the present first part of the paper the mechanism of the binding of electrons by a positive nucleus is discussed in relation to Planck's theory. It will be shown that it is possible from the point of view taken to account in a simple way for the law of the line spectrum of hydrogen. Further, reasons are given for a principal hypothesis on which the considerations contained in the following parts are based.

I wish here to express my thanks to Prof. Rutherford for his kind and encouraging interest in this work.

PART I.—BINDING OF ELECTRONS BY POSITIVE NUCLEI.

§ 1. *General Considerations.*

The inadequacy of the classical electrodynamics in accounting for the properties of atoms from an atom-model as Rutherford's, will appear very clearly if we consider a simple system consisting of a positively charged nucleus of very small dimensions and an electron describing closed orbits around it. For simplicity, let us assume that the mass of the electron is negligibly small in comparison with that of the nucleus, and further, that the velocity of the electron is small compared with that of light.

Let us at first assume that there is no energy radiation. In this case the electron will describe stationary elliptical orbits. The frequency of revolution ω and the major-axis of the orbit $2a$ will depend on the amount of energy W which must be transferred to the system in order to remove the electron to an infinitely great distance apart from the nucleus. Denoting the charge of the electron and of the nucleus by $-e$ and E respectively and the mass of the electron by m, we thus get

$$\omega = \frac{\sqrt{2}}{\pi}\,\frac{W^{\frac{3}{2}}}{eE\sqrt{m}}, \quad 2a = \frac{eE}{W}. \quad . \quad . \quad . \quad (1)$$

Further, it can easily be shown that the mean value of the kinetic energy of the electron taken for a whole revolution is equal to W. We see that if the value of W is not given, there will be no values of ω and a characteristic for the system in question.

Let us now, however, take the effect of the energy radiation into account, calculated in the ordinary way from the acceleration of the electron. In this case the electron will no longer describe stationary orbits. W will continuously increase, and the electron will approach the nucleus describing orbits of smaller and smaller dimensions, and with greater and greater frequency ; the electron on the average gaining in kinetic energy at the same time as the whole system loses energy. This process will go on until the dimensions of the orbit are of the same order of magnitude as the dimensions of the electron or those of the nucleus. A simple calculation shows that the energy radiated out during the process considered will be enormously great compared with that radiated out by ordinary molecular processes.

It is obvious that the behaviour of such a system will be very different from that of an atomic system occurring in nature. In the first place, the actual atoms in their permanent state seem to have absolutely fixed dimensions and frequencies. Further, if we consider any molecular process, the result seems always to be that after a certain amount of energy characteristic for the systems in question is radiated out, the systems will again settle down in a stable state of equilibrium, in which the distances apart of the particles are of the same order of magnitude as before the process.

Now the essential point in Planck's theory of radiation is that the energy radiation from an atomic system does not take place in the continuous way assumed in the ordinary electrodynamics, but that it, on the contrary, takes place in distinctly separated emissions, the amount of energy radiated out from an atomic vibrator of frequency ν in a single emission being equal to $\tau h\nu$, where τ is an entire number, and h is a universal constant*.

Returning to the simple case of an electron and a positive nucleus considered above, let us assume that the electron at the beginning of the interaction with the nucleus was at a great distance apart from the nucleus, and had no sensible velocity relative to the latter. Let us further assume that the electron after the interaction has taken place has settled down in a stationary orbit around the nucleus. We shall, for reasons referred to later, assume that the orbit in question is circular ; this assumption will, however, make no alteration in the calculations for systems containing only a single electron.

Let us now assume that, during the binding of the electron, a homogeneous radiation is emitted of a frequency ν, equal to half the frequency of revolution of the electron in its final

* See f. inst., M. Planck, *Ann. d. Phys.* xxxi. p. 758 (1910) ; xxxvii. p. 642 (1912) ; *Verh. deutsch. Phys. Ges.* 1911, p. 138.

orbit; then, from Planck's theory, we might expect that the amount of energy emitted by the process considered is equal to $\tau h\nu$, where h is Planck's constant and τ an entire number. If we assume that the radiation emitted is homogeneous, the second assumption concerning the frequency of the radiation suggests itself, since the frequency of revolution of the electron at the beginning of the emission is 0. The question, however, of the rigorous validity of both assumptions, and also of the application made of Planck's theory, will be more closely discussed in § 3.

Putting

$$W = \tau h \frac{\omega}{2}, \quad \ldots \ldots \quad (2)$$

we get by help of the formula (1)

$$W = \frac{2\pi^2 m e^2 E^2}{\tau^2 h^2}, \quad \omega = \frac{4\pi^2 m e^2 E^2}{\tau^3 h^3}, \quad 2a = \frac{\tau^2 h^2}{2\pi^2 m e E}. \quad (3)$$

If in these expressions we give τ different values, we get a series of values for W, ω, and a corresponding to a series of configurations of the system. According to the above considerations, we are led to assume that these configurations will correspond to states of the system in which there is no radiation of energy; states which consequently will be stationary as long as the system is not disturbed from outside. We see that the value of W is greatest if τ has its smallest value 1. This case will therefore correspond to the most stable state of the system, *i. e.* will correspond to the binding of the electron for the breaking up of which the greatest amount of energy is required.

Putting in the above expressions $\tau=1$ and $E=e$, and introducing the experimental values

$$e = 4{\cdot}7 \,.\, 10^{-10}, \quad \frac{e}{m} = 5{\cdot}31 \,.\, 10^{17}, \quad h = 6{\cdot}5 \,.\, 10^{-27},$$

we get

$$2a = 1{\cdot}1 \,.\, 10^{-8} \text{ cm.}, \quad \omega = 6{\cdot}2 \,.\, 10^{15} \frac{1}{\text{sec.}}, \quad \frac{W}{e} = 13 \text{ volt.}$$

We see that these values are of the same order of magnitude as the linear dimensions of the atoms, the optical frequencies, and the ionization-potentials.

The general importance of Planck's theory for the discussion of the behaviour of atomic systems was originally pointed out by Einstein*. The considerations of Einstein

* A. Einstein, *Ann. d. Phys.* xvii. p. 132 (1905); xx. p. 199 (1906); xxii. p. 180 (1907).

have been developed and applied on a number of different phenomena, especially by Stark, Nernst, and Sommerfield. The agreement as to the order of magnitude between values observed for the frequencies and dimensions of the atoms, and values for these quantities calculated by considerations similar to those given above, has been the subject of much discussion. It was first pointed out by Haas*, in an attempt to explain the meaning and the value of Planck's constant on the basis of J. J. Thomson's atom-model, by help of the linear dimensions and frequency of an hydrogen atom.

Systems of the kind considered in this paper, in which the forces between the particles vary inversely as the square of the distance, are discussed in relation to Planck's theory by J. W. Nicholson †. In a series of papers this author has shown that it seems to be possible to account for lines of hitherto unknown origin in the spectra of the stellar nebulæ and that of the solar corona, by assuming the presence in these bodies of certain hypothetical elements of exactly indicated constitution. The atoms of these elements are supposed to consist simply of a ring of a few electrons surrounding a positive nucleus of negligibly small dimensions. The ratios between the frequencies corresponding to the lines in question are compared with the ratios between the frequencies corresponding to different modes of vibration of the ring of electrons. Nicholson has obtained a relation to Planck's theory showing that the ratios between the wave-length of different sets of lines of the coronal spectrum can be accounted for with great accuracy by assuming that the ratio between the energy of the system and the frequency of rotation of the ring is equal to an entire multiple of Planck's constant. The quantity Nicholson refers to as the energy is equal to twice the quantity which we have denoted above by W. In the latest paper cited Nicholson has found it necessary to give the theory a more complicated form, still, however, representing the ratio of energy to frequency by a simple function of whole numbers.

The excellent agreement between the calculated and observed values of the ratios between the wave-lengths in question seems a strong argument in favour of the validity of the foundation of Nicholson's calculations. Serious

* A. E. Haas, *Jahrb. d. Rad. u. El.* vii. p. 261 (1910). See further, A. Schidlof, *Ann. d. Phys.* xxxv. p. 90 (1911); E. Wertheimer, *Phys. Zeitschr.* xii. p. 409 (1911), *Verh. deutsch. Phys. Ges.* 1912, p. 431; F. A. Lindemann, *Verh. deutsch. Phys. Ges.* 1911, pp. 482, 1107; F. Haber, *Verh. deutsch. Phys. Ges.* 1911, p. 1117.

† J. W. Nicholson, Month. Not. Roy. Astr. Soc. lxxii. pp. 49, 139, 677, 693, 729 (1912).

objections, however, may be raised against the theory. These objections are intimately connected with the problem of the homogeneity of the radiation emitted. In Nicholson's calculations the frequency of lines in a line-spectrum is identified with the frequency of vibration of a mechanical system in a distinctly indicated state of equilibrium. As a relation from Planck's theory is used, we might expect that the radiation is sent out in quanta; but systems like those considered, in which the frequency is a function of the energy, cannot emit a finite amount of a homogeneous radiation; for, as soon as the emission of radiation is started, the energy and also the frequency of the system are altered. Further, according to the calculation of Nicholson, the systems are unstable for some modes of vibration. Apart from such objections—which may be only formal (see p. 23)—it must be remarked, that the theory in the form given does not seem to be able to account for the well-known laws of Balmer and Rydberg connecting the frequencies of the lines in the line-spectra of the ordinary elements.

It will now be attempted to show that the difficulties in question disappear if we consider the problems from the point of view taken in this paper. Before proceeding it may be useful to restate briefly the ideas characterizing the calculations on p. 5. The principal assumptions used are:

(1) That the dynamical equilibrium of the systems in the stationary states can be discussed by help of the ordinary mechanics, while the passing of the systems between different stationary states cannot be treated on that basis.

(2) That the latter process is followed by the emission of a *homogeneous* radiation, for which the relation between the frequency and the amount of energy emitted is the one given by Planck's theory.

The first assumption seems to present itself; for it is known that the ordinary mechanics cannot have an absolute validity, but will only hold in calculations of certain mean values of the motion of the electrons. On the other hand, in the calculations of the dynamical equilibrium in a stationary state in which there is no relative displacement of the particles, we need not distinguish between the actual motions and their mean values. The second assumption is in obvious contrast to the ordinary ideas of electrodynamics, but appears to be necessary in order to account for experimental facts.

In the calculations on page 5 we have further made use

of the more special assumptions, viz. that the different stationary states correspond to the emission of a different number of Planck's energy-quanta, and that the frequency of the radiation emitted during the passing of the system from a state in which no energy is yet radiated out to one of the stationary states, is equal to half the frequency of revolution of the electron in the latter state. We can, however (see § 3), also arrive at the expressions (3) for the stationary states by using assumptions of somewhat different form. We shall, therefore, postpone the discussion of the special assumptions, and first show how by the help of the above principal assumptions, and of the expressions (3) for the stationary states, we can account for the line-spectrum of hydrogen.

§ 2. *Emission of Line-spectra.*

Spectrum of Hydrogen.—General evidence indicates that an atom of hydrogen consists simply of a single electron rotating round a positive nucleus of charge e*. The re-formation of a hydrogen atom, when the electron has been removed to great distances away from the nucleus—*e. g.* by the effect of electrical discharge in a vacuum tube—will accordingly correspond to the binding of an electron by a positive nucleus considered on p. 5. If in (3) we put $E=e$, we get for the total amount of energy radiated out by the formation of one of the stationary states,

$$W_\tau = \frac{2\pi^2 me^4}{h^2\tau^2}.$$

The amount of energy emitted by the passing of the system from a state corresponding to $\tau=\tau_1$ to one corresponding to $\tau=\tau_2$, is consequently

$$W_{\tau_2} - W_{\tau_1} = \frac{2\pi^2 me^4}{h^2}\left(\frac{1}{\tau_2^2} - \frac{1}{\tau_1^2}\right).$$

If now we suppose that the radiation in question is homogeneous, and that the amount of energy emitted is equal to $h\nu$, where ν is the frequency of the radiation, we get

$$W_{\tau_2} - W_{\tau_1} = h\nu,$$

* See f. inst. N. Bohr, Phil. Mag. xxv. p. 24 (1913). The conclusion drawn in the paper cited is strongly supported by the fact that hydrogen, in the experiments on positive rays of Sir J. J. Thomson, is the only element which never occurs with a positive charge corresponding to the loss of more than one electron (comp. Phil. Mag. xxiv. p. 672 (1912)).

and from this

$$\nu = \frac{2\pi^2 me^4}{h^3}\left(\frac{1}{\tau_2^2} - \frac{1}{\tau_1^2}\right). \quad . \quad . \quad . \quad . \quad . \quad . \quad (4)$$

We see that this expression accounts for the law connecting the lines in the spectrum of hydrogen. If we put $\tau_2 = 2$ and let τ_1 vary, we get the ordinary Balmer series. If we put $\tau_2 = 3$, we get the series in the ultra-red observed by Paschen* and previously suspected by Ritz. If we put $\tau_2 = 1$ and $\tau_2 = 4, 5, ..$, we get series respectively in the extreme ultra-violet and the extreme ultra-red, which are not observed, but the existence of which may be expected.

The agreement in question is quantitative as well as qualitative. Putting

$$e = 4{\cdot}7 \,.\, 10^{-10}, \quad \frac{e}{m} = 5{\cdot}31 \,.\, 10^{17}, \quad \text{and} \quad h = 6{\cdot}5 \,.\, 10^{-27},$$

we get

$$\frac{2\pi^2 me^4}{h^3} = 3{\cdot}1 \,.\, 10^{15}.$$

The observed value for the factor outside the bracket in the formula (4) is

$$3{\cdot}290 \,.\, 10^{15}.$$

The agreement between the theoretical and observed values is inside the uncertainty due to experimental errors in the constants entering in the expression for the theoretical value. We shall in § 3 return to consider the possible importance of the agreement in question.

It may be remarked that the fact, that it has not been possible to observe more than 12 lines of the Balmer series in experiments with vacuum tubes, while 33 lines are observed in the spectra of some celestial bodies, is just what we should expect from the above theory. According to the equation (3) the diameter of the orbit of the electron in the different stationary states is proportional to τ^2. For $\tau = 12$ the diameter is equal to $1{\cdot}6 \,.\, 10^{-6}$ cm., or equal to the mean distance between the molecules in a gas at a pressure of about 7 mm. mercury; for $\tau = 33$ the diameter is equal to $1{\cdot}2 \,.\, 10^{-5}$ cm., corresponding to the mean distance of the molecules at a pressure of about 0·02 mm. mercury. According to the theory the necessary condition for the appearance of a great number of lines is therefore a very small density of the gas; for simultaneously to obtain an

* F. Paschen, *Ann. d. Phys.* xxvii. p. 565 (1908).

intensity sufficient for observation the space filled with the gas must be very great. If the theory is right, we may therefore never expect to be able in experiments with vacuum tubes to observe the lines corresponding to high numbers of the Balmer series of the emission spectrum of hydrogen; it might, however, be possible to observe the lines by investigation of the absorption spectrum of this gas (see § 4).

It will be observed that we in the above way do not obtain other series of lines, generally ascribed to hydrogen; for instance, the series first observed by Pickering* in the spectrum of the star ζ Puppis, and the set of series recently found by Fowler† by experiments with vacuum tubes containing a mixture of hydrogen and helium. We shall, however, see that, by help of the above theory, we can account naturally for these series of lines if we ascribe them to helium.

A neutral atom of the latter element consists, according to Rutherford's theory, of a positive nucleus of charge $2e$ and two electrons. Now considering the binding of a single electron by a helium nucleus, we get, putting $E=2e$ in the expressions (3) on page 5, and proceeding in exactly the same way as above,

$$\nu=\frac{8\pi^2me^4}{h^3}\left(\frac{1}{\tau_2^2}-\frac{1}{\tau_1^2}\right)=\frac{2\pi^2me^4}{h^3}\left(\frac{1}{\left(\frac{\tau_2}{2}\right)^2}-\frac{1}{\left(\frac{\tau_1}{2}\right)^2}\right).$$

If we in this formula put $\tau_2=1$ or $\tau_2=2$, we get series of lines in the extreme ultra-violet. If we put $\tau_2=3$, and let τ_1 vary, we get a series which includes 2 of the series observed by Fowler, and denoted by him as the first and second principal series of the hydrogen spectrum. If we put $\tau_2=4$, we get the series observed by Pickering in the spectrum of ζ Puppis. Every second of the lines in this series is identical with a line in the Balmer series of the hydrogen spectrum; the presence of hydrogen in the star in question may therefore account for the fact that these lines are of a greater intensity than the rest of the lines in the series. The series is also observed in the experiments of Fowler, and denoted in his paper as the Sharp series of the hydrogen spectrum. If we finally in the above formula put $\tau_2=5, 6,..$, we get series, the strong lines of which are to be expected in the ultra-red.

The reason why the spectrum considered is not observed in

* E. C. Pickering, Astrophys. J. iv. p. 369 (1896); v. p. 92 (1897).
† A. Fowler, Month. Not. Roy. Astr. Soc. lxxiii. Dec. 1912.

ordinary helium tubes may be that in such tubes the ionization of helium is not so complete as in the star considered or in the experiments of Fowler, where a strong discharge was sent through a mixture of hydrogen and helium. The condition for the appearance of the spectrum is, according to the above theory, that helium atoms are present in a state in which they have lost both their electrons. Now we must assume that the amount of energy to be used in removing the second electron from a helium atom is much greater than that to be used in removing the first. Further, it is known from experiments on positive rays, that hydrogen atoms can acquire a negative charge ; therefore the presence of hydrogen in the experiments of Fowler may effect that more electrons are removed from some of the helium atoms than would be the case if only helium were present.

Spectra of other substances.—In case of systems containing more electrons we must—in conformity with the result of experiments—expect more complicated laws for the line-spectra than those considered. I shall try to show that the point of view taken above allows, at any rate, a certain understanding of the laws observed.

According to Rydberg's theory—with the generalization given by Ritz *—the frequency corresponding to the lines of the spectrum of an element can be expressed by

$$\nu = F_r(\tau_1) - F_s(\tau_2),$$

where τ_1 and τ_2 are entire numbers, and F_1, F_2, F_3, ... are functions of τ which approximately are equal to $\frac{K}{(\tau+a_1)^2}$, $\frac{K}{(\tau+a_2)^2}$, ... K is a universal constant, equal to the factor outside the bracket in the formula (4) for the spectrum of hydrogen. The different series appear if we put τ_1 or τ_2 equal to a fixed number and let the other vary.

The circumstance that the frequency can be written as a difference between two functions of entire numbers suggests an origin of the lines in the spectra in question similar to the one we have assumed for hydrogen; *i. e.* that the lines correspond to a radiation emitted during the passing of the system between two different stationary states. For systems containing more than one electron the detailed discussion may be very complicated, as there will be many different configurations of the electrons which can be taken into consideration as stationary states. This may account for the different sets of series in the line spectra emitted from the

* W. Ritz, *Phys. Zeitschr.* ix. p. 521 (1908).

substances in question. Here I shall only try to show how, by help of the theory, it can be simply explained that the constant K entering in Rydberg's formula is the same for all substances.

Let us assume that the spectrum in question corresponds to the radiation emitted during the binding of an electron; and let us further assume that the system including the electron considered is neutral. The force on the electron, when at a great distance apart from the nucleus and the electrons previously bound, will be very nearly the same as in the above case of the binding of an electron by a hydrogen nucleus. The energy corresponding to one of the stationary states will therefore for τ great be very nearly equal to that given by the expression (3) on p. 5, if we put $E=e$. For τ great we consequently get

$$\lim (\tau^2 . F_1(\tau)) = \lim (\tau^2 . F_2(\tau)) = \ldots = \frac{2\pi^2 me^4}{h^3},$$

in conformity with Rydberg's theory.

§ 3. *General Considerations continued.*

We shall now return to the discussion (see p. 7) of the special assumptions used in deducing the expressions (3) on p. 5 for the stationary states of a system consisting of an electron rotating round a nucleus.

For one, we have assumed that the different stationary states correspond to an emission of a different number of energy-quanta. Considering systems in which the frequency is a function of the energy, this assumption, however, may be regarded as improbable; for as soon as one quantum is sent out the frequency is altered. We shall now see that we can leave the assumption used and still retain the equation (2) on p. 5, and thereby the formal analogy with Planck's theory.

Firstly, it will be observed that it has not been necessary, in order to account for the law of the spectra by help of the expressions (3) for the stationary states, to assume that in any case a radiation is sent out corresponding to more than a single energy-quantum, $h\nu$. Further information on the frequency of the radiation may be obtained by comparing calculations of the energy radiation in the region of slow vibrations based on the above assumptions with calculations based on the ordinary mechanics. As is known, calculations on the latter basis are in agreement with experiments on the energy radiation in the named region.

Let us assume that the ratio between the total amount of

energy emitted and the frequency of revolution of the electron for the different stationary states is given by the equation $W = f(\tau) \cdot h\omega$, instead of by the equation (2). Proceeding in the same way as above, we get in this case instead of (3)

$$W = \frac{\pi^2 m e^2 E^2}{2h^2 f^2(\tau)}, \quad \omega = \frac{\pi^2 m e^2 E^2}{2h^3 f^3(\tau)}.$$

Assuming as above that the amount of energy emitted during the passing of the system from a state corresponding to $\tau = \tau_1$ to one for which $\tau = \tau_2$ is equal to $h\nu$, we get instead of (4)

$$\nu = \frac{\pi^2 m e^2 E^2}{2h^3}\left(\frac{1}{f^2(\tau_2)} - \frac{1}{f^2(\tau_1)}\right).$$

We see that in order to get an expression of the same form as the Balmer series we must put $f(\tau) = c\tau$.

In order to determine c let us now consider the passing of the system between two successive stationary states corresponding to $\tau = N$ and $\tau = N - 1$; introducing $f(\tau) = c\tau$, we get for the frequency of the radiation emitted

$$\nu = \frac{\pi^2 m e^2 E^2}{2c^2 h^3} \cdot \frac{2N - 1}{N^2 (N-1)^2}.$$

For the frequency of revolution of the electron before and after the emission we have

$$\omega_N = \frac{\pi^2 m e^2 E^2}{2c^3 h^3 N^3} \quad \text{and} \quad \omega_{N-1} = \frac{\pi^2 m e^2 E^2}{2c^3 h^3 (N-1)^3}.$$

If N is great the ratio between the frequency before and after the emission will be very near equal to 1; and according to the ordinary electrodynamics we should therefore expect that the ratio between the frequency of radiation and the frequency of revolution also is very nearly equal to 1. This condition will only be satisfied if $c = \frac{1}{2}$. Putting $f(\tau) = \frac{\tau}{2}$, we, however, again arrive at the equation (2) and consequently at the expression (3) for the stationary states.

If we consider the passing of the system between two states corresponding to $\tau = N$ and $\tau = N - n$, where n is small compared with N, we get with the same approximation as above, putting $f(\tau) = \frac{\tau}{2}$,

$$\nu = n\omega.$$

The possibility of an emission of a radiation of such a frequency may also be interpreted from analogy with the ordinary electrodynamics, as an electron rotating round a nucleus in an elliptical orbit will emit a radiation which according to Fourier's theorem can be resolved into homogeneous components, the frequencies of which are $n\omega$, if ω is the frequency of revolution of the electron.

We are thus led to assume that the interpretation of the equation (2) is not that the different stationary states correspond to an emission of different numbers of energy-quanta, but that the frequency of the energy emitted during the passing of the system from a state in which no energy is yet radiated out to one of the different stationary states, is equal to different multiples of $\frac{\omega}{2}$, where ω is the frequency of revolution of the electron in the state considered. From this assumption we get exactly the same expressions as before for the stationary states, and from these by help of the principal assumptions on p. 7 the same expression for the law of the hydrogen spectrum. Consequently we may regard our preliminary considerations on p. 5 only as a simple form of representing the results of the theory.

Before we leave the discussion of this question, we shall for a moment return to the question of the significance of the agreement between the observed and calculated values of the constant entering in the expressions (4) for the Balmer series of the hydrogen spectrum. From the above consideration it will follow that, taking the starting-point in the form of the law of the hydrogen spectrum and assuming that the different lines correspond to a homogeneous radiation emitted during the passing between different stationary states, we shall arrive at exactly the same expression for the constant in question as that given by (4), if we only assume (1) that the radiation is sent out in quanta $h\nu$, and (2) that the frequency of the radiation emitted during the passing of the system between successive stationary states will coincide with the frequency of revolution of the electron in the region of slow vibrations.

As all the assumptions used in this latter way of representing the theory are of what we may call a qualitative character, we are justified in expecting—if the whole way of considering is a sound one—an absolute agreement between the values calculated and observed for the constant in question, and not only an approximate agreement. The formula (4) may therefore be of value in the discussion of the results of experimental determinations of the constants e, m, and h.

While there obviously can be no question of a mechanical foundation of the calculations given in this paper, it is, however, possible to give a very simple interpretation of the result of the calculation on p. 5 by help of symbols taken from the ordinary mechanics. Denoting the angular momentum of the electron round the nucleus by M, we have immediately for a circular orbit $\pi M = \frac{T}{\omega}$, where ω is the frequency of revolution and T the kinetic energy of the electron; for a circular orbit we further have $T = W$ (see p. 3) and from (2), p. 5, we consequently get

$$M = \tau M_0,$$

where

$$M_0 = \frac{h}{2\pi} = 1{\cdot}04 \times 10^{-27}.$$

If we therefore assume that the orbit of the electron in the stationary states is circular, the result of the calculation on p. 5 can be expressed by the simple condition: that the angular momentum of the electron round the nucleus in a stationary state of the system is equal to an entire multiple of a universal value, independent of the charge on the nucleus. The possible importance of the angular momentum in the discussion of atomic systems in relation to Planck's theory is emphasized by Nicholson *.

The great number of different stationary states we do not observe except by investigation of the emission and absorption of radiation. In most of the other physical phenomena, however, we only observe the atoms of the matter in a single distinct state, *i. e.* the state of the atoms at low temperature. From the preceding considerations we are immediately led to the assumption that the "permanent" state is the one among the stationary states during the formation of which the greatest amount of energy is emitted. According to the equation (3) on p. 5, this state is the one which corresponds to $\tau = 1$.

§ 4. *Absorption of Radiation.*

In order to account for Kirchhoff's law it is necessary to introduce assumptions on the mechanism of absorption of radiation which correspond to those we have used considering the emission. Thus we must assume that a system consisting of a nucleus and an electron rotating round it under certain circumstances can absorb a radiation of a frequency equal to the frequency of the homogeneous radiation emitted during

* J. W. Nicholson, *loc. cit.* p. 679.

the passing of the system between different stationary states. Let us consider the radiation emitted during the passing of the system between two stationary states A_1 and A_2 corresponding to values for τ equal to τ_1 and τ_2, $\tau_1 > \tau_2$. As the necessary condition for an emission of the radiation in question was the presence of systems in the state A_1, we must assume that the necessary condition for an absorption of the radiation is the presence of systems in the state A_2.

These considerations seem to be in conformity with experiments on absorption in gases. In hydrogen gas at ordinary conditions for instance there is no absorption of a radiation of a frequency corresponding to the line-spectrum of this gas; such an absorption is only observed in hydrogen gas in a luminous state. This is what we should expect according to the above. We have on p. 9 assumed that the radiation in question was emitted during the passing of the systems between stationary states corresponding to $\tau \geqq 2$. The state of the atoms in hydrogen gas at ordinary conditions should, however, correspond to $\tau = 1$; furthermore, hydrogen atoms at ordinary conditions combine into molecules, *i. e.* into systems in which the electrons have frequencies different from those in the atoms (see Part III.). From the circumstance that certain substances in a non-luminous state, as, for instance, sodium vapour, absorb radiation corresponding to lines in the line-spectra of the substances, we may, on the other hand, conclude that the lines in question are emitted during the passing of the system between two states, one of which is the permanent state.

How much the above considerations differ from an interpretation based on the ordinary electrodynamics is perhaps most clearly shown by the fact that we have been forced to assume that a system of electrons will absorb a radiation of a frequency different from the frequency of vibration of the electrons calculated in the ordinary way. It may in this connexion be of interest to mention a generalization of the considerations to which we are led by experiments on the photo-electric effect, and which may be able to throw some light on the problem in question. Let us consider a state of the system in which the electron is free, *i. e.* in which the electron possesses kinetic energy sufficient to remove to infinite distances from the nucleus. If we assume that the motion of the electron is governed by the ordinary mechanics and that there is no (sensible) energy radiation, the total energy of the system—as in the above considered stationary states—will be constant. Further, there will be perfect continuity between the two kinds of states, as the difference between frequency and dimensions of the systems in successive stationary states will diminish without limit if τ increases. In the following considerations we shall for the sake of brevity refer to the two kinds of states in question as "mechanical" states; by this notation only emphasizing the assumption that the motion of the electron in both cases can be accounted for by the ordinary mechanics.

Tracing the analogy between the two kinds of mechanical states, we might now expect the possibility of an absorption of radiation, not only corresponding to the passing of the system between two different stationary states, but also corresponding to the passing between one of the stationary states and a state in which the electron is free; and as above, we might expect that the frequency of this radiation was determined by the equation $E=h\nu$, where E is the difference between the total energy of the system in the two states. As it will be seen, such an absorption of radiation is just what is observed in experiments on ionization by ultra-violet light and by Röntgen rays. Obviously, we get in this way the same expression for the kinetic energy of an electron ejected from an atom by photo-electric effect as that deduced by Einstein *, *i. e.* $T=h\nu-W$, where T is the kinetic energy of the electron ejected, and W the total amount of energy emitted during the original binding of the electron.

The above considerations may further account for the result of some experiments of R. W. Wood † on absorption of light by sodium vapour. In these experiments, an absorption corresponding to a very great number of lines in the principal series of the sodium spectrum is observed, and in addition a continuous absorption which begins at the head of the series and extends to the extreme ultra-violet. This is exactly what we should expect according to the analogy in question, and, as we shall see, a closer consideration of the above experiments allows us to trace the analogy still further. As mentioned on p. 9 the radii of the orbits of the electrons will for stationary states corresponding to high values for τ be very great compared with ordinary atomic dimensions. This circumstance was used as an explanation of the non-appearance in experiments with vacuum-tubes of lines corresponding to the higher numbers in the Balmer series of the hydrogen spectrum. This is also in conformity with experiments on the emission spectrum of sodium; in the principal series of the emission spectrum of this substance

* A. Einstein, *Ann. d. Phys.* xvii. p. 146 (1905).
† R. W. Wood, Physical Optics, p. 513 (1911).

rather few lines are observed. Now in Wood's experiments the pressure was not very low, and the states corresponding to high values for τ could therefore not appear; yet in the absorption spectrum about 50 lines were detected. In the experiments in question we consequently observe an absorption of radiation which is not accompanied by a complete transition between two different stationary states. According to the present theory we must assume that this absorption is followed by an emission of energy during which the systems pass back to the original stationary state. If there are no collisions between the different systems this energy will be emitted as a radiation of the same frequency as that absorbed, and there will be no true absorption but only a scattering of the original radiation; a true absorption will not occur unless the energy in question is transformed by collisions into kinetic energy of free particles. In analogy we may now from the above experiments conclude that a bound electron—also in cases in which there is no ionization—will have an absorbing (scattering) influence on a homogeneous radiation, as soon as the frequency of the radiation is greater than W/h, where W is the total amount of energy emitted during the binding of the electron. This would be highly in favour of a theory of absorption as the one sketched above, as there can in such a case be no question of a coincidence of the frequency of the radiation and a characteristic frequency of vibration of the electron. It will further be seen that the assumption, that there will be an absorption (scattering) of any radiation corresponding to a transition between two different mechanical states, is in perfect analogy with the assumption generally used that a free electron will have an absorbing (scattering) influence on light of any frequency. Corresponding considerations will hold for the emission of radiation.

In analogy to the assumption used in this paper that the emission of line-spectra is due to the re-formation of atoms after one or more of the lightly bound electrons are removed, we may assume that the homogeneous Röntgen radiation is emitted during the settling down of the systems after one of the firmly bound electrons escapes, *e. g.* by impact of cathode particles *. In the next part of this paper, dealing with the constitution of atoms, we shall consider the question more closely and try to show that a calculation based on this assumption is in quantitative agreement with the results of experiments: here we shall only mention briefly a problem with which we meet in such a calculation.

* Compare J. J. Thomson, Phil. Mag. xxiii. p. 456 (1912).

Experiments on the phenomena of X-rays suggest that not only the emission and absorption of radiation cannot be treated by the help of the ordinary electrodynamics, but not even the result of a collision between two electrons of which the one is bound in an atom. This is perhaps most clearly shown by some very instructive calculations on the energy of β-particles emitted from radioactive substances recently published by Rutherford *. These calculations strongly suggest that an electron of great velocity in passing through an atom and colliding with the electrons bound will loose energy in distinct finite quanta. As is immediately seen, this is very different from what we might expect if the result of the collisions was governed by the usual mechanical laws. The failure of the classical mechanics in such a problem might also be expected beforehand from the absence of anything like equipartition of kinetic energy between free electrons and electrons bound in atoms. From the point of view of the "mechanical" states we see, however, that the following assumption—which is in accord with the above analogy—might be able to account for the result of Rutherford's calculation and for the absence of equipartition of kinetic energy: two colliding electrons, bound or free, will, after the collision as well as before, be in mechanical states. Obviously, the introduction of such an assumption would not make any alteration necessary in the classical treatment of a collision between two free particles. But, considering a collision between a free and a bound electron, it would follow that the bound electron by the collision could not acquire a less amount of energy than the difference in energy corresponding to successive stationary states, and consequently that the free electron which collides with it could not lose a less amount.

The preliminary and hypothetical character of the above considerations needs not to be emphasized. The intention, however, has been to show that the sketched generalization of the theory of the stationary states possibly may afford a simple basis of representing a number of experimental facts which cannot be explained by help of the ordinary electrodynamics, and that the assumptions used do not seem to be inconsistent with experiments on phenomena for which a satisfactory explanation has been given by the classical dynamics and the wave theory of light.

* E. Rutherford, Phil. Mag. xxiv. pp. 453 & 893 (1912).

§ 5. *The permanent State of an Atomic System.*

We shall now return to the main object of this paper—the discussion of the "permanent" state of a system consisting of nuclei and bound electrons. For a system consisting of a nucleus and an electron rotating round it, this state is, according to the above, determined by the condition that the angular momentum of the electron round the nucleus is equal to $\frac{h}{2\pi}$.

On the theory of this paper the only neutral atom which contains a single electron is the hydrogen atom. The permanent state of this atom should correspond to the values of a and ω calculated on p. 5. Unfortunately, however, we know very little of the behaviour of hydrogen atoms on account of the small dissociation of hydrogen molecules at ordinary temperatures. In order to get a closer comparison with experiments, it is necessary to consider more complicated systems.

Considering systems in which more electrons are bound by a positive nucleus, a configuration of the electrons which presents itself as a permanent state is one in which the electrons are arranged in a ring round the nucleus. In the discussion of this problem on the basis of the ordinary electrodynamics, we meet—apart from the question of the energy radiation—with new difficulties due to the question of the stability of the ring. Disregarding for a moment this latter difficulty, we shall first consider the dimensions and frequency of the systems in relation to Planck's theory of radiation.

Let us consider a ring consisting of n electrons rotating round a nucleus of charge E, the electrons being arranged at equal angular intervals around the circumference of a circle of radius a.

The total potential energy of the system consisting of the electrons and the nucleus is

$$\mathrm{P} = -\frac{ne}{a}(\mathrm{E} - es_n),$$

where

$$s_n = \frac{1}{4}\sum_{s=1}^{s=n-1} \operatorname{cosec}\frac{s\pi}{n}.$$

For the radial force exerted on an electron by the nucleus and the other electrons we get

$$\mathrm{F} = -\frac{1}{n}\frac{d\mathrm{P}}{da} = -\frac{e}{a^2}(\mathrm{E} - es_n).$$

Denoting the kinetic energy of an electron by T and neglecting the electromagnetic forces due to the motion of the electrons (see Part II.), we get, putting the centrifugal force on an electron equal to the radial force,

$$\frac{2T}{a} = \frac{e}{a^2}(E - es_n),$$

or

$$T = \frac{e}{2a}(E - es_n).$$

From this we get for the frequency of revolution

$$\omega = \frac{1}{2\pi}\sqrt{\frac{e(E - es_n)}{ma^3}}.$$

The total amount of energy W necessary transferred to the system in order to remove the electrons to infinite distances apart from the nucleus and from each other is

$$W = -P - nT = \frac{ne}{2a}(E - es_n) = nT,$$

equal to the total kinetic energy of the electrons.

We see that the only difference in the above formula and those holding for the motion of a single electron in a circular orbit round a nucleus is the exchange of E for $E - es_n$. It is also immediately seen that corresponding to the motion of an electron in an elliptical orbit round a nucleus, there will be a motion of the n electrons in which each rotates in an elliptical orbit with the nucleus in the focus, and the n electrons at any moment are situated at equal angular intervals on a circle with the nucleus as the centre. The major axis and frequency of the orbit of the single electrons will for this motion be given by the expressions (1) on p. 3 if we replace E by $E - es_n$ and W by $\frac{W}{n}$. Let us now suppose that the system of n electrons rotating in a ring round a nucleus is formed in a way analogous to the one assumed for a single electron rotating round a nucleus. It will thus be assumed that the electrons, before the binding by the nucleus, were at a great distance apart from the latter and possessed no sensible velocities, and also that during the binding a homogeneous radiation is emitted. As in the case of a single electron, we have here that the total amount of energy emitted during the formation of the system is equal to the final kinetic energy of the electrons. If we now suppose that during the formation of the system the electrons at any moment are situated at equal angular intervals on the circumference of a circle with the nucleus in the centre, from analogy with the considerations on p. 5 we are here led to assume the existence of a series of stationary configurations in which the kinetic energy per electron is equal to $\tau h \frac{\omega}{2}$, where τ is an entire number, h Planck's constant, and ω the frequency of revolution. The configuration in which the greatest amount of energy is emitted is, as before, the one in which $\tau = 1$. This configuration we shall assume to be the permanent state of the system if the electrons in this state are arranged in a single ring. As for the case of a single electron we get that the angular momentum of each of the electrons is equal to $\frac{h}{2\pi}$. It may be remarked that instead of considering the single electrons we might have considered the ring as an entity. This would, however, lead to the same result, for in this case the frequency of revolution ω will be replaced by the frequency $n\omega$ of the radiation from the whole ring calculated from the ordinary electrodynamics, and T by the total kinetic energy nT.

There may be many other stationary states corresponding to other ways of forming the system. The assumption of the existence of such states seems necessary in order to account for the line-spectra of systems containing more than one electron (p. 11); it is also suggested by the theory of Nicholson mentioned on p. 6, to which we shall return in a moment. The consideration of the spectra, however, gives, as far as I can see, no indication of the existence of stationary states in which all the electrons are arranged in a ring and which correspond to greater values for the total energy emitted than the one we above have assumed to be the permanent state.

Further, there may be stationary configurations of a system of n electrons and a nucleus of charge E in which all the electrons are not arranged in a single ring. The question, however, of the existence of such stationary configurations is not essential for our determination of the permanent state, as long as we assume that the electrons in this state of the system are arranged in a single ring. Systems corresponding to more complicated configurations will be discussed on p. 24.

Using the relation $\mathrm{T} = h\frac{\omega}{2}$ we get, by help of the above expressions for T and ω, values for a and ω corresponding to the permanent state of the system which only differ from those given by the equations (3) on p. 5, by exchange of E for $E - es_n$.

The question of stability of a ring of electrons rotating round a positive charge is discussed in great detail by Sir J. J. Thomson *. An adaption of Thomson's analysis for the case here considered of a ring rotating round a nucleus of negligibly small linear dimensions is given by Nicholson †. The investigation of the problem in question naturally divides in two parts: one concerning the stability for displacements of the electrons in the plane of the ring; one concerning displacements perpendicular to this plane. As Nicholson's calculations show, the answer to the question of stability differs very much in the two cases in question. While the ring for the latter displacements in general is stable if the number of electrons is not great; the ring is in no case considered by Nicholson stable for displacements of the first kind.

According, however, to the point of view taken in this paper, the question of stability for displacements of the electrons in the plane of the ring is most intimately connected with the question of the mechanism of the binding of the electrons, and like the latter cannot be treated on the basis of the ordinary dynamics. The hypothesis of which we shall make use in the following is that the stability of a ring of electrons rotating round a nucleus is secured through the above condition of the universal constancy of the angular momentum, together with the further condition that the configuration of the particles is the one by the formation of which the greatest amount of energy is emitted. As will be shown, this hypothesis is, concerning the question of stability for a displacement of the electrons perpendicular to the plane of the ring, equivalent to that used in ordinary mechanical calculations.

Returning to the theory of Nicholson on the origin of lines observed in the spectrum of the solar corona, we shall now see that the difficulties mentioned on p. 7 may be only formal. In the first place, from the point of view considered above the objection as to the instability of the systems for displacements of the electrons in the plane of the ring may not be valid. Further, the objection as to the emission of the radiation in quanta will not have reference to the calculations in question, if we assume that in the coronal spectrum we are not dealing with a true emission but only with a scattering of radiation. This assumption seems probable if we consider

* *Loc. cit.* † *Loc. cit.*

the conditions in the celestial body in question; for on account of the enormous rarefaction of the matter there may be comparatively few collisions to disturb the stationary states and to cause a true emission of light corresponding to the transition between different stationary states; on the other hand there will in the solar corona be intense illumination of light of all frequencies which may excite the natural vibrations of the systems in the different stationary states. If the above assumption is correct, we immediately understand the entirely different form for the laws connecting the lines discussed by Nicholson and those connecting the ordinary line-spectra considered in this paper.

Proceeding to consider systems of a more complicated constitution, we shall make use of the following theorem, which can be very simply proved :—

"In every system consisting of electrons and positive nuclei, in which the nuclei are at rest and the electrons move in circular orbits with a velocity small compared with the velocity of light, the kinetic energy will be numerically equal to half the potential energy."

By help of this theorem we get—as in the previous cases of a single electron or of a ring rotating round a nucleus—that the total amount of energy emitted, by the formation of the systems from a configuration in which the distances apart of the particles are infinitely great and in which the particles have no velocities relative to each other, is equal to the kinetic energy of the electrons in the final configuration.

In analogy with the case of a single ring we are here led to assume that corresponding to any configuration of equilibrium a series of geometrically similar, stationary configurations of the system will exist in which the kinetic energy of every electron is equal to the frequency of revolution multiplied by $\frac{\tau}{2}h$ where τ is an entire number and h Planck's constant. In any such series of stationary configurations the one corresponding to the greatest amount of energy emitted will be the one in which τ for every electron is equal to 1. Considering that the ratio of kinetic energy to frequency for a particle rotating in a circular orbit is equal to π times the angular momentum round the centre of the orbit, we are therefore led to the following simple generalization of the hypotheses mentioned on pp. 15 and 22.

"*In any molecular system consisting of positive nuclei and electrons in which the nuclei are at rest relative to each other and the electrons move in circular orbits, the angular momentum* *of every electron round the centre of its orbit will in the permanent state of the system be equal to $\frac{h}{2\pi}$, where h is Planck's constant"* *.

In analogy with the considerations on p. 23, we shall assume that a configuration satisfying this condition is stable if the total energy of the system is less than in any neighbouring configuration satisfying the same condition of the angular momentum of the electrons.

As mentioned in the introduction, the above hypothesis will be used in a following communication as a basis for a theory of the constitution of atoms and molecules. It will be shown that it leads to results which seem to be in conformity with experiments on a number of different phenomena.

The foundation of the hypothesis has been sought entirely in its relation with Planck's theory of radiation ; by help of considerations given later it will be attempted to throw some further light on the foundation of it from another point of view.

April 5, 1913.

Reprint 4

Physik. Zeitschr. XVII, 1916. Sommerfeld, Theorie d. Zeeman-Effekts d. Wasserstofflinien. 491

Zur Theorie des Zeeman-Effekts der Wasserstofflinien, mit einem Anhang über den Stark-Effekt.

Von A. Sommerfeld.

§ 1. Einleitung.

In einer Annalen-Arbeit[1]), welche die Ergebnisse zweier in der Münchner Akademie veröffentlichter Aufsätze zusammenfaßt und erweitert, habe ich die Keplerschen Ellipsen, die ein Elektron um einen Wasserstoffkern herum beschreibt, nicht nur ihrer Größe und Gestalt, sondern auch ihrer räumlichen Lage nach „gequantelt". Um durch Quantenbedingungen gewisse Lagen der Bahnebene im Raume auszuzeichnen, ist allerdings erforderlich, daß durch irgendwelche physikalische Umstände eine Bezugsrichtung oder Bezugsebene im Raume gegeben ist, gegen welche die Lage der Bahnebene orientiert werden kann. A. a. O. habe ich mich beschränkt auf ein „magnetisches Feld von der Stärke Null", welches durch seine Ebene senkrecht zu den magnetischen Kraftlinien eine solche Bezugsebene definiert, ohne im übrigen die Bewegung des Elektrons zu beeinflussen. Macht man diese Ebene zur Äquatorebene $\vartheta = \pi/2$ eines räumlichen Polarkoordinatensystems r, ϑ, φ und unterwirft diese Koordinaten dem Quantenansatz

$$\int p_\varphi \, d\varphi = n_1 h, \quad \int p_\vartheta \, d\vartheta = n_2 h, \quad \int p_r \, dr = n' h,$$

unter p_φ, p_ϑ, p_r die zu den Koordinaten φ, ϑ, r gehörigen Impulse verstanden, wobei die Integrale über denjenigen Wertebereich der betr. Koordinate zu erstrecken sind, der zur Darstellung aller Phasenpunkte des Systems erforderlich ist, so ergibt sich das folgende bemerkenswerte Resultat: Während Größe und Gestalt der Ellipse dieselben bleiben wie ohne Magnetfeld, wird die Lage im Raum ganzzahlig beschränkt durch die Bedingung

$$\cos\alpha = \frac{n_1}{n_1 + n_2}. \tag{1}$$

Dabei bedeutet α den Winkel der Bahnebene gegen die Äquatorebene des Koordinatensystems. Die Energie der Bewegung lautet ebenso wie bei Abwesenheit des (als verschwindend angenommenen) Magnetfeldes:

$$W = -\frac{Nh}{(n+n')^2}, \quad n = n_1 + n_2. \tag{2}$$

Das radiale Quantum n' bestimmt mittels der Exzentrizität die Gestalt der Ellipse; das azimutale Quantum n bestimmt die Größe der Ellipse durch Vermittelung der für die Bahnebene berechneten Flächenkonstanten p.

1) Sommerfeld, Ann. d. Phys. **51**, 1, 1916.

Dieses azimutale Quantum n zerlegt sich in die beiden zu den Koordinaten φ und ϑ gehörigen Quanten n_1 und n_2. Das Quantum n_1 steht zu der Flächenkonstanten p_φ für die Äquatorebene in derselben Beziehung, wie das Quantum n zu der Flächenkonstanten p für die Bahnebene:

$$2\pi p = nh; \quad 2\pi p_\varphi = n_1 h. \tag{3}$$

Da $p_\varphi = p \cos\alpha$, folgt durch Division der beiden Gl. (3) unmittelbar die Bestimmungsgleichung (1) für $\cos\alpha$. Gl. (1) zeigt:

Für $n = 1$ gibt es nur zwei Lagen der Bahnebene, entweder senkrecht zur Äquatorebene ($n_1 = 0$, $n_2 = 1$) oder parallel dazu ($n_1 = 1$, $n_2 = 0$).

Für $n = 2$ gibt es vier Lagen, die senkrechte ($n_1 = 0$, $n_2 = 2$), die parallele ($n_1 = 2$, $n_2 = 0$) oder die beiden unter 60° geneigten Lagen ($n_1 = 1$, $n_2 = 1$, $\cos\alpha = 1/2$, $\alpha = \pm\pi/3$).

Allgemein beträgt bei beliebigem n die Anzahl der quantenhaft ausgezeichneten Lagen $2n$; unter ihnen ist stets die senkrechte und parallele Lage enthalten (vgl. Fig. 1).

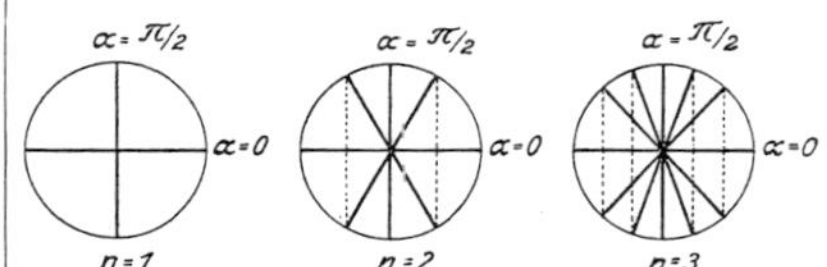

Fig. 1.

Wie ich a. a. O. bemerkt habe, versteht man auf Grund dieses Richtungseffekts nicht nur die qualitative Möglichkeit der scharfen Polarisation der Zeeman-Komponenten, sondern man gelangt auch zu gewissen quantitativen Analogien mit dem normalen Zeemaneffekt, allerdings nicht zu einer vollen Identität. Ich möchte dies im folgenden näher ausführen, da, wie ich höre, Beobachtungen über den Zeemaneffekt bei den Wasserstofflinien vorbereitet werden. Meine Ergebnisse decken sich, wie ich jetzt bei der Korrektur feststellen kann, mit Resultaten, die P. Debye inzwischen in den Göttinger Nachrichten veröffentlicht hat (vgl. auch unten Seite 507).

Bei den obigen Angaben war stillschweigend von dem Einfluß der Relativität abgesehen, welche, wie ich früher gezeigt habe, eine Feinstruktur der Wasserstofflinien bewirkt, insbesondere das bekannte Wasserstoff-Dublett erklärt. Da bei einfachen Linien der normale Zeemaneffekt, bei mehrfachen Linien aber zunächst die komplizierten Zeemaneffekte auftreten, die erst bei hinreichend starken

Reprinted from Physikalische Zeitschrift, Vol. 17, No. 20, pp. 491–507 (1916).

Magnetfeldern mittels des Paschen-Back-Effekts in das normale Lorentzsche Triplett übergehen, so versprach die relativistische Behandlung des Zeemaneffekts der Wasserstofflinien besonders interessante Resultate. Ich werde daher im folgenden das Problem des Zeemaneffektes auch mit Berücksichtigung der relativistischen Massenveränderlichkeit behandeln. Der Erfolg ist allerdings nicht der erwartete: es ergibt sich nämlich auch hier nur das normale Lorentzsche Triplett, und auch dieses nur in demselben Sinne (nämlich unter Absehung von gewissen überzähligen Komponenten), wie vom Standpunkte der gewöhnlichen Mechanik.

Ich beantworte bei dieser Gelegenheit zugleich eine Frage, die P. Epstein[1]) kürzlich gestellt hat, die Frage nach der kanonischen Form der relativistischen Bewegungsgleichungen mit Magnetfeld. Stellt man sich nämlich auf den Standpunkt, den Schwarzschild und Epstein bei der Behandlung des Starkeffekts mit so großem Erfolge durchgeführt haben, so hat man die Quantenansätze an die Hamilton-Jacobische Form der Bewegungsgleichungen anzuschließen. Die von Jacobi eingeführte Wirkungsfunktion ist nichts anderes als die Summe der von mir benutzten, aber mit unbestimmter oberer Grenze geschriebenen Phasenintegralen, genommen über alle Freiheitsgrade; die Periodizitätsmoduln der Jacobischen Wirkungsfunktion werden auf diese Weise direkt Vielfache des Planckschen Wirkungsquantums. Wegen dieses naturgemäßen und engen Zusammenhangs zwischen Quantentheorie und Hamilton-Jacobischer Mechanik wird man wünschen, auch die Differentialgleichungen des Zeemaneffekts in kanonischer Form zu schreiben. Man gelangt dazu einfach, wenn man statt der gewöhnlichen Zeit die relativistische Eigenzeit einführt.

Auch vom Standpunkte der analytischen Einfachheit empfiehlt sich dieser Weg. Es ist nicht nötig, die Bewegungsgleichungen vollständig auszuintegrieren; man kann vielmehr schon bei den ersten Integralgleichungen Halt machen, da diese bereits die uns interessierende Energiekonstante W liefern. Sodann aber ergibt sich eine bemerkenswerte Vereinfachung, wenn man die Periodizitätsmoduln der Wirkungsfunktion — unsere Quantenzahlen — nach der Methode der komplexen Integration ausrechnet. Da stets die Grenzen der Integration durch die Verzweigungspunkte des Integranden gegeben werden, so handelt es sich bei unseren Phasenintegralen notwendig um „vollständige" Integrale auf geschlossenem Wege, die durch Deformation des Integrationsweges vereinfacht und in den wichtigsten Fällen nach der Residuenmethode direkt ausgewertet werden können.

In einem Anhang wende ich die hier entwickelten Methoden auf den Starkeffekt an, indem ich erstens die Berechnung durch komplexe Integration vereinfache und zweitens die kanonische Form der Bewegungsgleichungen mit Rücksicht auf die relativistische Massenveränderlichkeit entwickle, ohne dieselben übrigens zu integrieren. Auf das Interesse, welches die relativistische Behandlung des Starkeffekts bieten würde, hat bereits P. Epstein[1]) hingewiesen.

1) P. Epstein, Ann. d. Phys. 51, 198, 1916.

§ 2. Zeemaneffekt ohne Relativität.

Es ist seit langem bekannt, daß ein Elektron unter dem gleichzeitigen Einfluß einer Coulombschen Kraft und eines homogenen Magnetfeldes eine Keplerellipse beschreibt, die um die Richtung der magnetischen Kraftlinien eine Präzession ausführt, sofern man die Quadrate der magnetischen Feldstärke vernachlässigen darf. Wir beweisen dieses Ergebnis, welches uns auch als Grundlage für die relativistische Behandlung dienen wird, in Kürze folgendermaßen[2]).

Die Bewegungsgleichungen lauten, wenn man die Richtung des homogenen Magnetfeldes zur z-Achse macht und den Kern als unbeweglich (Masse ∞) ansieht:

$$\left.\begin{aligned} \frac{d}{dt}(m\dot{x}) &= -\frac{e^2}{r^3}x - \frac{e}{c}H\dot{y}, \\ \frac{d}{dt}(m\dot{y}) &= -\frac{e^2}{r^3}y + \frac{e}{c}H\dot{x}, \\ \frac{d}{dt}(m\dot{z}) &= -\frac{e^2}{r^3}z. \end{aligned}\right\} \qquad (1)$$

Neben den rechtwinkligen Koordinaten benutzen wir das schon in der Einleitung genannte Polarkoordinatensytem r, ϑ, φ, indem wir setzen

$$x = r\sin\vartheta\cos\varphi\,,\quad y = r\sin\vartheta\sin\varphi\,,\quad z = r\cos\vartheta. \qquad (2)$$

Das zur Koordinate φ gehörige Impulsmoment p_φ ist

$$p_\varphi = m(x\dot{y} - y\dot{x}) = mr^2\sin^2\vartheta\,\dot{\varphi}. \qquad (3)$$

Dasselbe ist nicht konstant; vielmehr folgt aus (1)

$$\frac{dp_\varphi}{dt} = \frac{mf}{2}\frac{d}{dt}(x^2+y^2) = \frac{mf}{2}\frac{d}{dt}(r^2\sin^2\vartheta). \qquad (4)$$

1) Ann. d. Phys. 50, 500 und 507, 1916.
2) In Übereinstimmung mit P. Epstein, Ann. d. Phys. 51, 198, 1916.

Physik. Zeitschr. XVII, 1916. Sommerfeld, Theorie d. Zeeman-Effekts d. Wasserstofflinien. 493

Hierbei ist die für den Zeemaneffekt charakteristische Abkürzung eingeführt:

$$f = \frac{e}{m}\,\frac{H}{c}\,. \tag{5}$$

Schreiben wir für (4)

$$\frac{d}{dt}\left(m r^2 \sin^2\vartheta\,\left(\dot\varphi - \frac{f}{2}\right)\right) = 0\,, \tag{6}$$

so liegt es nahe, einen Hilfswinkel χ einzuführen durch

$$\dot\chi = \dot\varphi - \frac{f}{2}\,,\quad \varphi = \chi + \frac{f}{2}\,t\,. \tag{7}$$

Die Integration von (6) liefert dann

$$m r^2 \sin^2\vartheta\,\dot\chi = p_\chi\,, \tag{8}$$

wo p_χ eine Konstante ist.

Wir zeigen, daß auch die ursprünglichen Bewegungsgleichungen vereinfacht werden, wenn wir ein System $\xi\eta\zeta$ einführen, welches gegen das System xyz mit der Winkelgeschwindigkeit $f/2$ um die z-Achse gedreht wird, indem wir setzen:

$$\xi = r\sin\vartheta\cos\chi,\ \eta = r\sin\vartheta\sin\chi,\ \zeta = r\cos\vartheta. \tag{9}$$

Wegen (2) und (7) ergibt sich

$$x = \xi\cos\frac{f}{2}t - \eta\sin\frac{f}{2}t$$

$$y = \xi\sin\frac{f}{2}t + \eta\cos\frac{f}{2}t$$

$$z = \zeta\,.$$

Hieraus folgt:

$$\left.\begin{aligned}\dot x + f y &= (\dot\xi + \frac{f}{2}\eta)\cos\frac{f}{2}t - (\dot\eta - \frac{f}{2}\xi)\sin\frac{f}{2}t\\ \dot y - f x &= (\dot\xi + \frac{f}{2}\eta)\sin\frac{f}{2}t + (\dot\eta - \frac{f}{2}\xi)\cos\frac{f}{2}t\end{aligned}\right| \tag{10}$$

und

$$\begin{aligned}\frac{d}{dt}(\dot x + f y) &= (\ddot\xi + \frac{f^2}{4}\xi)\cos\frac{f}{2}t - (\ddot\eta + \frac{f^2}{4}\eta)\sin\frac{f}{2}t,\\ \frac{d}{dt}(\dot y + f x) &= (\ddot\xi + \frac{f^2}{4}\xi)\sin\frac{f}{2}t + (\ddot\eta + \frac{f^2}{4}\eta)\cos\frac{f}{2}t.\end{aligned} \tag{10a}$$

Die Gl. (1) gehen daher über in

$$\left.\begin{aligned}\ddot\xi &= -\frac{e^2\xi}{mr^3} - \frac{f^2}{4}\xi\,,\\ \ddot\eta &= -\frac{e^2\eta}{mr^3} - \frac{f^2}{4}\eta\,,\\ \ddot\zeta &= -\frac{e^2\zeta}{mr^3}\,.\end{aligned}\right| \tag{11}$$

Dies sind im $\xi\eta\zeta$-System die Bewegungsgleichungen eines Elektrons unter dem Einfluß des Potentials:

$$V = -\frac{e^2}{r} + \frac{mf^2}{8}(\xi^2 + \eta^2). \tag{11a}$$

Vernachlässigt man aber Glieder mit f^2, so hat man genau die Bewegungsgleichungen ohne Magnetfeld und die Keplersche Ellipse; ihre Bahnebene liegt im $\xi\eta\zeta$-System fest, beschreibt aber im xyz-System wegen (7) eine gleichförmige Präzession um die z-Achse mit der Winkelgeschwindigkeit $f/2$.

Die Vernachlässigung der Glieder mit f^2 läßt sich numerisch rechtfertigen; denn es ist für den $n$ten Bohrschen Kreis

$$\frac{e^2}{mr^3} = \frac{(2\pi)^6 m^2 e^8}{h^6 n^6} = \frac{1{,}7\cdot 10^{33}}{n^6}\,,$$

dagegen für ein Magnetfeld $H = 10000$ Gauß

$$\frac{f^2}{4} = 7{,}8\cdot 10^{21}\,.$$

Das erste Glied auf der rechten Seite von (11) überwiegt also das zweite auch bei höheren Werten von n um einen Faktor 10^{10}.

Durch unsere Reduktion auf das $\xi\eta\zeta$-System sind nun genau diejenigen Verhältnisse geschaffen, von denen in der Einleitung die Rede war: ein Magnetfeld von verschwindender Stärke und eine physikalisch ausgezeichnete Äquator-Ebene $\zeta = 0$. In diesem System gelten daher die dort angegebenen Resultate: Die Bahnebene ist gegen die Äquatorebene um einen Winkel α geneigt, welcher nur die durch (1) § 1 ganzzahlig bestimmten Werte annehmen darf. Die Flächenkonstante für die Äquatorebene (die wir jetzt sinngemäß nicht p_φ sondern p_χ nennen werden) ist durch (3) § 1 bestimmt:

$$2\pi p_\chi = n_1 h\,, \tag{12}$$

die Energie im $\xi\eta\zeta$-System (die wir entsprechend mit W_χ bezeichnen) durch (2) § 1:

$$W_\chi = \frac{Nh}{(n+n')^2}\,.$$

Wegen des Beweises dieser Resultate sei auf die eingangs zitierte Annalenarbeit verwiesen; übrigens enthalten die späteren relativistischen Rechnungen den vorliegenden Fall als Spezialfall.

Wir fragen nun nach der Energie der Bewegung im xyz-System. Sie ist

$$W = \frac{m}{2}(\dot r^2 + r^2\dot\vartheta^2 + r^2\sin^2\vartheta\,\dot\varphi^2) - \frac{e^2}{r}\,.$$

Benutzen wir die Beziehung (7) und vernachlässigen das Glied mit f^2, so entsteht

$$W = \frac{m}{2}(\dot r^2 + r^2\dot\vartheta^2 + r^2\sin^2\vartheta\,\dot\chi^2) - \frac{e^2}{r} + m r^2\sin^2\vartheta\,\dot\chi\,\frac{f}{2}\,.$$

Die beiden ersten Terme rechts sind identisch mit W_χ, der Faktor von $\frac{f}{2}$ im dritten Term ist nach (8) gleich p_χ. Benutzen wir für p_χ Gl. (12), so entsteht

$$W = W_\chi + \frac{f}{4\pi} n_1 h, \; \Delta W = W - W_0 = \frac{f}{4\pi} n_1 h.$$

W_χ ist identisch mit der Energie W_0, die unsere Bahn ohne Magnetfeld besitzen würde;

$$\Delta W = W - W_\chi = W - W_0 \qquad (13)$$

stellt also diejenige Energie dar, die in der magnetischen Präzession der Bahn steckt. Nach dem Bohrschen Frequenzansatz ist die Schwingungszahl der Spektrallinie gegeben durch

$$h\nu = W_m - W_n.$$

W_m bedeutet die Energie der Anfangsbahn, welche durch die Quantenzahlen $m = m_1 + m_2$ und m' bestimmt ist, W_n die Energie der Endbahn, die zu den Quantenzahlen $n = n_1 + n_2$ und n' gehört. Die Änderung der Schwingungszahl durch das Magnetfeld beträgt also

$$\Delta \nu = \frac{\Delta W_m - \Delta W_n}{h} = \frac{f}{4\pi}(m_1 - n_1). \qquad (14)$$

Die normale Lorentzsche Verschiebung (Abstand der s-Komponenten im Lorentzschen Triplett von der p-Komponente) ist in unseren Bezeichnungen

$$\Delta \nu_{\text{norm}} = \frac{f}{4\pi}.$$

Es ergibt sich also:

$$\Delta \nu = \Delta \nu_{\text{norm}} \cdot Z, \; Z = m_1 - n_1 = \text{ganze Zahl.} \qquad (15)$$

§ 3. Diskussion des Resultats.

Unser Resultat (15) fällt mit dem Ergebnis der klassischen Theorie nicht zusammen, aber dicht daneben. Unser $\Delta\nu$ ist, allgemein gesprochen, ein ganzes Vielfaches Z des normalen Lorentzschen $\Delta\nu$.

Naturgemäß bilden bei Bohrscher Auffassung die Polarisationen der Komponenten im Zeeman- und Starkeffekt eine besondere Schwierigkeit. Indessen zeigt unser Resultat auch in dieser Hinsicht eine Analogie mit der Wirklichkeit. Beim Starkeffekt hat Epstein die folgende merkwürdige Regel durchweg bestätigt gefunden: Es findet parallele (senkrechte) Polarisation der Starkkomponente statt, wenn die azimutale Quantenzahl beim Übergange um eine gerade (ungerade) Anzahl von Einheiten springt. Bei Epstein bezieht sich die azimutale Quantenzahl auf einen Winkel, der in der Ebene senkrecht zu den elektrischen Kraftlinien gezählt wird; ihr entspricht bei uns (in gewissem Sinne) die Quantenzahl n_1 bzw. m_1. Übertragen wir also die Epsteinsche Regel auf unseren Fall, so haben wir zu erwarten: Bei geradem Z, insbesondere für $Z = 0$, haben wir parallele, bei ungeradem Z, insbesondere für $Z = 1$, senkrechte Polarisation, wie es sein soll.

Nach der von mir früher aufgestellten Regel der „Quantenungleichungen", nach der bei dem positiven Charakter der Quantenzahlen diese nur (oder vorzugsweise) von größeren zu kleineren Werten übergehen, würde $m_1 \geq n_1$, also Z eine positive ganze Zahl (einschl. der Null) sein. Dies scheint dem rechts und links symmetrischen Verhalten der Zeeman-Aufspaltung zu widersprechen. Indessen erhält man offenbar die Komponenten auf der negativen Seite, wenn man den Sinn von H, also das Vorzeichen von f umkehrt. Umkehrung des Umlaufsinnes in der Keplerellipse ist aber gleichbedeutend mit Umkehrung von H. Da der Umlaufsinn bei uns unbestimmt bleibt, ist also auch unser Resultat rechts und links symmetrisch.

Hiernach ergibt sich das Bild der Fig. 2, ein normales Lorentzsches Triplett mit

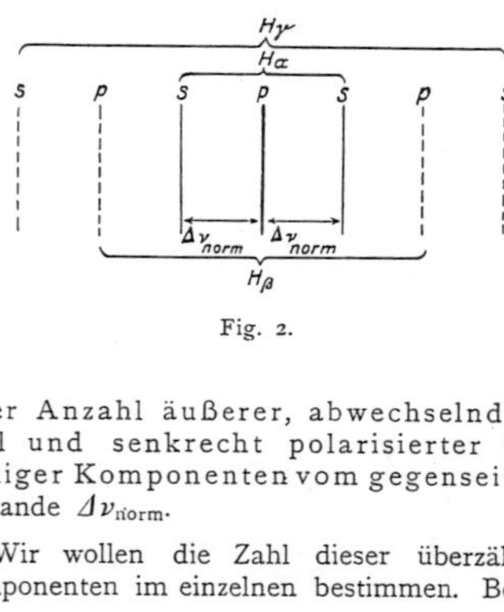

Fig. 2.

einer Anzahl äußerer, abwechselnd parallel und senkrecht polarisierter überzähliger Komponenten vom gegenseitigen Abstande $\Delta\nu_{\text{norm}}$.

Wir wollen die Zahl dieser überzähligen Komponenten im einzelnen bestimmen. Bei H_α

$$\frac{\nu}{N} = \frac{1}{2^2} - \frac{1}{3^2}$$

ist $m + m' = 3$, $n + n' = 2$. Der Fall n bzw. m gleich Null ist auszuschließen, da derselbe einer ausgearteten Ellipse von der Exzentrizität 1 entsprechen würde, die das Elektron durch den Kern hindurch führen würde. Es sind also die folgenden Fälle möglich, wenn man nur Übergänge von abnehmenden Quantenzahlen in allen drei Zahlenpaaren $m_1 n_1$, $m_2 n_2$, $m' n'$ betrachtet:

$m' = 0$	$m_2 = 3$	$m_1 = 0$	$n' = 0$	$n_2 = 2$	$n_1 = 0$	$Z = 0$
$m' = 0$	$m_2 = 2$	$m_1 = 1$	$n' = 0$	$n_2 = 2$	$n_1 = 0$	$Z = 1$
$m' = 0$	$m_2 = 2$	$m_1 = 1$	$n' = 0$	$n_2 = 1$	$n_1 = 1$	$Z = 0$
$m' = 0$	$m_2 = 1$	$m_1 = 2$	$n' = 0$	$n_2 = 1$	$n_1 = 1$	$Z = 1$
$m' = 0$	$m_2 = 1$	$m_1 = 2$	$n' = 0$	$n_2 = 0$	$n_1 = 2$	$Z = 0$
$m' = 0$	$m_2 = 0$	$m_1 = 3$	$n' = 0$	$n_2 = 0$	$n_1 = 2$	$Z = 1$
$m' = 1$	$m_2 = 2$	$m_1 = 0$	$n' = 1$	$n_2 = 1$	$n_1 = 0$	$Z = 0$
$m' = 1$	$m_2 = 2$	$m_1 = 0$	$n' = 0$	$n_2 = 2$	$n_1 = 0$	$Z = 0$
$m' = 1$	$m_2 = 1$	$m_1 = 1$	$n' = 1$	$n_2 = 1$	$n_1 = 0$	$Z = 1$
$m' = 1$	$m_2 = 1$	$m_1 = 1$	$n' = 1$	$n_2 = 0$	$n_1 = 1$	$Z = 0$
$m' = 1$	$m_2 = 1$	$m_1 = 1$	$n' = 0$	$n_2 = 1$	$n_1 = 1$	$Z = 0$
$m' = 1$	$m_2 = 0$	$m_1 = 2$	$n' = 1$	$n_2 = 0$	$n_1 = 1$	$Z = 1$
$m' = 1$	$m_2 = 0$	$m_1 = 2$	$n' = 0$	$n_2 = 0$	$n_1 = 2$	$Z = 0$
$m' = 2$	$m_2 = 1$	$m_1 = 0$	$n' = 1$	$n_2 = 1$	$n_1 = 0$	$Z = 0$
$m' = 2$	$m_2 = 1$	$m_1 = 0$	$n' = 0$	$n_2 = 2$	$n_1 = 0$	$Z = 0$
$m' = 2$	$m_2 = 0$	$m_1 = 1$	$n' = 1$	$n_2 = 0$	$n_1 = 1$	$Z = 0$

Hier treten also noch keine überzähligen Komponenten auf; Z ist wegen unserer „Quantenungleichungen" auf die Werte 0 und 1 eingeschränkt.

Wohl aber wird bei H_β neben $Z = 0$ und 1 auch der Wert $Z = 2$ möglich, so daß wir eine überzählige Komponente haben, wie die folgenden Beispiele zeigen:

$m' = 0$	$m_2 = 0$	$m_1 = 4$	$n' = 0$	$n_2 = 0$	$n_1 = 2$	$Z = 2$
$m' = 0$	$m_2 = 1$	$m_1 = 3$	$n' = 0$	$n_2 = 1$	$n_1 = 1$	$Z = 2$
$m' = 1$	$m_2 = 0$	$m_1 = 3$	$n' = 1$	$n_2 = 0$	$n_1 = 1$	$Z = 2$
$m' = 1$	$m_2 = 1$	$m_1 = 2$	$n' = 1$	$n_2 = 1$	$n_1 = 0$	$Z = 2$

Bei H_γ tritt mit $Z = 3$ eine zweite überzählige Komponente hinzu usf. Allgemein haben wir mit $n + n' = N$, $m + m' = M > N$ als Höchstwert von Z

$$Z = M - N,$$

also $M - N - 1$ überzählige Komponenten.

Es ist interessant zu bemerken, daß eine ältere Theorie von Ritz[1]) überzählige Komponenten derselben Lage ergeben hat. Daß diese Komponenten real und bisher nur wegen geringer Intensität nicht bemerkt wären, ist recht unwahrscheinlich.

Wenn somit das Ergebnis auch keineswegs befriedigend ist, so überrascht doch die Leistungsfähigkeit der Quantentheorie im Verein mit dem Bohrschen Modelle auch in diesem Falle. Diese Leistungsfähigkeit übertrifft sogar die Erwartungen, die Herr Bohr selbst hinsichtlich des Zeemaneffekts geäußert hat. Er meinte: die elektrische Zerlegung der Spektrallinien werde durch eine Änderung der Elektronenbahnen und ihrer Energie bewirkt, die magnetische Zerlegung dagegen durch eine magnetische Beeinflussung des Übergangs von der Anfangs- in die Endbahn, die zu einer Abänderung des Frequenzansatzes führe. Wir sahen dagegen: Unter wörtlicher Beibehaltung dieses Ansatzes ergibt sich eine Energieänderung der Elektronenbahnen, die der Hauptsache nach auf das Lorentzsche Triplett führt, ohne Einführung irgendwelcher neuer Annahmen.

1) W. Ritz, Gesammelte Werke, Paris 1911, pag. 120, Ann. d. Phys. **25**, 660, 1908.

§ 4. Zeemaneffekt mit Relativität; die Bewegungsgleichungen.

Die Differentialgleichungen behalten die Form (1) § 2 bei; die rechten Seiten bleiben völlig ungeändert, auf der linken Seite ist unter m zu verstehen

$$m = \frac{m_0}{\sqrt{1 - \beta^2}} = \frac{m_0}{\dot{\tau}},$$

wo m_0 die Ruhemasse und τ die Eigenzeit bedeuten

$$\tau = \int \sqrt{1 - \beta^2}\, dt, \quad \dot{\tau} = \sqrt{1 - \beta^2} = \frac{m_0}{m}. \tag{1}$$

Bei Benutzung der Polarkoordinaten r, ϑ, φ wird das Impulsmoment für die φ-Koordinate:

$$\begin{aligned} p_\varphi = m r^2 \sin^2 \vartheta\, \dot{\varphi} &= m_0 r^2 \sin^2 \vartheta \frac{\dot{\varphi}}{\dot{\tau}} \\ &= m_0 r^2 \sin^2 \vartheta \frac{d\varphi}{d\tau}. \end{aligned} \tag{2}$$

496 Sommerfeld, Theorie d. Zeeman-Effekts d. Wasserstofflinien. Physik. Zeitschr. XVII, 1916.

Die Stärke des Magnetfeldes ist jetzt zu messen durch das Produkt

$$f = \frac{e}{m_0}\frac{H}{c}. \tag{3}$$

Daraufhin folgt aus den Bewegungsgleichungen durch die übliche Rechnung an Stelle von Gl. (4) § 2:

$$\frac{dp_\varphi}{dt} = \frac{m_0 f}{2}\frac{d}{dt}(x^2+y^2) = \frac{m_0 f}{2}\frac{d}{dt}(r^2\sin^2\vartheta)$$

oder wegen (2) und (1)

$$\frac{d}{dt}\left(mr^2\sin^2\vartheta\,(\dot\varphi - \frac{m_0}{m}\frac{f}{2})\right) = \frac{d}{dt}\left((mr^2\sin^2\vartheta\,(\dot\varphi - \frac{f}{2}\dot\tau)\right) = 0. \tag{4}$$

Ein Hilfswinkel χ ist jetzt offenbar folgendermaßen zu definieren:

$$\dot\chi = \dot\varphi - \frac{f}{2}\dot\tau,\quad \varphi = \chi + \frac{f}{2}\tau. \tag{5}$$

Nach Einführung desselben schreibt sich der Flächensatz (4) einfach so:

$$\frac{dp_\chi}{dt} = 0,\quad p_\chi = mr^2\sin^2\vartheta\,\dot\chi = m_0 r^2\sin^2\vartheta\frac{\dot\chi}{\dot\tau} = m_0 r^2\sin^2\vartheta\frac{d\chi}{d\tau}. \tag{6}$$

Während p_φ unter dem Einfluß des Magnetfeldes variabel ist, wird wieder das auf den Hilfswinkel χ bezogene Impulsmoment p_χ konstant. Dabei unterscheiden sich φ und χ nicht wie früher um den Betrag $\frac{f}{2}t$, der proportional mit der Zeit wächst, sondern um den Betrag $\frac{f}{2}\tau$, der proportional der Eigenzeit anwächst. Dementsprechend ist jetzt ein $\xi\eta\zeta$-System einzuführen, welches gegen das xyz-System um die z-Achse gleichförmig in der Eigenzeit τ, aber ungleichförmig in der Zeit t gedreht wird.

Wir schreiben die Gl. (1) § 2 zunächst so:

$$\left.\begin{aligned}
&\frac{d}{dt}(m\dot x + m_0 f y) = \frac{d}{dt}(m(\dot x + f\dot\tau y)) = -\frac{e^2}{r^3}x,\\
&\frac{d}{dt}(m\dot y - m_0 f x) = \frac{d}{dt}(m(\dot y - f\dot\tau x)) = -\frac{e^2}{r^3}y,\\
&\frac{d}{dt}(m\dot z) = -\frac{e^2}{r^3}z.
\end{aligned}\right\} \tag{7}$$

Aus

$x = r\sin\vartheta\cos\varphi,\quad y = r\sin\vartheta\sin\varphi,\quad z = r\cos\vartheta$
$\xi = r\sin\vartheta\cos\chi,\quad \eta = r\sin\vartheta\sin\chi,\quad \zeta = r\cos\vartheta$

ergibt sich wegen (5)

$$\left.\begin{aligned}
x &= \xi\cos\frac{f}{2}\tau - \eta\sin\frac{f}{2}\tau,\\
y &= \xi\sin\frac{f}{2}\tau + \eta\cos\frac{f}{2}\tau,\\
z &= \zeta,
\end{aligned}\right\} \tag{8}$$

ferner

$$\begin{aligned}
\dot x + f\dot\tau y &= \left(\dot\xi + \frac{f}{2}\dot\tau\eta\right)\cos\frac{f}{2}\tau - \left(\dot\eta - \frac{f}{2}\dot\tau\xi\right)\sin\frac{f}{2}\tau,\\
\dot y - f\dot\tau x &= \left(\dot\xi + \frac{f}{2}\dot\tau\eta\right)\sin\frac{f}{2}\tau + \left(\dot\eta - \frac{f}{2}\dot\tau\xi\right)\cos\frac{f}{2}\tau,
\end{aligned}$$

endlich mit Rücksicht auf die Beziehung

$$m\dot\tau = m_0,\ \text{Gl. (1)}$$

$$\begin{aligned}
\frac{d}{dt}(m(\dot x + f\dot\tau y)) &= \left(\frac{d}{dt}(m\dot\xi) + m_0\left(\frac{f}{2}\right)^2\dot\tau\xi\right)\cos\frac{f}{2}\tau - \left(\frac{d}{dt}(m\dot\eta) + m_0\left(\frac{f}{2}\right)^2\dot\tau\eta\right)\sin\frac{f}{2}\tau\\
\frac{d}{dt}(m(\dot y - f\dot\tau y)) &= \left(\frac{d}{dt}(m\dot\xi) + m_0\left(\frac{f}{2}\right)^2\dot\tau\eta\right)\sin\frac{f}{2}\tau + \left(\frac{d}{dt}(m\dot\eta) + m_0\left(\frac{f}{2}\right)^2\dot\tau\eta\right)\cos\frac{f}{2}\tau.
\end{aligned}$$

Für das $\xi\eta\zeta$-System folgt daher aus (7) und (8)

$$\left.\begin{aligned}
\frac{d}{dt}(m\dot\xi) &= -\frac{e^2}{r^3}\xi - m_0\left(\frac{f}{2}\right)^2\dot\tau\xi,\\
\frac{d}{dt}(m\dot\eta) &= -\frac{e^2}{r^3}\eta - m_0\left(\frac{f}{2}\right)^2\dot\tau\eta,\\
\frac{d}{dt}(m\dot\zeta) &= -\frac{e^2}{r^3}\zeta.
\end{aligned}\right\} \tag{9}$$

Vernachlässigt man wie in § 2 Glieder mit f^2, so hat man wieder genau die Bewegungsgleichungen ohne Magnetfeld, also eine Keplersche Ellipse, deren Bahnebene im $\xi\eta\zeta$-System festliegt, im xyz-System aber eine Präzession um die z-Achse ausführt, u. zw. eine in der Zeit t ungleichförmige, in der Eigenzeit τ gleichförmige Präzession.

Will man aber die Glieder mit f^2 zunächst beibehalten, so lassen sich die auf den rechten Seiten von (9) stehenden Kräfte nicht mehr wie früher aus einem Potential ableiten und die Bewegungsgleichungen haben dementsprechend nicht die kanonische Form. Um hierzu zu gelangen, muß man zweierlei tun: Man muß einerseits die Eigenzeit τ als unabhängige Variable einführen, andererseits Gebrauch machen von dem Satze der lebendigen Kraft.

Dividiert man die Gl. (9) rechts und links mit $\dot\tau$, so erhält man zunächst:

$$\left.\begin{aligned}
m_0\frac{d^2\xi}{d\tau^2} &= -\frac{e^2}{r^3}\frac{\xi}{\dot\tau} - m_0\left(\frac{f}{2}\right)^2\xi\\
m_0\frac{d^2\eta}{d\tau^2} &= -\frac{e^2}{r^3}\frac{\eta}{\dot\tau} - m_0\left(\frac{f}{2}\right)^2\eta\\
m_0\frac{d^2\zeta}{d\tau^2} &= -\frac{e^2}{r^3}\frac{\zeta}{\dot\tau}.
\end{aligned}\right\} \tag{10}$$

Man beachte dabei, daß τ nicht die Eigenzeit im $\xi\eta\zeta$-System, sondern diejenige im xyz-System ist. Daß man die Bewegungsgleichungen des Massenpunktes durch Einführung seiner Eigenzeit vereinfacht, ist seit Minkowski wohl bekannt. Mit der hier gewonnenen Vereinfachung hat es dagegen eine andere Bewandtnis, da es sich hier sozusagen um eine „fremde Eigenzeit" handelt.

Der Zusammenhang zwischen der Eigenzeit im xyz-System τ und der Eigenzeit im $\xi\eta\zeta$-System — wir wollen sie etwa T nennen — folgt aus der Gl. (8) durch Differentiation. Es ist nämlich

$$\dot{x} = \left(\dot{\xi} - \frac{f}{2}\dot{\tau}\eta\right)\cos\frac{f}{2}\tau - \left(\dot{\eta} + \frac{f}{2}\dot{\tau}\xi\right)\sin\frac{f}{2}\tau,$$

$$\dot{y} = \left(\dot{\xi} - \frac{f}{2}\dot{\tau}\eta\right)\sin\frac{f}{2}\tau + \left(\dot{\eta} + \frac{f}{2}\dot{\tau}\xi\right)\cos\frac{f}{2}\tau,$$

daher

$$\left.\begin{aligned} \dot{x}^2+\dot{y}^2+\dot{z}^2 &= \left(\dot{\xi} - \frac{f}{2}\dot{\tau}\eta\right)^2 + \left(\dot{\eta} + \frac{f}{2}\dot{\tau}\xi\right)^2 + \dot{\zeta}^2 = \\ & \dot{\xi}^2+\dot{\eta}^2+\dot{\zeta}^2 + f\dot{\tau}(\xi\dot{\eta}-\eta\dot{\xi}) + \left(\frac{f}{2}\right)^2(\xi^2+\eta^2)\dot{\tau}^2. \end{aligned}\right\} \tag{11}$$

Nun gilt aber

$$\dot{x}^2+\dot{y}^2+\dot{z}^2 = c^2(1-\dot{\tau}^2),\quad \dot{\xi}^2+\dot{\eta}^2+\dot{\zeta}^2 = c^2(1-\dot{T}^2),$$

$$\xi\dot{\eta}-\eta\dot{\xi} = \frac{p_\chi}{m} = \frac{p_\chi\dot{\tau}}{m_0}. \tag{12}$$

Also

$$c^2(1-\dot{\tau}^2) = c^2(1-\dot{T}^2) + \left(\frac{fp_\chi}{m_0} + \left(\frac{f}{2}\right)^2(\xi^2+\eta^2)\right)\dot{\tau}^2,$$

$$\dot{T}^2 = \dot{\tau}^2\left\{1 + \frac{1}{c^2}\left(\frac{fp_\chi}{m_0} + \left(\frac{f}{2}\right)^2(\xi^2+\eta^2)\right)\right\}. \tag{13}$$

Was sodann den Satz der lebendigen Kraft betrifft, so beträgt die kinetische Energie im xyz-System bekanntlich

$$m_0c^2\left(\frac{1}{\sqrt{1-\beta^2}} - 1\right) = \frac{m_0c^2}{\dot{\tau}} - m_0c^2.$$

Die potentielle Energie ist, da das magnetische Feld wattlos wirkt:

$$V = -\frac{e^2}{r}.$$

Bedeutet W die bei der Bewegung konstante Gesamtenergie, so gilt also

$$\frac{1}{\dot{\tau}} = \frac{1}{m_0c^2}\left(W + \frac{e^2}{r}\right) + 1. \tag{14}$$

Bei Benutzung dieser Form der Energiegleichung nehmen die Bewegungsgleichungen (10) eine wesentlich einfachere Form an.

§ 5. Kanonische Form der Bewegungsgleichungen.

Durch Eintragen von (14) in (10) des vorigen Paragraphen ergibt sich

$$\begin{aligned} m_0\frac{d^2\xi}{d\tau^2} &= -\frac{e^2\xi}{r^3}\left(1+\frac{W}{m_0c^2}\right) - \frac{e^4\xi}{m_0c^2r^4} - m_0\left(\frac{f}{2}\right)^2\xi, \\ m_0\frac{d^2\eta}{d\tau^2} &= -\frac{e^2\eta}{r^3}\left(1+\frac{W}{m_0c^2}\right) - \frac{e^4\eta}{m_0c^2r^4} - m_0\left(\frac{f}{2}\right)^2\eta, \\ m_0\frac{d^2\zeta}{d\tau^2} &= -\frac{e^2\zeta}{r^3}\left(1+\frac{W}{m_0c^2}\right) - \frac{e^4\zeta}{m_0c^2r^4}. \end{aligned} \tag{1}$$

Die rechten Seiten sind die negativen Ableitungen nach $\xi\eta\zeta$ des Potentials

$$V = -\frac{e^2}{r}\left(1+\frac{W}{m_0c^2}\right) - \frac{e^4}{2m_0c^2r^2} + m_0\frac{f^2}{8}(\xi^2+\eta^2). \tag{2}$$

Führt man noch die Impulskomponenten

$$p_\xi = m_0\frac{d\xi}{d\tau},\quad p_\eta = m_0\frac{d\eta}{d\tau},\quad p_\zeta = m_0\frac{d\zeta}{d\tau} \tag{3}$$

(die wegen Benutzung der „fremden Eigenzeit" τ nicht mit den Impulskomponenten im $\xi\eta\zeta$-Sytem identisch sind) sowie die Hamiltonsche Funktion

$$H = \frac{p_\xi^2+p_\eta^2+p_\zeta^2}{2m_0} + V \tag{4}$$

ein, so schreiben sich die Gl. (1) und (3) in die folgende kanonische Form um:

$$\left.\begin{aligned} \frac{dp_\xi}{d\tau} &= -\frac{\partial H}{\partial\xi}, & \frac{dp_\eta}{d\tau} &= -\frac{\partial H}{\partial\eta}, & \frac{dp_\zeta}{d\tau} &= -\frac{\partial H}{\partial\zeta}, \\ \frac{d\xi}{d\tau} &= +\frac{\partial H}{\partial p_\xi}, & \frac{d\eta}{d\tau} &= +\frac{\partial H}{\partial p_\eta}, & \frac{d\zeta}{d\tau} &= +\frac{\partial H}{\partial p_\zeta}. \end{aligned}\right\} \tag{5}$$

Aus dieser Form der Gleichungen schließt man sofort, daß gilt:

$$\frac{dH}{d\tau} = 0, \quad \text{also } H = H_0 = \text{konst.} \tag{6}$$

Allerdings ist die Konstante H_0 der Hamiltonschen Funktion nicht, wie in den gewöhnlichen Problemen der Mechanik, identisch mit der Energiekonstanten W. Wir finden aber den Ausdruck von H_0 durch W folgendermaßen: Zunächst ist

$$\frac{p_\xi^2+p_\eta^2+p_\zeta^2}{2m_0} = \frac{m_0}{2}\,\frac{\dot{\xi}^2+\dot{\eta}^2+\dot{\zeta}^2}{\dot{\tau}^2}.$$

Aus den Gl. (11) und (12) des vorigen Paragraphen entnimmt man:

$$\begin{aligned} \frac{\dot{\xi}^2+\dot{\eta}^2+\dot{\zeta}^2}{\dot{\tau}^2} &= \frac{\dot{x}^2+\dot{y}^2+\dot{z}^2}{\dot{\tau}^2} - \frac{fp_\chi}{m_0} - \left(\frac{f}{2}\right)^2(\xi^2+\eta^2) \\ &= c^2\left(\frac{1}{\dot{\tau}^2} - 1\right) - \frac{fp_\chi}{m_0} - \left(\frac{f}{2}\right)^2(\xi^2+\eta^2). \end{aligned}$$

Nimmt man noch Gl. (14) des vorigen Paragraphen hinzu, so wird

$$\begin{aligned} \frac{p_\xi^2+p_\eta^2+p_\zeta^2}{2m_0} &= \frac{1}{2m_0c^2}\left(W+\frac{e^2}{r}\right)^2 \\ &+ \left(W+\frac{e^2}{r}\right) - \frac{fp_\chi}{2} - \frac{m_0f^2}{8}(\xi^2+\eta^2). \end{aligned} \tag{7}$$

Addiert man hierzu den Ausdruck von V in Gl. (2), so fallen alle veränderlichen Terme heraus und es entsteht als Konstante der Hamiltonschen Funktion:

$$H_0 = W\left(1 + \frac{W}{2m_0c^2}\right) - \frac{fp_\chi}{2}. \tag{8}$$

Bekanntlich besteht Jacobis Integrationsmethode der kanonischen Gleichungen der Mechanik darin, eine Wirkungsfunktion S einzuführen, die in unserem Falle der drei Variabeln $\xi\eta\zeta$ folgendermaßen zu definieren ist

$$S = \int_{\xi_0}^{\xi} p_\xi d\xi + \int_{\eta_0}^{\eta} p_\eta d\eta + \int_{\zeta_0}^{\zeta} p_\zeta d\zeta, \tag{9}$$

also als Summe der drei „Phasenintegrale" mit unbestimmter oberer und beliebiger unterer Grenze. Da hiernach

$$p_\xi = \frac{\partial S}{\partial \xi},\ p_\eta = \frac{\partial S}{\partial \eta},\ p_\zeta = \frac{\partial S}{\partial \zeta}, \tag{10}$$

so folgt aus (7) als Jacobische partielle Differentialgleichung

$$\left(\frac{\partial S}{\partial \xi}\right)^2 + \left(\frac{\partial S}{\partial \eta}\right)^2 + \left(\frac{\partial S}{\partial \zeta}\right)^2 = \frac{1}{c^2}\left(W + \frac{e^2}{r}\right)^2 + 2m_0\left(W + \frac{e^2}{r}\right) - m_0 f p_\chi - \left(\frac{m_0 f}{2}\right)^2(\xi^2 + \eta^2) \tag{11}$$

oder, in Polarkoordinaten $r\vartheta\chi$ geschrieben:

$$\left(\frac{\partial S}{\partial r}\right)^2 + \frac{1}{r^2}\left(\frac{\partial S}{\partial \vartheta}\right)^2 + \frac{1}{r^2\sin^2\vartheta}\left(\frac{\partial S}{\partial \chi}\right) = \frac{1}{c^2}\left(W + \frac{e^2}{r}\right)^2 + 2m_0\left(W + \frac{e^2}{r}\right) - m_0 f p_\chi - \left(\frac{m_0 f}{2}\right)^2 r^2 \sin^2\vartheta. \tag{12}$$

Setzt man hierin $c = \infty$, so erhält man natürlich die entsprechende partielle Differentialgleichung für den Zeemaneffekt ohne Relativität. Die zugehörigen gewöhnlichen Differentialgleichungen sind durch (5) gegeben, wenn man darin τ durch t und H ersetzt durch

$$H = \frac{p_\xi^2 + p_\eta^2 + p_\zeta^2}{2m_0} - \frac{e^2}{r} + \frac{m_0 f^2}{8}(\xi^2 + \eta^2),$$

wie sich aus (2) mit $c = \infty$ ergibt.

Setzt man andererseits $f = 0$, so fällt man auf das Keplersche Problem ohne Magnetfeld, aber mit Relativität zurück. Die Hamiltonsche Funktion lautet in diesem Falle nach unserer Methode, bei der als unabhängige Variable die Eigenzeit benutzt wird:

$$H = \frac{p_\xi^2 + p_\eta^2 + p_\zeta^2}{2m_0} - \frac{e^2}{r}\left(1 + \frac{W}{m_0c^2}\right) - \frac{e^4}{2m_0c^2r^2}$$

und unterscheidet sich scheinbar wesentlich von der gewöhnlich angegebenen Form der Hamiltonschen Funktion, bei der als unabhängige Variable in den kanonischen Differentialgleichungen die Zeit t benutzt wird:

$$H = m_0c^2\left(\sqrt{1 + \frac{p_\xi^2 + p_\eta^2 + p_\zeta^2}{m_0^2c^2}} - 1\right) - \frac{e^2}{r}. \tag{13}$$

Dieser letzte Ausdruck ist nichts anderes als die Summe von kinetischer und potentieller Energie in relativistischer Form, ausgedrückt durch die rechtwinkligen Impulskoordinaten, welche im Anschluß an das Vorhergehende mit p_ξ, p_η, p_ζ bezeichnet sind. Bildet man hiermit die kanonischen Differentialgleichungen

$$\frac{dp_\xi}{dt} = -\frac{\partial H}{\partial \xi}, \ldots \frac{d\xi}{dt} = \frac{\partial H}{\partial p_\xi}, \ldots$$

so unterscheiden sich dieselben formal von denjenigen Gleichungen, die aus (5) mit $f = 0$ entstehen. Bildet man aber die Jacobische partielle Differentialgleichung, indem man in (13) p_ξ, p_η, p_ζ durch (10) ersetzt und H gleich W macht, so entsteht

$$1 + \frac{1}{m_0^2c^2}\left(\left(\frac{\partial S}{\partial \xi}\right)^2 + \left(\frac{\partial S}{\partial \eta}\right)^2 + \left(\frac{\partial S}{\partial \zeta}\right)^2\right) = \left(1 + \frac{1}{m_0c^2}\left(W + \frac{e^2}{r}\right)\right)^2.$$

Dies erweist sich als völlig identisch mit der Gl. (11), wenn man darin $f = 0$ setzt. Es ist auch von vornherein klar, daß die Einführung der Eigenzeit, die nur ein analytisches Hilfsmittel zur Vereinfachung der Gleichungen war, keinen Einfluß haben kann auf die Jacobische Differentialgleichung, aus der sie herausfällt, und auf die weiteren aus ihr zu ziehenden Schlüsse.

Schließlich läßt sich auch für den Zeemaneffekt mit Relativität eine Form der Hamiltonschen Funktion angegeben, die der Gl. (13) analog gebaut ist und zu kanonischen Gleichungen mit der unabhängigen Variabeln t statt τ gehört. Sie lautet nach einer freundlichen Mitteilung von G. Herglotz:

$$H = m_0c^2\left(\sqrt{1 + \frac{w^2}{m_0^2c^2}} - 1\right) - \frac{e^2}{r}$$

mit der Abkürzung

$$w^2 = \left(p_\xi - \frac{m_0 f}{2}\eta\right)^2 + \left(p_\eta + \frac{m_0 f}{2}\xi\right)^2 + p_\zeta^2.$$

Die Konstante H_0 wird alsdann mit der Energie W identisch; die zugehörige Jacobische Gleichung stimmt mit (11) überein.

§ 6. Aufstellung der Quantenbedingungen.

Ebenso wie die rechtwinkligen Impulskomponenten in Gl. (10) des vorigen Paragraphen sind bei Einführung beliebiger Koordinaten q_i die zugehörigen Impulskomponenten die Ableitungen der Wirkungsfunktion nach den betr. q:

$$p_i = \frac{\partial S}{\partial q_i}.$$

Die von mir aufgestellte allgemeine Quantenbedingung für die Koordinate q_i

Physik. Zeitschr. XVII, 1916. Sommerfeld, Theorie d. Zeeman-Effekts d. Wasserstofflinien. 499

$$\int p_i dq_i = n_i \cdot h \; (n_i = \text{ganze Zahl})$$

schreibt sich daher einfach

$$\int \frac{\partial S}{\partial q_i} dq_i = [S]_i = n_i h \, .$$

Hier bedeutet $[S]_i$ die Änderung, welche die Wirkungsfunktion erfährt, wenn die Koordinate q_i denjenigen Bereich durchläuft, der zur Darstellung aller „Phasen" des betrachteten Systems erforderlich ist. Wir nennen diese Änderung den i^{ten} Periodizitätsmodul von S und haben den in der Einleitung genannten Satz: Die Periodizitätsmoduln der Jacobischen Wirkungsfunktion sind ganze Vielfache des Planckschen Wirkungsquantums.

Wenden wir dieses Schema auf unsere Polarkoordinaten r, ϑ, χ an, so lautet es:

$$[S]_\chi = n_1 h, \; [S]_\vartheta = n_2 h, \; [S]_r = n' h \, . \qquad (1)$$

Aus der Differentialgleichung (12) des vorigen Paragraphen, in der χ rechter Hand nicht vorkommt, ergibt sich

$$\frac{\partial S}{\partial \chi} = p_\chi = \text{konst.}; \qquad (2)$$

da der Wertebereich des Winkels χ sich von 0 bis 2π erstreckt, hat man

$$[S]_\chi = \int_0^{2\pi} p_\chi \, d\chi = 2\pi \, p_\chi \, . \qquad (3)$$

Setzt man den aus (1), (2) und (3) folgenden Wert

$$p_\chi = \frac{\partial S}{\partial \chi} = \frac{n_1 h}{2\pi}$$

in Gl. (12) des vorigen Paragraphen ein, so ergibt sich:

$$\left(\frac{\partial S}{\partial r}\right)^2 + \frac{1}{r^2}\left(\frac{\partial S}{\partial \vartheta}\right)^2 = \frac{1}{c^2}\left(W + \frac{e^2}{r}\right)^2 + 2m_0\left(W + \frac{e^2}{r}\right) - m_0 f \cdot \frac{n_1 h}{2\pi} - \left(\frac{m_0 f}{2}\right)^2 r^2 \sin^2\vartheta - \frac{1}{r^2 \sin^2\vartheta}\left(\frac{n_1 h}{2\pi}\right)^2, \qquad (4)$$

wofür man auch schreiben kann:

$$\left(\frac{\partial S}{\partial r}\right)^2 + \frac{1}{r^2}\left(\frac{\partial S}{\partial \vartheta}\right)^2 = \left(\frac{1}{c}\left(W + \frac{e^2}{r}\right) + m_0 c\right)^2 - \left(\frac{m_0 f}{2} r \sin\vartheta + \frac{n_1 h}{2\pi r \sin\vartheta}\right)^2 - m_0^2 c^2 \, . \qquad (4a)$$

Es wäre nun prinzipiell erforderlich, diese Gleichung in Strenge zu integrieren. Dies ist aber nach der gewöhnlichen Methode der Separation der Variabeln jedenfalls nicht möglich. Wir vernachlässigen daher das Glied mit f^2, indem wir uns auf die Abschätzung seiner Größe in § 2 berufen. Dann läßt sich (4) unter Hinzufügung des Gliedes

$$\pm \frac{1}{r^2 \sin^2\vartheta_0}\left(\frac{n_1 h}{2\pi}\right)^2$$

separieren in die beiden folgenden Gleichungen:

$$\left(\frac{\partial S}{\partial \vartheta}\right)^2 = \left(\frac{1}{\sin^2\vartheta_0} - \frac{1}{\sin^2\vartheta}\right)\left(\frac{n_1 h}{2\pi}\right)^2 . \qquad (5)$$

$$\left(\frac{\partial S}{\partial r}\right)^2 = \frac{1}{c^2}\left(W + \frac{e^2}{r}\right)^2 + 2m_0\left(W + \frac{e^2}{r}\right) - m_0 f \frac{n_1 h}{2\pi} - \frac{1}{r^2 \sin^2\vartheta_0}\left(\frac{n_1 h}{2\pi}\right)^2 . \qquad (6)$$

ϑ_0 ist eine durch unsere Zerfällung neu eingeführte Integrationskonstante, deren Bedeutung aus (5) hervorgeht. Es muß nämlich $\partial S / \partial \vartheta$ reell sein und ϑ seiner geometrischen Bedeutung nach zwischen 0 und π liegen. Aus beidem folgt:

$$\sin\vartheta \geqq \sin\vartheta_0, \; \vartheta_0 \leqq \vartheta \leqq \pi - \vartheta_0 \, . \qquad (7)$$

ϑ_0 bestimmt also den Wertebereich, in dem der Winkel ϑ variieren kann. Daß ϑ die Grenzen ϑ_0 und $\pi - \vartheta_0$ wirklich erreicht, folgt aus Stetigkeitsgründen; eine Umkehr des Fortschreitungssinnes von ϑ kann nämlich nur an solchen Stellen erfolgen, wo $\frac{\partial S}{\partial \vartheta} = 0$, also $\sin\vartheta = \sin\vartheta_0$ ist. Infolgedessen wächst ϑ dauernd von ϑ_0 bis $\pi - \vartheta_0$, nimmt dauernd ab bis ϑ_0 usf. Der volle Wertebereich von ϑ entspricht dem Hin- und Rückwege von ϑ_0 bis $\pi - \vartheta_0$ und von $\pi - \vartheta_0$ bis ϑ_0.

Hiernach wird, das Integral über den eben genannten Wertebereich genommen:

$$[S]_\vartheta = \int \frac{\partial S}{\partial \vartheta} d\vartheta = \frac{n_1 h}{2\pi} \int \sqrt{\frac{1}{\sin^2\vartheta_0} - \frac{1}{\sin^2\vartheta}} \, d\vartheta \, .$$

Nach (1) lautet also unsere Quantenbedingung für ϑ:

$$\int \sqrt{\frac{1}{\sin^2\vartheta_0} - \frac{1}{\sin^2\vartheta}} \, d\vartheta = 2\pi \frac{n_2}{n_1} . \qquad (8)$$

Wir wenden uns zur Quantenbedingung in r und schreiben Gl. (6) zunächst abkürzend

$$\frac{\partial S}{\partial r} = \sqrt{A + 2\frac{B}{r} + \frac{C}{r^2}}, \qquad (9)$$

$$\left.\begin{aligned} A &= \frac{W^2}{c^2} + 2m_0 W - m_0 f \frac{n_1 h}{2\pi}, \\ B &= \frac{We^2}{c^2} + m_0 e^2, \; C = \frac{e^4}{c^2} - \frac{1}{\sin^2\vartheta_0}\left(\frac{n_1 h}{2\pi}\right)^2 . \end{aligned}\right\} \qquad (9a)$$

Aus denselben Gründen wie vorher kann r im Verlauf der Bewegung seinen Fortschreitungssinn nur an solchen Stellen umkehren, für welche $dS / \partial r$ verschwindet. Die Grenzen $r_{\min}$ und $r_{\max}$, innerhalb deren r variiert, sind daher gegeben durch die quadratische Gleichung

$$A + \frac{2B}{r} + \frac{C}{r^2} = 0 \, .$$

Der volle Wertebereich von r erstreckt sich von der einen Wurzel dieser Gleichung bis zur anderen und wieder zurück zur ersten. Infolge-

500 Sommerfeld, Theorie d. Zeeman-Effekts d. Wasserstofflinien. Physik. Zeitschr. XVII, 1916.

dessen wird nach (1) unsere dritte Quantenbedingung:

$$[S]_r = \int \sqrt{A + \frac{2B}{r} + \frac{C}{r^2}}\, dr = n' h. \qquad (10)$$

In beiden Fällen (8) und (10) haben wir es also, wie in der Einleitung bemerkt, mit sog. vollständigen Integralen zu tun, das Wort entlehnt aus der Theorie der elliptischen Integrale und in dem Sinne verstanden, daß die Integration von einem Verzweigungspunkte des Integranden (Verschwindungspunkte der Quadratwurzel) bis zum anderen und wieder zurück zum ersten geführt werden soll. Da ein vollständiges Integral analytisch ein wesentlich einfacheres Ding ist als ein unvollständiges, werden wir bei der Ausrechnung unserer Quantenbedingungen im nächsten Paragraphen von denjenigen Methoden Nutzen ziehen, die sich für vollständige, d. h. im Komplexen geschlossene Integrale darbieten. Die Allgemeinheit unserer Überlegung, die eigentlich nur auf der Stetigkeit von S und dem quadratischen Charakter des Differentialparameters

$$\left(\frac{\partial S}{\partial \xi}\right)^2 + \left(\frac{\partial S}{\partial \eta}\right)^2 + \left(\frac{\partial S}{\partial \zeta}\right)^2$$

beruhte, zeigt, daß auch ganz allgemein bei beliebigen Freiheitsgraden und bei beliebigen Kraftfeldern (äußeren oder inneratomistischen) unsere Quantenbedingungen auf vollständige Integrale führen und der Methode der komplexen Integration zugänglich sein müssen.

§ 7. Ausrechnung der Quantenbedingungen durch komplexe Integration.

Als Typus unserer Phasenintegrale betrachten wir die Form (10) des vorigen Paragraphen

$$J = \oint \sqrt{A + 2\frac{B}{r} + \frac{C}{r^2}}\, dr. \qquad (1)$$

Durch den am Integralzeichen angebrachten Kreis wird wie üblich der geschlossene Charakter des Integrationsweges angedeutet. Er

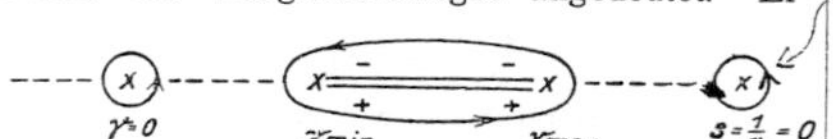

Fig. 3.

erstreckt sich (vgl. Fig. 3) um die beiden Verzweigungspunkte $r_{\min}$ und $r_{\max}$ herum. Bezüglich des Vorzeichens der Quadratwurzel ist festzusetzen, daß dieses auf der reellen Achse zwischen $r_{\min}$ und $r_{\max}$ positiv sein soll bei positivem dr, negativ bei negativem dr. Wir führen demnach von $r_{\min}$ bis $r_{\max}$ einen „Verzweigungsschnitt", sprechen von einem „negativen" und einem „positiven Ufer" desselben und bestimmen den Wert der Quadratwurzel an allen anderen Stellen der komplexen Ebene unserer Figur (des „oberen Blattes" einer zweiblättrigen Riemannschen Fläche) im Anschluß an die Vorzeichen Festsetzung auf den beiden Ufern unseres Verzweigungsschnittes. Z. B. gilt in der Nähe des Nullpunktes die Entwicklung

$$\sqrt{A + \frac{2B}{r} + \frac{C}{r^2}} = \frac{\sqrt{C}}{r}\left(1 + \frac{B}{C} r + \ldots\right), \qquad (2)$$

wo $\sqrt{C}$ eine negativ-imaginäre Größe sein muß, und in der Nähe des Unendlichkeitspunktes, den wir in der Figur der Deutlichkeit wegen ins Endliche hineintransformiert haben

$$\sqrt{A + \frac{2B}{r} + \frac{C}{r^2}} = \sqrt{A}\left(1 + \frac{B}{A}\frac{1}{r} + \ldots\right), \qquad (3)$$

wo $\sqrt{A}$ eine positiv-imaginäre Größe sein muß. In der Tat gelangen wir von dem positiven Ufer des Verzweigungschnittes nach dem Nullpunkt (bzw. Unendlichkeitspunkt) durch einen Umgang um den Verzweigungspunkt $r_{\min}$ (bzw. $r_{\max}$) vom Winkel 180^0 im negativen (bzw. positiven) Sinne, welcher Umgang den Faktor

$$e^{-\frac{i\pi}{2}} = -i \quad (\text{bzw. } e^{+\frac{i\pi}{2}} = +i)$$

mit sich bringt. Hierdurch bestätigen sich unsere Angaben bei (2) und (3) über den negativ-imaginären bzw. positiv-imaginären Charakter von $\sqrt{C}$ und $\sqrt{A}$.

Wir formen nun unseren Integrationsweg, der ursprünglich um die Verzweigungspunkte herum führte, um in einen Umlauf um die singulären Punkte des Integranden; es sind dieses der Nullpunkt $r = 0$ und der Unendlichkeitspunkt $s = \frac{1}{r} = 0$. Beide Umläufe lassen sich nach der Residuenregel unmittelbar berechnen. Da sie beide (vgl. die Figur) im negativen Sinne erfolgen, ist ihr Beitrag bzw.

$$-2\pi i N \text{ bzw. } -2\pi i U,$$

wo N und U die Residuen bei $r = 0$ und $s = 0$ sind. N ergibt sich aus (2) als Koeffizient von $1/r$ zu

$$N = \sqrt{C}.$$

Um U zu erhalten, haben wir (3) mit $dr = -ds/s^2$ zu multiplizieren und den Koeffizienten von ds/s zu nehmen. Dieser ist

$$U = -\sqrt{A}\frac{B}{A} = -\frac{B}{\sqrt{A}}.$$

Unser Integral (1) wird daher

$$J = -2\pi i (N+U) = -2\pi i\left(\sqrt{C} - \frac{B}{\sqrt{A}}\right). \quad (4)$$

Offenbar hat diese Berechnungsweise von J den Vorzug der Verallgemeinerungsfähigkeit.

Ebenso behandeln wir wegen einer späteren Anwendung das Integral

$$J' = \oint \frac{r\,dr}{\sqrt{A + 2\frac{B}{r} + \frac{C}{r^2}}}, \quad (5)$$

genommen auf demselben geschlossenen, in Fig 3 gezeichneten Wege. Auch jetzt ergibt sich:

$$J' = -2\pi i (N + U).$$

Indessen ist jetzt das Residuum N im Nullpunkte gleich Null, da hier der Nullpunkt als singulärer Punkt ausscheidet. Für den Unendlichkeitspunkt $s = 1/r = 0$ entwickelt, lautet der Integrand

$$-\frac{ds}{s^3}\frac{1}{\sqrt{A}}\left(1 + \frac{2B}{A}s + \frac{C}{A}s^2\right)^{-1/2} =$$
$$-\frac{ds}{s^3}\frac{1}{\sqrt{A}}\left(1 - \frac{B}{A}s + \left(\frac{3}{2}\frac{B^2}{A^2} - \frac{C}{2A}\right)s^2 + \ldots\right)$$

Der Faktor von ds/s ist also

$$U = -\frac{1}{2A^{3/2}}\left(\frac{3B^2}{A} - C\right).$$

Somit wird

$$J' = \frac{\pi i}{A^{3/2}}\left(\frac{3B^2}{A} - C\right). \quad (6)$$

Ähnlich verfahren wir mit dem Integral (8) des vorigen Paragraphen

$$J_1 = \sqrt{\frac{1}{\sin^2\vartheta_0} - \frac{1}{\sin^2\vartheta}}\,d\vartheta.$$

Wir setzen

$$x = \sin^2\vartheta, \quad a = \sin^2\vartheta_0 \quad (7)$$

und erhalten

$$\sqrt{a}\,J_1 = \oint\sqrt{\frac{x-a}{1-x}}\,\frac{dx}{x}.$$

Die Integration, die in ϑ von ϑ_0 bis $\pi - \vartheta_0$ und zurück zu erstrecken war, geht in der x-Ebene (vgl. Fig. 4) von $x = a$ bis $x = 1$ und nach Umlaufung dieses Verzweigungspunktes

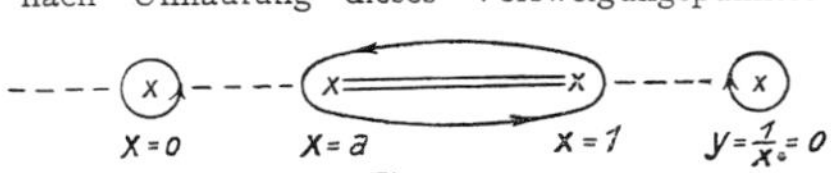

Fig. 4.

zurück bis $x = a$. Wir haben also auch hier ein in der x-Ebene geschlossenes Integral um die beiden Verzweigungspunkte $x = a$ und $x = 1$, welches wir auf die Umläufe um $x = 0$ und $x = \infty$ zusammenziehen können. Wiederum ergibt sich

$$\sqrt{a}\,J_1 = -2\pi i (N + U) \quad (8)$$

Das Residuum im Nullpunkt wird

$$N = -i\sqrt{a}. \quad (9)$$

Die Entwicklung des Integranden im Unendlichkeitspunkte $y = \frac{1}{x} = 0$ lautet

$$-\frac{dy}{y}\left(-1 - (1-a)y + \ldots\right)^{-1/2}.$$

Da man von dem positiven Ufer des Verzweigungsschnittes der Fig. 4 nach dem Unendlichkeitspunkt durch einen positiven Umgang um den Verzweigungspunkt $x = 1$ vom Winkel 180^0 gelangt, wird das Residuum im Unendlichkeitspunkte

$$U = -\frac{1}{e^{+\frac{i\pi}{2}}} = +i. \quad (9a)$$

Aus (8), (9), und (9a) ergibt sich also

$$\sqrt{a}\,J_1 = 2\pi(1 - \sqrt{a}). \quad (10)$$

Da aber nach Gl. (8) des vorigen Paragraphen war

$$J_1 = 2\pi\frac{n_2}{n_1},$$

so ergibt sich aus (7) und (10) dieses Paragraphen

$$\sqrt{a} = \sin\vartheta_0 = \frac{n_1}{n_1 + n_2} = \frac{n_1}{n} \quad (11)$$

mit der früheren Abkürzung

$$n = n_1 + n_2.$$

Der Inhalt der Gl. (11) wurde bereits durch Gl. (1) und Fig. 1 der Einleitung dargestellt, welche also bei der relativistischen Behandlung ihre Gültigkeit behalten. In der Tat besteht zwischen unserem jetzigen Winkel ϑ_0 und dem früheren Neigungswinkel α der Bahnebene gegen die Äquatorebene die geometrische Beziehung

$$\sin\vartheta_0 = \cos\alpha.$$

Das auf den Winkel ϑ bezügliche Phasenintegral bestimmt also, wie in der Einleitung betont, die Lage der Bahnebene im Raume, und zwar im Raume des $\xi\eta\zeta$-Systems.

§ 8. Die Energie der Bewegung. Zeemaneffekt mit Relativität.

Wir überzeugen uns weiter, daß das auf den Radiusvektor r bezügliche Phasenintegral entsprechenderweise die Energie der Bewegung bestimmt. Nach Gl. (10) des vorvorigen und Gl. (4) des vorvorigen Paragraphen hat man

$$n'h = -2\pi i\left(\sqrt{C} - \frac{B}{\sqrt{A}}\right). \quad (1)$$

Wegen Gl. (11) des vorigen Paragraphen wird der Wert von C (vgl. Gl. (9a) des vorvorigen Paragraphen)

$$C = \frac{e^4}{c^2} - \left(\frac{nh}{2\pi}\right)^2 = -\left(\frac{h}{2\pi}\right)^2 (n^2 - \alpha^2), \qquad (2)$$

Hier bedeutet α die von mir früher eingeführte „Konstante der Feinstruktur"[1])

$$\alpha = \frac{2\pi e^2}{hc} = 7{,}3 \cdot 10^{-3}.$$

Somit nach der der Gl. (2) des vorigen Paragraphen hinzugefügten Vorzeichenbestimmung

$$-2\pi i \sqrt{C} = -h\sqrt{n^2 - \alpha^2}$$

und nach Gl. (1)

$$\frac{Ah^2}{(2\pi B)^2} = -\frac{1}{(n' + \sqrt{n^2 - \alpha^2})^2}. \qquad (3)$$

Nach Gl. (9a) des vorvorigen Paragraphen ist

$$2\pi B = \frac{2\pi e^2}{c}\left(\frac{W}{c} + m_0 c\right) = \alpha h \left(\frac{W}{c} + m_0 c\right),$$

$$A = \left(\frac{W}{c} + m_0 c\right)^2 - m_0{}^2 c^2 - m_0 f \frac{n_1 h}{2\pi},$$

$$\frac{Ah^2}{(2\pi B)^2} = \frac{1}{\alpha^2}\left(1 - \frac{m_0{}^2 c^2 + m_0 f \frac{n_1 h}{2\pi}}{\left(\frac{W}{c} + m_0 c\right)^2}\right),$$

also nach (3)

$$\frac{m_0{}^2 c^2 + m_0 f \frac{n_1 h}{2\pi}}{\left(\frac{W}{c} + m_0 c\right)^2} = 1 + \frac{\alpha^2}{(n' + \sqrt{n^2 - \alpha^2})^2}$$

$$\frac{W}{c} + m_0 c = \left(m_0{}^2 c^2 + m_0 f \frac{n_1 h}{2\pi}\right)^{1/2} \left(1 + \frac{\alpha^2}{(n' + \sqrt{n^2 - \alpha^2})^2}\right)^{-1/2}.$$

Durch Division mit $m_0 c$ ergibt sich

$$1 + \frac{W}{m_0 c^2} = \left(1 + \frac{f}{2\pi} \frac{n_1 h}{m_0 c^2}\right)^{1/2} \left(1 + \frac{\alpha^2}{(n' + \sqrt{n^2 - \alpha^2})^2}\right)^{-1/2} \qquad (4)$$

Setzt man hierin $f = 0$, so folgt die Energie der Keplerschen Ellipse ohne Magnetfeld[2]), die wir W_0 nennen wollen

$$1 + \frac{W_0}{m_0 c^2} = \left(1 + \frac{\alpha^2}{(n' + \sqrt{n^2 - \alpha^2})^2}\right)^{-1/2}. \qquad (4a)$$

Man kann daher die Gl. (4) auch so schreiben

$$1 + \frac{W}{m_0 c^2} = \left(1 + \frac{f}{2\pi} \frac{n_1 h}{m_0 c^2}\right)^{1/2} \left(1 + \frac{W_0}{m_0 c^2}\right)$$

oder bei konsequenter Vernachlässigung höherer Potenzen von f

$$1 + \frac{W}{m_0 c^2} = \left(1 + \frac{f}{4\pi} \frac{n_1 h}{m_0 c^2}\right)\left(1 + \frac{W_0}{m_0 c^2}\right).$$

Durch Ausrechnung folgt:

$$W - W_0 = \frac{f}{4\pi} n_1 h \left(1 + \frac{W_0}{m_0 c^2}\right). \qquad (5)$$

Aus Gl. (4a) ersieht man aber, daß

$\frac{W_0}{m_0 c^2}$ von der Ordnung $\alpha^2 = 5{,}3 \cdot 10^{-5}$

ist. Man kann daher ohne Fehler die vorige Gleichung einfacher schreiben:

$$W - W_0 = \frac{f}{4\pi} n_1 h. \qquad (6)$$

Damit sind wir auf Gl. (13) des § 2 zurückgefallen, aus welcher in gewissem Sinne (d. h. abgesehen von den überzähligen Komponenten) das normale Lorentzsche Triplett folgte. Unser Resultat lautet also: Der Zeemaneffekt mit Relativität ist nicht verschieden von dem Zeemaneffekt ohne Relativität. Oder anders ausgedrückt: Der Zeemaneffekt wird durch die aus der Relativitätstheorie folgende Feinstruktur der Wasserstofflinien nicht beeinflußt.

Dieses Resultat ist höchst verdächtig. Zwar kennt man den Zeemaneffekt der Wasserstofflinien wegen ihrer Diffusität nicht hinlänglich gut. Daß er aber für jede Komponente der H-Linien aus einem normalen Triplett bestehen sollte, ist sehr unwahrscheinlich. Tatsächlich zeigen den normalen Zeemaneffekt bei kleinen Feldern doch nur die einfachen Spektrallinien. Zweifellos ist der Paschen-Back-Effekt die allgemeinste und wichtigste magnetooptische Erfahrung, die wir haben; sie erklärt erst das normale Triplett der sog. einfachen Linien, die in Wirklichkeit wohl stets nur hinreichend enge mehrfache Linien sind. Von dem Paschen-Back-Effekt zeigt sich aber in unserer Berechnung keine Andeutung. Vielmehr würde nach unserer Gl. (6) die Mehrfachheit der H-Linien, die in der Mehrwertigkeit von W_0 steckt, auch in dem Werte von W erhalten bleiben, unabhängig von der Größe von f. Wir hätten also so viele sich überlagernde und sich nicht störende Tripletts (nebst überzähligen äußeren Linien) als wir ohne Magnetfeld Komponenten der Feinstruktur haben. Der Paschen-Back-Effekt tritt bekanntlich in Erscheinung, wenn f nicht mehr klein ist gegen $\Delta\nu_0$, d. h. gegen den ursprünglichen Abstand zweier Komponenten der mehrfachen Spektrallinie. In unserer Formel tritt aber das Verhältnis $f : \Delta\nu_0$ überhaupt nicht auf.

Wir müssen also unser Resultat (6) für

1) Vgl. Ann d. Phys. l. c. Teil II. § 3, Gl. (12a); die ursprünglich in der Münchn. Akademie benutzte und ebenso bezeichnete Konstante war gleich $\alpha^2/4$.

2) Vgl. Ann. d. Phys. l. c. Teil II, Gl. (17), wo für Wasserstoff $E = e$ zu setzen ist.

unzureichend erklären und müssen nach dem Grunde hierfür fragen.

1. In meiner ursprünglichen Darstellung machte die Wahl der Koordinaten, in denen die Quantenbedingungen angesetzt wurden, Schwierigkeit. In den Arbeiten von Schwarzschild und Epstein scheint diese Schwierigkeit behoben durch die Anweisung: Man gehe von kanonischen Gleichungen aus und benutze solche Koordinaten, in denen sich die Jacobische Differentialgleichung separieren läßt. Nach dieser Anweisung sind wir in den letzten Paragraphen verfahren.

Indessen gelang die Separierung der Variabeln nicht in Strenge, sondern nur unter Vernachlässigung höherer Potenzen von f. Trotz der Kleinheit der fraglichen Glieder ist es möglich, daß sie einen prinzipiellen Einfluß auf den Ansatz der Quantenbedingungen haben könnten. Jedenfalls sind unsere Variabeln r, ϑ, χ nicht in Strenge die richtigen Variabeln, die der Quantelung unterworfen werden sollten. Bei dem diskreten Charakter der Quantenbedingungen wäre es nicht ausgeschlossen, daß diese Bedingungen für die in Strenge richtigen Koordinaten wesentlich anders lauten und sich nicht stetig anschließen an diejenigen Bedingungen, die für die nur angenähert und scheinbar richtigen Koordinaten r, ϑ, χ berechnet wurden.

2. Man könnte andererseits die Schuld auf die Bohrsche Frequenzgleichung

$$h\nu = W_m - W_n$$

schieben. Sie ist aus verschiedenen Gründen unzureichend: als skalare Gleichung kann sie niemals die Polarisationen erklären; da sie nur die Anfangs- und Endbahn berücksichtigt, nicht aber die näheren Umstände des Überganges, verzichtet sie auf den Schwingungscharakter der Spektrallinien; ihr Ursprung aus der Vorstellung der Energiequanten steht mit der neueren Entwicklung der Quantentheorie im Widerspruch, welche nur mehr Wirkungsquanten kennt.

Um die Entscheidung zu treffen, ob mutmaßlich der Einwand 1 oder 2 für das negative Ergebnis unserer Untersuchung verantwortlich zu machen ist, werden wir uns am besten durch die Erfahrungen beim Starkeffekt leiten lassen. Hier wird die Bohrsche Frequenzgleichung ungeändert übernommen und der Einfluß des elektrischen Feldes auf die Art des Übergangs zwischen Anfangs- und Endbahn durchaus nicht in Rechnung gesetzt.

Andererseits aber wird das mechanische Problem der Anfangs- und Endbewegung streng gelöst und der Quantenansatz in solchen Koordinaten vorgenommen, die zur genauen Separierung der Jacobischen Differentialgleichung führen. Die nachträgliche Beschränkung auf die erste Potenz des elektrischen Feldes ist hier nur Sache der praktischen Ausrechnung, hat aber keinen prinzipiellen Einfluß auf Koordinatenwahl oder Quantenansatz.

Hiernach werden wir auch beim Zeemaneffekt geneigt sein, den Einwand 1 als maßgebend anzusehen. Es ergibt sich also als Hauptresultat unserer Untersuchung die Forderung: die für den Quantenansatz strenge richtigen Koordinaten durch genaue Integration der Bewegungsgleichungen zu ermitteln.

§ 9. Anhang über den Stark-Effekt.

Das Elektron (Ladung $-e$) befinde sich unter dem Einfluß des Wasserstoffkernes (Ladung $+e$) und eines homogenen elektrischen Feldes von der Stärke E und der Richtung x. Seine potentielle Energie ist

$$V = -\frac{e^2}{r} + eEx,\ r^2 = x^2 + \varrho^2,\ \varrho^2 = y^2 + z^2. \quad (1)$$

a) Stark-Effekt ohne Relativität.

Anwendung der Methode der komplexen Integration.

Mit Schwarzschild und Epstein führen wir parabolische Koordinaten ξ, η, φ ein durch die Gleichungen

$$\left.\begin{aligned} &y = \varrho\cos\varphi,\ z = \varrho\sin\varphi,\ x + i\varrho = \frac{1}{2}(\xi + i\eta)^2 \\ &x = \frac{1}{2}(\xi^2 - \eta^2),\ \varrho = \xi\eta,\ r = \frac{1}{2}(\xi^2 + \eta^2). \end{aligned}\right\} \quad (2)$$

Das Linienelement wird

$$ds^2 = dx^2 + d\varrho^2 + \varrho^2 d\varphi^2 = (\xi^2 + \eta^2)(d\xi^2 + d\eta^2) + \xi^2\eta^2 d\varphi^2.$$

Daher die lebendige Kraft

$$T = \frac{m}{2}\left\{(\xi^2 + \eta^2)(\dot\xi^2 + \dot\eta^2) + \xi^2\eta^2\dot\varphi^2\right\}$$

oder, in den Impulsen

$$p_\xi = \frac{\partial T}{\partial \xi},\ p_\eta = \frac{\partial T}{\partial \eta},\ p_\varphi = \frac{\partial T}{\partial \varphi}$$

geschrieben

$$T = \frac{1}{2m(\xi^2 + \eta^2)}\left\{p_\xi^2 + p_\eta^2 + \left(\frac{1}{\xi^2} + \frac{1}{\eta^2}\right)p_\varphi^2\right\}. \quad (3)$$

Die potentielle Energie wird nach (1) und (2)

$$V = \frac{1}{2(\xi^2 + \eta^2)}\left\{-4e^2 + eE(\xi^4 - \eta^4)\right\}. \quad (4)$$

Daher lautet die Hamiltonsche Funktion:

$$H = T + V = W:$$

$$H = \frac{1}{2m(\xi^2+\eta^2)}\Big\{p_\xi^2 + p_\eta^2 + \left(\frac{1}{\xi^2}+\frac{1}{\eta^2}\right)p_\varphi^2 - 4me^2 + meE(\xi^4-\eta^4)\Big\}$$

und die Jacobische Differentialgleichung:

$$\left(\frac{\partial S}{\partial \xi}\right)^2 + \left(\frac{\partial S}{\partial \eta}\right)^2 + \left(\frac{1}{\xi^2}+\frac{1}{\eta^2}\right)\left(\frac{\partial S}{\partial \varphi}\right)^2 = 2mW(\xi^2+\eta^2) + 4me^2 - meE(\xi^4-\eta^4). \tag{5}$$

Ihre Integration ergibt

$$\frac{\partial S}{\partial \varphi} = p_\varphi = \text{Konst.}, \quad \frac{\partial S}{\partial \xi} = \sqrt{f_1(\xi)}, \quad \frac{\partial S}{\partial \eta} = \sqrt{f_2(\eta)}. \tag{6}$$

$$\left.\begin{aligned} f_1(\xi) &= 2mW\xi^2 + 2m(e^2+\beta) - meE\xi^4 - \frac{p_\varphi^2}{\xi^2} \\ f_2(\eta) &= 2mW\eta^2 + 2m(e^2-\beta) + meE\eta^4 - \frac{p_\varphi^2}{\eta^2} \end{aligned}\right\} \tag{7}$$

β ist eine bei der Separation der Variabeln ξ η eingeführte Integrationskonstante.

Die Quantenbedingungen lauten in der Schreibweise der Gl. (1) § 6:

$$[S]_\xi = n_1 h, \quad [S]_\eta = n_2 h, \quad [S]_\varphi = n_3 h. \tag{8}$$

Die Integration in $[S]_\varphi$ ist von 0 bis 2π zu erstrecken; die letzte Bedingung (8) liefert daher

$$p_\varphi = \frac{n_3 h}{2\pi}.$$

Die Integrationen in $[S]_\xi$ und $[S]_\eta$ erstrecken sich hin und her zwischen ξ_1 ξ_2 bzw. η_1 η_2, wo ξ_1 ξ_2 bzw. η_1 η_2 diejenigen beiden positiven Wurzeln von $f_1(\xi) = 0$ bzw. $f_2(\eta) = 0$ sind, die für $E = 0$ nicht ∞ werden. Die beiden ersten Gl. (8) lassen sich daher zusammenfassend schreiben:

$$\oint \sqrt{A + 2\frac{B}{x} + \frac{C}{x^2} + Dx}\, dx = 2nh. \tag{9}$$

Hier ist das eine Mal $\xi^2 = x$, das andere Mal $\eta^2 = x$ gesetzt und berücksichtigt, daß dann $d\xi/\xi$ und $d\eta/\eta$ gleich $dx/2x$ wird, wobei der Faktor 2 nach rechts gesetzt wurde und n die Zahl n_1 bzw. n_2 vertritt. Die Koeffizienten A, B, C, D bedeuten das einemal

$$A_1 = 2mW, \; B_1 = m(e^2+\beta), \; C_1 = -\left(\frac{n_3 h}{2\pi}\right)^2, \; D_1 = -meE. \tag{9a}$$

das anderemal

$$A_2 = 2mW, \; B_2 = m(e^2-\beta), \; C_2 = -\left(\frac{n_3 h}{2\pi}\right)^2, \; D_2 = +meE. \tag{9b}$$

Die Integration ist in der komplexen x-Ebene ein Umgang um diejenigen beiden Verzweigungspunkte x_1 x_2 des Radikanden von (9), die mit $D = 0$ nicht ins Unendliche rücken. Außer diesen beiden Verzweigungspunkten gibt es noch zwei weitere, x_3 und ∞, so daß (9) ein vollständiges elliptisches Integral darstellt.

Nachdem so der Quantenansatz an die genaue Integration der Bewegungsgleichungen angeschlossen ist (vgl. den Schluß des vorigen Paragraphen), darf man für alle praktischen Zwecke das elliptische Integral nach Potenzen von D entwickeln, um es auf elementare Integrale zurückzuführen. Man schreibe:

$$\sqrt{A + 2\frac{B}{x} + \frac{C}{x^2} + Dx} = \sqrt{A + 2\frac{B}{x} + \frac{C}{x^2}} + \frac{1}{2}\frac{Dx}{\sqrt{A + 2\frac{B}{x} + \frac{C}{x^2}}}$$

Dann ergibt sich statt (9)

$$J + \frac{D}{2}J' = 2nh. \tag{10}$$

J und J' haben genau dieselbe Bedeutung wie in § 7 Gl. (1) und (5). Dabei ist folgender Umstand für unsere Rechnungsmethode charakteristisch. Die Integration in (9) ist ursprünglich um die Verzweigungspunkte x_1 x_2 des unverkürzten Radikanden in (9) zu erstrecken. Diese Verzweigungspunkte stimmen nicht genau überein mit den Verzweigungspunkten, sagen wir x^1_1, x^1_2, des verkürzten Radikanden $A + \frac{2B}{x} + \frac{C}{x^2}$. Jedenfalls aber kann der ursprüngliche Integrationsweg so gewählt werden, daß er auch die Verzweigungspunkte x^1_1, x^1_2 umschließt. Nach Vornahme unserer Potenzentwicklung ist er daher für die weitere Behandlung der Integrale J und J' völlig äquivalent mit demjenigen Integrationswege, der in § 7 zugrunde gelegt wurde. Es ist daher bei unserer Methode gar nicht nötig, die Verschiedenheit der Wurzeln x^1_1, x^1_2 einerseits und x_1, x_2 andrerseits zu berücksichtigen. Nicht so, wenn man an den ursprünglichen reellen Integralen festhält und auf die Freiheit in der Verfügung über den Integrationsweg, die die komplexe Auffassung mit sich bringt, verzichtet. Dementsprechend bemerkt Epstein[1]) daß man beim Übergang zur „zweiten Näherung" (d. h. bei Berücksichtigung der ersten Potenz der Feldstärke E) auf die abweichenden Werte der Integrationsgrenzen (bei Epstein u_1, u_2) zu achten habe, daß diese aber nur „Effekte zweiter Ordnung" mit sich bringen. Dagegen zeigt unsere Methode unmittelbar, daß diese Effekte in jeder Ordnung exakt verschwinden.

Setzt man nun aus § 7 Gl. (4) und (6) die

1) Ann. d. Phys. **50**, 507, 1916.

Werte von J und J' in (10) ein, so entsteht

$$\sqrt{C}-\frac{B}{\sqrt{A}}-\frac{D}{4A^{3/2}}\left(\frac{3B^2}{A}-C\right)=\frac{nhi}{\pi}. \quad (11)$$

Diese Gleichung ist in A, B, C, D und n mit den Indizes 1 und 2 zu versehen und vertritt daher eigentlich zwei Gleichungen. Wir eliminieren aus ihnen die in B_1 und B_2 vorkommende Integrationskonstante β (vgl. Gl. (9a) und (9b)), da wir uns für sie nicht interessieren, und berechnen zu dem Ende zunächst B aus (11). Es ergibt sich

$$B\left(1+\frac{3DB}{4A^2}\right)=\sqrt{A}\left(\sqrt{C}-\frac{nhi}{\pi}\right)+\frac{CD}{4A}. \quad (12)$$

In dem wegen D kleinen Gliede linker Hand kann man für B den Näherungswert $B=\sqrt{A}\,K$ einführen, wo zur Abkürzung geschrieben wird

$$K=\sqrt{C}-\frac{nhi}{\pi}. \quad (13)$$

Dann folgt aus (12)

$$B=\sqrt{A}\,K+\frac{CD}{4A}-\frac{3}{4}\frac{DK^2}{A}$$

oder ausführlicher geschrieben:

$$\left.\begin{aligned} B_1&=\sqrt{A}\,K_1-\frac{meEC}{4A}+\frac{3}{4}\frac{meEK_1^2}{A}\\ &=m(e^2+\beta)\\ B_2&=\sqrt{A}\,K_2+\frac{meEC}{4A}-\frac{3}{4}\frac{meEK_2^2}{A}\\ &=m(e^2-\beta)\end{aligned}\right\} \quad (14)$$

Es unterscheiden sich nämlich nach (9a) und (9b) A_1, A_2 und C_1, C_2 nicht voneinander, weshalb für diese Größen die Buchstaben A und C beibehalten werden können. Die nur dem Vorzeichen nach verschiedenen Werte $\mp meE$ von D_1 und D_2 sind eingesetzt, ebenso die Werte von B_1 und B_2. K_1 und K_2 sind nach (13) verschieden wegen der verschiedenen Bedeutung von $n=n_1$ bzw. $n=n_2$.

Nehmen wir die halbe Summe der beiden Gl. (14)

$$\frac{K_1+K_2}{2}\left(\sqrt{A}+\frac{3}{4}\frac{meE}{A}\left(K_1-K_2\right)\right)=me^2, \quad (15)$$

so ist die beabsichtigte Elimination von β erreicht und wir können zur Berechnung von $A=2mW$ übergehen. In erster Näherung folgt

$$A=\left(\frac{2me^2}{K_1+K_2}\right)^2,$$

welchen Wert wir in dem mit E behafteten kleinen Gliede links eintragen können. Somit

$$\begin{aligned}\sqrt{A}&=\frac{2me^2}{K_1+K_2}\\ &\quad-\frac{3}{4}\frac{meE}{(2me^2)^2}(K_1-K_2)(K_2+K_1)^2,\\ A&=\frac{(2me^2)^2}{(K_1+K_2)^2}-\frac{3}{4}\frac{E}{e}(K_1-K_2)(K_2+K_1),\\ W&=\frac{2me^4}{(K_1+K_2)^2}-\frac{3}{8}\frac{E}{me}(K_1-K_2)(K_2+K_1).\end{aligned} \quad (16)$$

Nach (13) ist aber

$$K_1-K_2=\frac{hi}{\pi}(n_2-n_1),$$

$$K_2+K_1=2\sqrt{C}-\frac{hi}{\pi}(n_1+n_2)$$

und nach (9a) oder (9b) mit Rücksicht auf die Vorzeichenbestimmung von $\sqrt{C}$ bei Gl. (2) in § 7

$$\sqrt{C}=-\frac{n_3hi}{2\pi},$$

also

$$\begin{aligned}K_1+K_2&=-\frac{hi}{\pi}(n_1+n_2+n_3),(K_1-K_2)(K_2+K_1)\\ &=\frac{h^2}{\pi^2}(n_2-n_1)(n_1+n_2+n_3).\end{aligned} \quad (17)$$

Setzt man dieses in (16) ein, so ergibt sich als endgültiges Resultat

$$\begin{aligned}-W&=\frac{2\pi^2me^4}{(n_1+n_2+n_3)^2h^2}\\ &\quad+\frac{3h^2E}{8\pi^2me}(n_2-n_1)(n_1+n_2+n_3).\end{aligned} \quad (18)$$

Dies stimmt genau mit Epsteins Gl. (62) überein und unterscheidet sich von Schwarzschilds Resultat nur unwesentlich aus Gründen, die Epstein inzwischen aufgeklärt hat. Wie man weiß, liegt in dem zweiten Term rechts in Gl. (18) die Erklärung für die ganze Mannigfaltigkeit der Erscheinungen, die Stark bei den verschiedenen Balmer-Linien entdeckt hat. Man darf wohl ohne Übertreibung sagen, daß die formale Erklärung dieser Erscheinungen nicht einfacher gegeben werden kann, als in den vorstehenden Zeilen geschehen.

b) Stark-Effekt mit Relativität.

Die relativistische Bewegungsgleichung für die x-Komponente lautet mit dem Werte (1) für die potentielle Energie V:

$$\frac{dp_x}{dt}=-\frac{\partial V}{\partial x},\quad p_x=m\frac{dx}{dt}=m_0\frac{dx}{d\tau}, \quad (19)$$

wo τ die Eigenzeit ist. Die Energiegleichung liefert ähnlich wie in § 4, Gl. (14) für $\dot{\tau}=d\tau/dt=\sqrt{1-\beta^2}$:

$$\frac{1}{\tau} = \frac{1}{m_0 c^2}(W - V) + 1. \tag{20}$$

Durch Multiplikation dieser Gleichung mit der vorhergehenden erhält man ähnlich wie in § 5:

$$\frac{dp_x}{d\tau} = -\frac{\partial V}{\partial x}\left(1 + \frac{W}{m_0 c^2}\right) + \frac{1}{m_0 c^2} V \frac{\partial V}{\partial x}$$
$$= -\frac{\partial}{\partial x}\left[V\left(1 + \frac{W}{m_0 c^2}\right) - \frac{V^2}{2 m_0 c^2}\right]. \tag{21}$$

Führt man also eine Hamiltonsche Funktion ein

$$H = \frac{p_x^2 + p_y^2 + p_z^2}{2 m_0} + V\left(1 + \frac{W}{m_0 c^2}\right) - \frac{V^2}{2 m_0 c^2}, \tag{22}$$

so lassen sich die Bewegungsgleichungen nach (19) und (21) in die folgende kanonische Form bringen:

$$\frac{dp_x}{d\tau} = -\frac{\partial H}{\partial x}, \ldots, \frac{dx}{d\tau} = +\frac{\partial H}{\partial p_x}, \ldots \tag{23}$$

Die Konstante der Hamiltonschen Funktion ist nicht gleich der Energiekonstanten W sondern, wie man nach der Methode des § 5 leicht nachrechnet, gleich

$$H_0 = W\left(1 + \frac{W}{2 m_0 c^2}\right). \tag{24}$$

Will man andrerseits in (22) die gewöhnliche Zeit t statt τ benutzen, so muß man statt (23) und (24) schreiben:

$$H = m_0 c^2 \left(\sqrt{1 + \frac{p_x^2 + p_y^2 + p_z^2}{m_0^2 c^2}} - 1\right) + V, \tag{23a}$$

$$H_0 = W \tag{24a}$$

Aus $H = H_0$ folgt nach (23) 24) ebenso wie nach (23a), (24a) die Jacobische partielle Differentialgleichung unseres Problems zu

$$\left(\frac{\partial S}{\partial x}\right)^2 + \left(\frac{\partial S}{\partial y}\right)^2 + \left(\frac{\partial S}{\partial z}\right)^2 = 2 m_0 W\left(1 + \frac{W}{2 m_0 c^2}\right)$$
$$- 2 m_0 V\left(1 + \frac{W}{m_0 c^2}\right) + \frac{V^2}{c^2}.$$

Diese Gleichung wird aber durch Einführung der parabolischen Koordinaten ξ, η, φ nicht mehr in eine separierbare Form übergeführt. Nach den Rechnungen vom Anfange dieses Paragraphen ergibt sich nämlich:

$$\left(\frac{\partial S}{\partial \xi}\right)^2 + \left(\frac{\partial S}{\partial \eta}\right)^2 + \left(\frac{1}{\xi^2} + \frac{1}{\eta^2}\right)\left(\frac{\partial S}{\partial \varphi}\right)^2$$
$$= 2 m_0 W\left(1 + \frac{W}{2 m_0 c^2}\right)(\xi^2 + \eta^2)$$
$$+ m_0\left(1 + \frac{W}{m_0 c^2}\right)(4 e^2 - eE(\xi^4 - \eta^4))$$
$$+ \frac{(4 e^2 - eE(\xi^4 - \eta^4))^2}{4 c^2(\xi^2 + \eta^2)}.$$

Hier steht das letzte Glied, welches aus V^2/c^2 entstanden ist, der Separierung der Variabeln $\xi\,\eta$ im Wege. Infolgedessen können wir diese Variabeln nicht mehr als die für den relativistischen Stark-Effekt richtigen anerkennen. In den richtigen Koordinaten angesetzt, mögen die Quantenbedingungen wesentlich und unstetig von den obigen Gl. (8) abweichen, ähnlich wie dies am Schluß des vorigen Paragraphen für die richtigen und die scheinbar angenähert richtigen Koordinaten beim Zeeman-Effekt vermutet wurde.

Offenbar wird es auch im Gebiete des Stark-Effektes ein Analogon zu dem magnetischen Paschen-Back-Effekte geben. Die ohne Rücksicht auf die Multiplizität gewonnene, nicht relativistische Formel (18) hat sich bei „starken" Feldern vollkommen bewährt. Dieselbe Formel kann aber bei „schwachen" Feldern gewiß nicht genau gelten, da sie mit $E = 0$ nicht den Übergang in die wirkliche Feinstruktur der Wasserstofflinien, sondern in die gröbere Balmersche Formel für einfache Linien vermittelt. Man wird hier offenbar ebenso wie beim Zeeman-Effekt sagen können: Die Feinstruktur wird gegenstandslos für elektrische Felder von solcher Stärke, daß ihre Aufspaltung groß ist gegenüber den Komponentenabständen der ursprünglichen Feinstruktur. Gegenüber solchen „starken Feldern" verhalten sich die Wasserstofflinien wie einfache Linien. Dagegen muß sich bei „schwachen" Feldern, deren Aufspaltung kleiner ist als die Feinstruktur, die letztere bemerklich machen.

Zusatz bei der Korrektur. Eine ganz elementare Betrachtung bestätigt die Ergebnisse der allgemeinen Behandlung des Zeeman-Effektes und gestattet einen Einblick in den Ursprung der magnetischen Aufspaltung.

Wir beschränken uns auf Kreisbahnen in der Ebene senkrecht zum Felde H. a sei der Radius des Kreises, w die Winkelgeschwindigkeit des Elektrons. Die magnetische Energie des unserer Kreisbahn äquivalenten Elementarmagneten im magnetischen Felde ist

$$W_m = H \frac{a^2 w}{2} \frac{e}{c}.$$

Mit Rücksicht auf die Quantenbedingung $m a^2 w = \frac{nh}{2\pi}$ und die frühere Abkürzung $f = \frac{e}{m}\frac{H}{c}$ wird dieses

$$W_m = \frac{f}{4\pi} nh.$$

Rechnet man andererseits die kinetische und potentielle elektrische Energie aus, auf Grund derselben Quantenbedingung und der durch das Feld abgeänderten dynamischen Bedingung, so ergibt sich als Summe beider bis auf Glieder mit f^2 derselbe Wert wie ohne Feld. Daraus erkennt man:

Die magnetische Aufspaltung rührt von der magnetischen Energie W_m her (paramagnetischer Effekt). Die Änderung der Kreisbahn durch das Feld ruft nur eine Korrektion zweiter Ordnung hervor (einen diamagnetischen Effekt). Die frühere Rechnung, bei welcher in der Differentialgleichung das Glied mit f^2 vernachlässigt wurde, ist exakt in dem Gliede erster Ordnung. Dasselbe gilt auch im relativistischen Falle. Die am Ende von § 8 ins Auge gefaßte Möglichkeit, daß das Auftreten überzähliger Linien und der Widerspruch gegen den Paschen-Back-Effekt von der ungenauen Integration der partiellen Differentialgleichung herrühren möchte, ist daher fallen zu lassen.

(Eingegangen 7. September 1916.)

Reprint 5

264 NATURE [FEBRUARY 20, 1926

Letters to the Editor.

[*The Editor does not hold himself responsible for opinions expressed by his correspondents. Neither can he undertake to return, nor to correspond with the writers of, rejected manuscripts intended for this or any other part of* NATURE. *No notice is taken of anonymous communications.*]

Spinning Electrons and the Structure of Spectra.

So far as we know, the idea of a quantised spinning of the electron was put forward for the first time by A. K. Compton (*Journ. Frankl. Inst.*, Aug. 1921, p. 145), who pointed out the possible bearing of this idea on the origin of the natural unit of magnetism. Without being aware of Compton's suggestion, we have directed attention in a recent note (*Naturwissenschaften*, Nov. 20, 1925) to the possibility of applying the spinning electron to interpret a number of features of the quantum theory of the Zeeman effect, which were brought to light by the work especially of van Lohuizen, Sommerfeld, Landé and Pauli, and also of the analysis of complex spectra in general. In this letter we shall try to show how our hypothesis enables us to overcome certain fundamental difficulties which have hitherto hindered the interpretation of the results arrived at by those authors.

To start with, we shall consider the effect of the spin on the manifold of stationary states which corresponds to motion of an electron round a nucleus. On account of its magnetic moment, the electron will be acted on by a couple just as if it were placed at rest in a magnetic field of magnitude equal to the vector product of the nuclear electric field and the velocity of the electron relative to the nucleus divided by the velocity of light. This couple will cause a slow precession of the spin axis, the conservation of the angular momentum of the atom being ensured by a compensating precession of the orbital plane of the electron. This complexity of the motion requires that, corresponding to each stationary state of an imaginary atom, in which the electron has no spin, there shall in general exist a set of states which differ in the orientation of the spin axis relative to the orbital plane, the other characteristics of the motion remaining unchanged. If the spin corresponds to a one-quantum rotation there will be in general two such states. Further, the energy difference of these states will, as a simple calculation shows, be proportional to the fourth power of the nuclear charge. It will also depend on the quantum numbers which define the state of motion of the non-spinning electron in a way very similar to the energy differences connected with the rotation of the orbit in its own plane arising from the relativity variation of the electronic mass. We are indebted to Dr. Heisenberg for a letter containing some calculations on the quantitative side of the problem.

This result suggests an essential modification of the explanation hitherto given of the fine structure of the hydrogen-like spectra. As an illustration we may consider the energy levels corresponding to electronic orbits for which the principal quantum number is equal to three. The scheme on the left side of the accompanying figure (Fig. 1) corresponds to the results to be expected from Sommerfeld's theory. The so-called azimuthal quantum number k is defined by the quantity of moment of momentum of the electron about the nucleus, $kh/2\pi$, where $k = 1, 2, 3$. According to the new theory, depicted in the scheme[1] on the right, this moment of momentum is given by $Kh/2\pi$, where $K = \frac{1}{2}, \frac{3}{2}, \frac{5}{2}$. The total angular momentum of the atom is $Jh/2\pi$, where $J = 1, 2, 3$. The symbols K and J correspond to those used by Landé in his classification of the Zeeman effects of the optical multiplets. The letters S, P, D also relate to the analogy with the structure of optical spectra which we consider below. The dotted lines represent the position of the energy levels to be expected in the absence of the spin of the electron. As the arrows indicate, this spin now splits each level into two, with the exception of the level $K = \frac{1}{2}$, which is only displaced.

In order to account for the experimental facts, the resulting levels must fall in just the same places as the levels given by the older theory. Nevertheless, the two schemes differ fundamentally. In particular, the new theory explains at once the occurrence of certain components in the fine structure of the hydrogen spectrum and of the helium spark spectrum

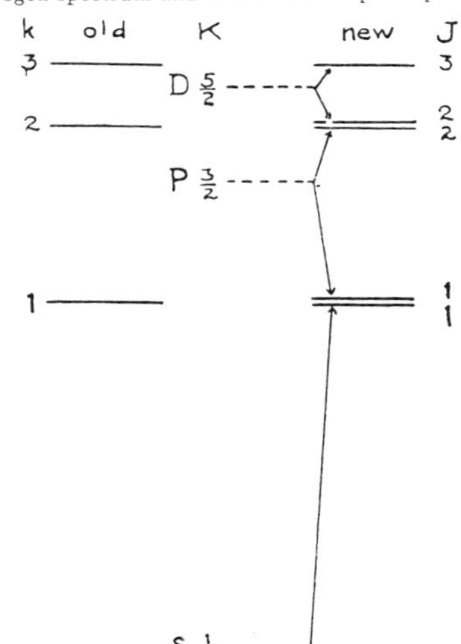

FIG. 1.

which according to the old scheme would correspond to transitions where K remains unchanged. Unless these transitions could be ascribed to the action of electric forces in the discharge which would perturb the electronic motion, their occurrence would be in disagreement with the correspondence principle, which only allows transitions in which the azimuthal quantum number changes by one unit. In the new scheme we see that, in the transitions in question, K will actually change by one unit and only J will remain unchanged. Their occurrence is, therefore, quite in conformity with the correspondence principle.

The modification proposed is specially important for explaining the structure of X-ray spectra. These spectra differ from the hydrogen-like spectra by the appearance of the so-called "screening" doublets, which are ascribed to the interaction of the electrons within the atom, effective mainly through reducing the effect of the nuclear attraction. In our view, these screening doublets correspond to pairs of levels which have the same angular momentum J but different azimuthal quantum numbers K. Consequently, the orbits will penetrate to different distances

[1] Quite independently of the ideas discussed here, a scheme of levels corresponding to this figure has been previously proposed by the writers (*Physica*, 5, 266, 1925), on the ground of the formal analogy between spectral structures. From similar formal considerations, this scheme has recently also been arrived at by J. C. Slater (*Proc. Washington Acad.*, December 1925).

Reprinted from Nature, Vol. 117, No. 2938, pp. 264–265 (1926).

from the nucleus, so that the screening of the nuclear charge by the other electrons in the atom will have different effects. This screening effect will, however, be the same for a pair of levels which have the same K but different J's and correspond to the same orbital shape. Such pairs of levels were, on the older theory, labelled with values of k differing by one unit, and it was quite impossible to understand why these so-called "relativity" doublets should appear separately from the screening doublets. On our view, the doublets in question may more properly be termed "spin" doublets, since the sole reason for their appearance is the difference in orientation of the spin axis relative to the orbital plane. It should be emphasised that our interpretation is in complete accordance with the correspondence principle as regards the rules of combination of X-ray levels.

The assumption of the spinning electron leads to a new insight into the remarkable analogy between the multiplet structure of the optical spectra and the structure of X-ray spectra, which was emphasised especially by Landé and Millikan. While the attempt to refer this analogy to a relativity effect common to all the structures was most unsatisfactory, it obtains an immediate explanation on the hypothesis of the spin electron. If, for example, we consider the spectra of the alkaline type, we are led to recognise in the well-known doublets regular spin doublets of the character described above. In fact, this enables us to explain the dependence of the doublet width on the effective nuclear charge and the quantum numbers describing the orbit, as well as the rules of combination.

The simplicity of the alkaline spectra is due to the fact that the atom consists of an electron revolving round an atomic residue which contains only completed electronic groups, which are magnetically inert. When we pass to atoms in which several electrons revolve round a residue of this kind we meet with new features, since we have to take account of other directing influences on the spin axis of each electron besides the couple due to its own motion in the electric field. Not only does this enable us to account for the appearance of multiplets of higher complexity, but it also seems to throw light on the so-called "branching" of spectra, which usually accompanies the adding of a further electron to the atom, and for which hitherto no satisfactory explanation has been given. In fact, it seems that the introduction of the concept of the spinning electron makes it possible throughout to maintain the principle of the successive building up of atoms utilised by Bohr in his general discussion of the relations between spectra and the natural system of the elements. Above all, it may be possible to account for the important results arrived at by Pauli, without having to assume an unmechanical "duality" in the binding of the electrons.

So far we have not mentioned the Zeeman effect, although the introduction of the spinning electron was primarily suggested by the analysis of the anomalous Zeeman effects shown by the components of multiplet structures. From the point of view of the correspondence principle, this effect shows that the influence of a magnetic field on the motion of the atom differs considerably from that to be expected if the electron had no spin. In fact, from the well-known theorem of Larmor we would expect the effect on any spectral line to be of the simple Lorentz type, quite independently of the character of the multiplet structure. Therefore the appearance of the anomalous Zeeman effects has hitherto presented very grave difficulties. However, these difficulties disappear at once when, as assumed, the electron has a spin and the ratio between magnetic moment and angular momentum of this spin is different from that corresponding to the revolution of the electron in an orbit large compared with its own size. On this assumption the spin axis of an electron not affected by other forces would precess with a frequency different from the Larmor rotation. It is easily shown that the resultant motion of the atom for magnetic fields of small intensity will be of just the type revealed by Landé's analysis. If the field is so strong that its influence on the precession of the spin axis is comparable with that due to the orbital motion in the atom, this motion will be changed in a way which directly explains the gradual transformation of the multiplet structure for increasing fields known as the Paschen-Back effect.

It seems possible on these lines to develop a quantitative theory of the Zeeman effect, if it is assumed that the ratio between magnetic moment and angular momentum due to the spin is twice the ratio corresponding to an orbital revolution. At present, however, it seems difficult to reconcile this assumption with a quantitative analysis of our explanation of the fine structure of levels. In fact it leads, in a preliminary calculation, to widths of the spin doublets just twice as large as those required by observation. It must be remembered, however, that we are here dealing with problems which for their final solution require a closer study of quantum mechanics and perhaps also of questions concerning the structure of the electron.

In conclusion, we wish to acknowledge our indebtedness to Prof. Niels Bohr for an enlightening discussion, and for criticisms which helped us distinguish between the essential points and the more technical details of the new interpretation.

G. E. UHLENBECK.
S. GOUDSMIT.
Instituut voor Theoretische Natuurkunde,
Leyden, December 1925.

HAVING had the opportunity of reading this interesting letter by Mr. Goudsmit and Mr. Uhlenbeck, I am glad to add a few words which may be regarded as an addition to my article on atomic theory and mechanics, which was published as a supplement to NATURE of December 5, 1925. As stated there, the attempts which have been made to account for the properties of the elements by applying the quantum theory to the nuclear atom have met with serious difficulties in the finer structure of spectra and the related problems. In my article expression was given to the view that these difficulties were inherently connected with the limited possibility of representing the stationary states of the atom by a mechanical model. The situation seems, however, to be somewhat altered by the introduction of the hypothesis of the spinning electron which, in spite of the incompleteness of the conclusions that can be derived from models, promises to be a very welcome supplement to our ideas of atomic structure. In fact, as Mr. Goudsmit and Mr. Uhlenbeck have described in their letter, this hypothesis throws new light on many of the difficulties which have puzzled the workers in this field during the last few years. Indeed, it opens up a very hopeful prospect of our being able to account more extensively for the properties of elements by means of mechanical models, at least in the qualitative way characteristic of applications of the correspondence principle. This possibility must be the more welcomed at the present time, when the prospect is held out of a quantitative treatment of atomic problems by the new quantum mechanics initiated by the work of Heisenberg, which aims at a precise formulation of the correspondence between classical mechanics and the quantum theory.

N. BOHR.
Copenhagen, January 1926.

336

Reprint 6

Über das Wasserstoffspektrum vom Standpunkt der neuen Quantenmechanik.

Von **W. Pauli jr.** in Hamburg.

(Eingegangen am 17. Januar 1926.)

Es wird gezeigt, daß sich die Balmerterme eines Atoms mit einem einzigen Elektron aus der neuen Quantenmechanik richtig ergeben und daß die in der bisherigen Theorie aus den Zusatzverboten von singulären Bewegungen entstehenden Schwierigkeiten, die insbesondere im Falle der gekreuzten Felder zutage treten, in der neuen Theorie verschwinden. Auch die Beeinflussung des Wasserstoffspektrums durch äußere elektrische und magnetische Kraftfelder wird vom Standpunkt der neuen Quantenmechanik aus diskutiert. Die Berücksichtigung der Relativitätskorrektionen sowie die Berechnung der Übergangswahrscheinlichkeiten (Intensitäten) bleibt jedoch zunächst noch außer Betracht.

§ 1. Die Grundlagen der neuen Quantenmechanik. Kürzlich ist von Heisenberg[1]) eine Formulierung der Prinzipien der Quantentheorie gegeben worden, die gegenüber der bisherigen Theorie der mehrfach periodischen Systeme einen großen Fortschritt darstellt. Gemäß dieser Heisenbergschen Fassung der Quantentheorie wird auf eine mechanisch-kinematische Veranschaulichung der Bewegung der Elektronen in den stationären Zuständen des Atoms gänzlich verzichtet, und es werden außer den zeitlichen Mittelwerten der klassisch-kinematischen Größen nur harmonische Partialschwingungen eingeführt, die jedem Übergang zwischen zwei stationären Zuständen zugeordnet werden und in unmittelbarem Zusammenhang mit den spontanen Übergangswahrscheinlichkeiten des Systems stehen. Ist

$$x_m^n = a_m^n \, e^{2\pi i \left(\nu_m^n t + \delta_m^n\right)}$$

die den Übergang von einem Zustand n nach einem anderen Zustand m charakterisierende Partialschwingung der kartesischen Koordinate x eines bestimmten Elektrons im Atom, so liefert diese den Beitrag

$$\frac{1}{h\,\nu_m^n}\,\frac{2}{3}\,\frac{e^2}{c^3}\left(2\pi\,\nu_m^n\right)^4 \left|x_m^n\right|^2 \cdot 2$$

zum Werte des Koeffizienten der spontanen Emissionswahrscheinlichkeit A_m^n des zugehörigen Überganges. Während in der früheren Theorie diese Beziehung gemäß dem Korrespondenzprinzip nur als im Grenzfall großer Quantenzahlen asymptotisch gültig anzusehen war, ist sie nunmehr als

[1]) W. Heisenberg, ZS. f. Phys. **33**, 879, 1925.

Reprinted from Zeitschrift für Physik, Vol. 36, No. 5, pp. 336–363 (1926).

die stets gültige Definition der Amplituden x^n_m aufzufassen, und es sind allgemeiner die den einzelnen Übergangsprozessen zugeordneten Partialschwingungen durch Intensität und Polarisation der emittierten Strahlung physikalisch definiert. Diese Partialschwingungen können jedoch nicht mehr zu bestimmten „Bahnen" der Elektronen des Atoms zusammengefaßt werden, da sie Übergangsprozessen, nicht aber stationären Zuständen zugeordnet werden.

Die Heisenbergsche Fassung der Quantentheorie wurde sodann von Born und Jordan [1]), Dirac [2]), sowie von Born, Jordan und Heisenberg [3]) weiter ausgebaut und in ein konsequentes mathematisches System gebracht, bei der alle früher der klassischen Mechanik entnommenen Beziehungen durch analog gebildete quantentheoretische Beziehungen zwischen den zu den Koordinaten der einzelnen Atomteilchen gehörigen zeitlichen Mittelwerten x^n_n und Partialschwingungen x^n_m ersetzt werden. Um diese Beziehungen zu formulieren, hat es sich als zweckmäßig erwiesen, jeder klassisch kinematischen Größe x eine Matrix zuzuordnen; in deren Diagonalgliedern stehen die zeitlichen Mittelwerte x^n_n, die zu den einzelnen stationären Zuständen gehören, an den Plätzen (n, m) [n-te Zeile, m-te Kolonne] und (m, n) [m-te Zeile, n-te Kolonne] stehen konjugiert komplexe Schwingungen

$$x^n_m = a^n_m \, e^{2\pi i \left(\nu^n_m t + \delta^n_m\right)} \quad \text{und} \quad x^m_n = a^m_n \, e^{2\pi i \left(\nu^m_n t + \delta^m_n\right)}, \tag{1}$$

worin $a^n_m = a^m_n$ positiv reell und

$$\nu^m_n = -\nu^n_m, \quad \delta^m_n = -\delta^n_m. \tag{2}$$

Die harmonische Schwingung x^m_n gehört zum Übergang von m nach n, die harmonische Schwingung x^m_n zum umgekehrten Übergang von n nach m, so daß der eine dieser Übergänge ein Emissions-, der andere ein Absorptionsübergang ist.

Der zeitlichen Ableitung $\dot{x}$ wird die Matrix zugeordnet, deren einzelne Elemente durch Differentiation nach der Zeit aus den entsprechenden Elementen der Matrix x gebildet werden, also

$$\dot{x}^n_m = 2\pi i \nu^n_m x^n_m, \tag{3}$$

insbesondere ist $\dot{x}^n_n = 0$, die Diagonalglieder von $\dot{x}$ verschwinden. Wegen $\nu^m_n = -\nu^n_m$ gilt auch hier, daß x^n_m und $\dot{x}^m_n$ konjugiert komplex sind (Hermitescher Charakter der Matrizen). Der Energie E ist eine Diagonal-

[1]) ZS. f. Phys. **34**, 858, 1925.

[2]) Proc. Roy. Soc. **109**, 642, 1925.

[3]) ZS. f. Phys. **35**, 557, 1926. Im folgenden zitiert als „Quantenmechanik II".

matrix zuzuordnen, das heißt eine solche, deren Nichtdiagonalglieder verschwinden. Es stellt $E_n = E_n^n$ den Energiewert in dem mit dem Index n bezeichneten Quantenzustand dar, und man hat zufolge der Frequenzbedingung

$$h\,\nu_m^n = E_n^n - E_m^m, \tag{I}$$

was mit der obigen Vorschrift $\nu_m^m = -\nu_m^n$, $\nu_n^n = 0$ im Einklang ist.

Das Wesentliche ist nun der von Heisenberg hervorgehobene Umstand, daß der Multiplikation zweier Matrizen $\boldsymbol{x}$ und $\boldsymbol{y}$ mit Rücksicht auf die Frequenzbedingung eine sinngemäße Bedeutung zukommt. Das Produkt $\boldsymbol{x\,y}$ der beiden Matrizen x und y ist definiert durch

$$(x\,y)_m^n = \sum_l x_l^n\,y_m^l\,. \tag{4}$$

Da nun aus (I) die Kombinationsbeziehung

$$\nu_l^n + \nu_m^l = \nu_m^n \tag{5}$$

hervorgeht, stellt $(x\,y)_m^n$ in der Tat wieder eine harmonische Schwingung der Frequenz ν_m^n dar, wenn x_l^n und y_m^l harmonische Schwingungen mit den Frequenzen ν_l^m und ν_m^l sind. Es muß dann auch für die Phasen δ_m^n eine Kombinationsrelation

$$\delta_l^n + \delta_m^l = \delta_m^n \tag{6}$$

angenommen werden.

Für die Multiplikation zweier Matrizen gelten alle gewöhnlichen Rechenregeln mit Ausnahme des kommutativen Gesetzes; es ist $\boldsymbol{x\,y}$ von $\boldsymbol{y\,x}$ im allgemeinen verschieden. So kann z. B. die Differenz $\boldsymbol{E\,x} - \boldsymbol{x\,E}$, in der $\boldsymbol{E}$ die Diagonalmatrix der Energie bedeutet und die Produkte gemäß der allgemeinen Vorschrift (4) auszuführen sind, gemäß (3) und der Frequenzbedingung (I) in einfache Verbindung mit der Matrix $\dot{\boldsymbol{x}}$ der zeitlichen Ableitung von $\boldsymbol{x}$ gebracht werden:

$$\boldsymbol{E\,x} - \boldsymbol{x\,E} = \frac{h}{2\pi i}\,\dot{\boldsymbol{x}}, \tag{7}$$

welche Relation für jede Matrix $\boldsymbol{x}$ gültig ist.

Die zur Berechnung der Matrizen $\boldsymbol{x}$ bei einem bestimmten mechanischen System erforderlichen Relationen, die physikalischen Grundgesetze der neuen Quantenmechanik, haben nun Born und Jordan auf folgende Form gebracht, die wir gleich für Systeme mit beliebig vielen Freiheitsgeraden anschreiben. Es seien mit $\boldsymbol{q}_\varrho$ und $\boldsymbol{p}_\varrho$ ($\varrho = 1 \cdots f$) die kartesischen Koordinaten der Atomteilchen und die zugehörigen Impulse

($p_x = m\,\dot{x}$) bezeichnet. Dann gelten außer der Frequenzbedingung (I) die „Quantenbedingungen"

$$\left.\begin{aligned} p_\varrho p_\sigma - p_\sigma p_\varrho = 0, \quad q_\varrho q_\sigma - q_\sigma q_\varrho = 0, \\ p_\varrho q_\sigma - q_\sigma p_\varrho = \begin{cases} 0 \text{ für } \varrho \neq \sigma, \\ \dfrac{h}{2\pi i}\cdot \mathbf{1} \text{ für } \varrho = \sigma. \end{cases} \end{aligned}\right\} \quad \text{(II)}$$

Hierin bedeutet **1** die „Einheitsmatrix" (deren Nichtdiagonalglieder verschwinden und deren Diagonalglieder gleich 1 sind). Wie Kramers[1]) gezeigt hat, können diese Relationen auf Grund der quantentheoretischen Dispersionsformeln von Ladenburg, Kramers und Kramers und Heisenberg durch die Forderung interpretiert werden, daß sich die einzelnen Atomteilchen gegenüber kurzperiodischen äußeren Kräften wie freie Teilchen verhalten. Endlich gilt als letztes Quantengesetz der Energiesatz

$$H(p, q) = E \text{ (Diagonalmatrix)}. \quad \text{(III)}$$

Die Matrixfunktion $H(p, q)$ charakterisiert ein bestimmtes mechanisches System, und die nächstliegende Annahme ist die, daß sie bei Verwendung von kartesischen Koordinaten mit der klassischen Funktion formal zusammenfällt. Wir können uns mit dem Fall begnügen, daß sie entsprechend der kinetischen und der potentiellen Energie aus zwei Teilen besteht, von denen der eine nur von den p, der andere nur von den q abhängt. Zunächst sind gemäß der Multiplikationsvorschrift (4) nur solche Matrixfunktionen definiert, die als Potenzreihen der p und q (mit positiven und negativen Potenzen) geschrieben werden können. Für diesen Fall haben Born, Jordan und Heisenberg gezeigt, daß die Grundgesetze (I), (II) und (III) solche Matrixrelationen zur Folge haben, die den Bewegungsgleichungen der klassischen Mechanik vollkommen analog sind und bei sinngemäßer Definition der auf der rechten Seite auftretenden partiellen Differentialquotienten geschrieben werden können:

$$\dot{q}_\varrho = \frac{\partial H(p, q)}{\partial p_\varrho}, \quad \dot{p}_\varrho = -\frac{\partial H(p, q)}{\partial q_\varrho}, \quad (8)$$

Es sei noch hervorgehoben, daß die Reihenfolge, in welcher in den Matrizen die stationären Zustände des betrachteten Systems angeordnet werden, gleichgültig ist und der Begriff „Quantenzahl" in der neuen Theorie in die Grundgesetze nicht eingeht. Im Gegensatz zu der bisherigen Theorie sind ferner in der neuen Theorie die Werte der Über-

[1]) H. A. Kramers, Physica **5**, 369, 1925.

gangswahrscheinlichkeiten auch für kleine Quantenzahlen im Prinzip quantitativ bestimmt.

§ 2. **Allgemeine Übersicht über Methoden und Ergebnisse der folgenden Berechnungen.** In der vorliegenden Arbeit haben wir es uns zur Aufgabe gestellt, die neue Theorie auf ein Atom mit einem einzigen Elektron anzuwenden. Jedoch ist es uns noch nicht gelungen, alle auf ein solches wasserstoffähnliches Atom bezüglichen Folgerungen aus den Grundgesetzen der neuen Theorie zu entwickeln, insbesondere wurde die Berechnung der Übergangswahrscheinlichkeiten von wasserstoffähnlichen Spektren noch nicht versucht, sondern es wurde nur die Berechnung der Energiewerte der stationären Zustände des Wasserstoffatoms im ungestörten Falle sowie bei Anwesenheit von äußeren elektrischen und magnetischen Feldern (unter Elimination der Übergangswahrscheinlichkeiten) durchgeführt. Die relativistischen Korrektionsglieder blieben zunächst außer Betracht. Als Resultat ergaben sich die Balmerterme sowie der Starkeffekt im Einklang mit der Erfahrung. Es fallen ferner die Schwierigkeiten fort, die infolge zusätzlicher Verbote von singulären Bewegungen, bei denen das Elektron dem Kern beliebig nahe kommt, in der bisherigen Theorie auftraten und sich besonders im Falle gekreuzter elektrischer und magnetischer Felder bemerkbar machten. Es mögen daher diese Schwierigkeiten hier näher erläutert werden.

Wir gehen zunächst aus vom Falle paralleler elektrischer und magnetischer Felder. Sind e und m_0 Ladung und Masse des Elektrons, Ze die Kernladung, a die Halbachse der Bahn, F und H die Feldstärken des elektrischen und magnetischen Feldes, so ist die Larmorfrequenz

$$o_H = \frac{eH}{4\pi m_0 c}, \tag{9}$$

die säkulare Starkeffektfrequenz o_F gegeben durch

$$o_F = \frac{3}{4\pi}\sqrt{\frac{a}{Zm_0}}\,F = \frac{3}{2}\,eF\,\frac{a_1}{h}\,n, \tag{10}$$

wenn

$$a_1 = \frac{h^2}{4\pi^2 Z e^2 m_0}$$

den Radius der einquantigen Kreisbahn des Atoms bedeutet. Im Felde treten dann die beiden Quantenbedingungen hinzu, welche die Projektion z des Abstandes des elektrischen Mittelpunkts der Bahn vom Kern auf die Feldrichtung gemäß

$$z = \frac{3}{2}\,a\,\frac{s}{n} \tag{11}$$

und das Impulsmoment P_z parallel zum Felde gemäß

$$P_z = m \frac{h}{2\pi} \tag{12}$$

festlegen. Bei gegebenem n und $|m| \leqq n$ durchläuft dabei, von Zusatzverboten zunächst abgesehen, die Starkeffektquantenzahl s die symmetrisch zu Null liegenden, in Schritten von zwei Einheiten aufeinanderfolgenden Werte [1])

$$\begin{gathered} s = -(n - |m|), \quad -(n - |m| - 2), \ldots \\ (n - |m| - 2), \quad n - |m|, \quad \text{wobei} \quad |m| \leqq n. \end{gathered} \tag{13}$$

Die zusätzliche Energie im Felde ist dann gegeben durch

$$E_1 = (s\, o_F + m\, o_H)\, h. \tag{14}$$

Für die Verallgemeinerung auf den Fall gekreuzter Felder ist es zweckmäßig, an Stelle von o_F und o_H die Frequenzen

$$\omega_1 = o_H + o_F, \quad \omega_2 = |o_H - o_F|$$

einzuführen. Es ist dann (13) und (14) gleichbedeutend mit

$$E_1 = \left(\frac{n}{2} - n_1\right) \omega_1 h + \left(\frac{n}{2} - n_2\right) \omega_2 h, \tag{15}$$

worin

$$0 \leqq n_1 \leqq n, \quad 0 \leqq n_2 \leqq n. \tag{16}$$

Da ω_1 und ω_2 stets als positive Größen definiert werden, so daß

$$\omega_2 = o_H - o_F \quad \text{für} \quad o_H > o_F,$$

$$\omega_2 = o_F - o_H \quad \text{für} \quad o_H > o_F,$$

ist der Zusammenhang der Zahlen s und m mit den Zahlen n_1 und n_2 beschrieben durch

$$\left.\begin{aligned} m &= n - (n_1 + n_2), & s &= n_2 - n_1 & \text{für} \quad o_H > o_F, \\ m &= n_2 - n_1, & s &= n - (n_1 + n_2) & \text{für} \quad o_H < o_F \end{aligned}\right\} \tag{17}$$

(im Falle $o_H = o_F$ wird $\omega_2 = 0$ und das System ist entartet).

Im allgemeinen Falle gekreuzter elektrischer und magnetischer Felder bleibt nun gemäß den Resultaten von Klein [2]) und Lenz [3]) der

---

[1]) Dies folgt unter anderem aus dem Zusammenhang von s und n mit den Quantenzahlen n_ξ, n_η der parabolischen Koordinaten ξ, η gemäß

$$n = n_\xi + n_\eta + |m|, \quad s = n_\xi - n_\eta,$$
$$0 \leqq n_\xi \leqq n, \quad 0 \leqq n_\eta \leqq n.$$

[2]) O. Klein, ZS. f. Phys. **22**, 109, 1924.

[3]) W. Lenz, ZS. f. Phys. **24**, 197, 1924. Die in dieser Arbeit mit n_1 und n_2 bezeichneten Zahlen sind für ungerades n nicht ganz, sondern laufen (falls man sich an die Quantenregeln für Periodizitätssysteme hält) in Schritten von einer Einheit von $-\frac{n}{2}$ bis $\frac{n}{2}$, die Grenzen eingeschlossen.

Ausdruck (15) für die Störungsenergie in den Quantenzuständen des Systems bestehen, wenn die Frequenzen ω_1 und ω_2 folgendermaßen definiert werden. Es seien jetzt $\mathfrak{o}_F$ und $\mathfrak{o}_H$ Vektoren parallel zu den Richtungen des äußeren elektrischen bzw. magnetischen Feldes, deren Beträge mit den säkularen Frequenzen (10) bzw. (9) übereinstimmen, welche jedes dieser Felder allein hervorrufen würde. Dann addiere und subtrahiere man $\mathfrak{o}_F$ und $\mathfrak{o}_H$ vektoriell und bilde die Beträge der resultierenden Vektoren. Auf diese Weise erhält man

$$\omega_1 = |\mathfrak{o}_H + \mathfrak{o}_F|, \quad \omega_2 = |\mathfrak{o}_H - \mathfrak{o}_F|. \tag{18}$$

Für parallele elektrische und magnetische Felder stimmt dies mit der früheren Vorschrift überein.

Dieses Resultat bringt nun beträchtliche Schwierigkeiten mit sich, wenn man es in Beziehung bringt mit dem Ausschluß derjenigen Bahnen, bei welchen das Elektron in den Kern fällt oder diesem im Laufe seiner Bewegung beliebig nahe kommt. Das erste derartige Zusatzverbot trat bereits in der Sommerfeldschen Theorie der relativistischen Feinstruktur auf, wo die Zustände mit verschwindender Impulsquantenzahl k, bei denen das Elektron auf einer durch den Kern gehenden Geraden hin und her pendeln würde, als zu stationären Zuständen ungeeignet ausgeschlossen werden mußten:

$$k \neq 0. \tag{19}$$

Im Falle des Starkeffekts entspricht der Wert $|s| = n$ der Starkeffektquantenzahl ebenfalls solchen geradlinigen Pendelbahnen und es kann als empirisch sicher gelten, daß er in Wahrheit nicht auftritt:

$$|s| \neq n. \tag{20}$$

Allgemein zeigte Bohr durch einen Vergleich der Anzahl der stationären Zustände der relativistischen Feinstruktur in schwachen achsensymmetrischen Kraftfeldern mit derjenigen beim Starkeffekt, daß infolge des Zusatzverbotes (19) in achsensymmetrischen Kraftfeldern stets auch alle Bahnen mit $m = 0$ ausgeschlossen werden müssen, bei denen übrigens im Falle des Starkeffektes das Elektron dem Kern beliebig nahe kommt:

$$m \neq 0. \tag{20'}$$

Es ist (20) in (20') als Spezialfall enthalten, da gemäß (13) für $s = n$ die Zahl m nur den Wert Null zur Verfügung hat. In gekreuzten Feldern ist es nun möglich, als stationäre Zustände zugelassene Bahnen in gemäß (20) oder (20') ausgeschlossene Bahnen stetig überzuführen. Man braucht hierzu nur folgenden adiabatischen Prozeß auszuführen. Es seien zunächst die beiden Felder parallel und es sei o_H von o_F verschieden, etwa $o_H > o_F$. Nachdem man die Richtungen der Felder gegeneinander langsam verdreht hat, vermindere man die Intensität des Magnetfeldes so lange, bis $|o_H| < |o_F|$ geworden ist; endlich richte man die Felder wieder parallel. Bei diesem Prozeß bleiben gemäß (18) ω_1 und ω_2 stets von Null verschieden, die Quantenzahlen n_1 und n_2 behalten daher stets denselben Wert. Da nun vor dem Prozeß $o_H > o_F$, nach dem Prozeß $o_H < o_F$ gilt, ist zufolge von (17) das Resultat des Prozesses dieses, daß solche Zustände ineinander übergeführt werden, bei denen die Werte der elektrischen Quantenzahl s und der magnetischen Quantenzahl m miteinander vertauscht sind. Insbesondere wird die Pendelbahn $s = n$, $m = 0$ in die zum Felde senkrecht stehende Kreisbahn $s = 0$, $m = n$ übergeführt. Es zeigt sich also, daß die Zusatzverbote, denen zufolge die geradlinigen Pendelbahnen auszuschließen sind, innerhalb des Rahmens der Quantentheorie mehrfach periodischer Systeme nicht widerspruchsfrei durchgeführt werden können.

Die im folgenden durchgeführte Rechnung zeigt nun (§ 5), daß in der neuen Quantenmechanik, in die eine Veranschaulichung der stationären Zustände durch bestimmte Elektronenbahnen nicht eingeführt wird, besondere Zusatzverbote überflüssig werden und die geschilderten Schwierigkeiten von selbst fortfallen. Es ergeben sich nämlich zum n-quantigen Zustand des ungestörten Atoms mit der Energie

$$E_n = -\frac{R h Z^2}{n^2} \tag{21}$$

(R = Rydbergkonstante) in äußeren parallelen bzw. gekreuzten elektrischen und magnetischen Feldern wieder die Werte (14) bzw. (15) der Zusatzenergie, worin o_H und o_F wieder durch (9) und (10), ferner ω_1 und ω_2 durch (18) gegeben sind. In den Formeln (13), (16) und (17) ist jedoch gemäß der neuen Mechanik n stets durch

$$n^* = n - 1 \tag{22}$$

zu ersetzen, welche Zahl jetzt als Maximum der Werte von s, m und n_1, n_2 fungiert, so daß nunmehr gilt:

$$\left.\begin{array}{c} s = -(n^* - |m|), \quad -(n^* - |m| - 2), \ldots \quad (n^* - |m| - 2), \\ n^* - |m|, \quad \text{mit} \quad |m| \leqq n^*, \end{array}\right\} \tag{13*}$$

$$0 \leqq n_1 \leqq n^*, \quad 0 \leqq n_2 \leqq n^*, \tag{16*}$$

$$\left.\begin{array}{lll} m = n^* - (n_1 + n_2), & s = n_2 - n_1, & \text{für} \quad o_H > o_F, \\ m = n_2 - n_1, & s = n^* - (n_1 + n_2), & \text{für} \quad o_H < o_F. \end{array}\right\} \tag{17*}$$

Insbesondere folgt für den Starkeffekt gemäß (10) und (14)

$$\text{mit} \quad \left.\begin{aligned} E_1 &= \tfrac{3}{2} e F a_1 n s \\ 0 &\leqq s \leqq n^* \end{aligned}\right\} \tag{23}$$

wie es die Erfahrung verlangt. Man sieht ferner, daß die Wertemannigfaltigkeit der Zahlen m und s nunmehr vollkommen symmetrisch ist, wie es der angegebene adiabatische Prozeß in gekreuzten Feldern erfordert.

Bei Aufhebung der Entartung des ungestörten Atoms durch ein zusätzliches Zentralkraftfeld (wie es z. B. durch die Relativitätskorrektionen geliefert wird) und ein äußeres Magnetfeld zerfällt der n-quantige Zustand des Atoms mit der Energie (21) gemäß der neuen Quantenmechanik in Zustände, die durch Quantenzahlen k und m klassifiziert werden können, welche den bekannten Auswahlregeln

$$\Delta k = \pm 1, \quad \Delta m = 0, \quad \pm 1$$

genügen. Die ganze Zahl m bestimmt wieder gemäß (12) die Impulskomponente des Atoms parallel dem Felde, während der den Wert der Störungsenergie des Zentralfeldes bestimmenden Zahl k keine so unmittelbare dynamische Bedeutung zukommt. Sie nimmt beim n-quantigen Zustand $n - 1$ aufeinanderfolgende Werte an, die sich je um eine Einheit unterscheiden, so daß sich jedenfalls die Anzahl der Feinstrukturniveaus richtig ergibt, ohne daß Zusatzverbote eingeführt werden (über ihre Energiewerte kann noch nichts ausgesagt werden). Wir wollen die Quantenzahl k speziell so normieren, daß im äußeren Magnetfeld die Quantenzahl m für jeden durch n und k charakterisierten Zustand die (ganzzahligen) Werte annimmt:

$$-k \leqq m \leqq k. \tag{24}$$

Die so normierte Zahl k kann im n-quantigen Zustand die Werte

$$k = 0, 1, 2, \ldots, n^* \tag{25}$$

annehmen. Zwischen den durch (24) und (25) und den durch (13*) klassifizierten Zuständen kann eine eindeutige Zuordnung hergestellt werden. Das Gewicht des n-quantigen Zustandes ist (in jedem Falle) gleich n^2.

Die angegebene, aus der neuen Theorie sich ergebende Termmannigfaltigkeit des Wasserstoffatoms in äußeren Feldern hat insbesondere zur Folge, daß im Normalzustand dieses Atoms, wo $n = 1$, $n^* = 0$ ist, die Quantenzahl m keinen anderen Wert als $m = 0$ annehmen kann, dieser Zustand also unmagnetisch ist. Diese Folgerung wird namentlich in Analogie zu dem Verhalten der Alkaliatome befremdlich erscheinen.

In diesem Zusammenhang ist zu betonen, daß die jetzige Fassung der neuen Quantenmechanik vom anomalen Zeemaneffekt (Versagen des Larmortheorems) anscheinend noch nicht Rechenschaft gibt und daß daher noch Modifikationen derselben erforderlich sein dürften. Es erscheint nicht ausgeschlossen, daß diese Modifikationen der Theorie bereits bei Atomen mit einem einzigen Elektron in Erscheinung treten werden. Auf diese Fragen kommen wir am Schlusse dieser Arbeit zurück (§ 6).

Was die im folgenden benutzte Methode zur Lösung der Matrixgleichungen der neuen Theorie im Falle eines Atoms mit einem Elektron betrifft, so müssen zunächst im folgenden § 3 die nötigen Rechenregeln für das gleichzeitige Operieren mit den Matrizen $\boldsymbol{x}, \boldsymbol{y}, \boldsymbol{z}$ der kartesischen Koordinaten des Elektrons (zu einer Vektormatrix $\mathfrak{r}$ zusammengefaßt) und der Matrix $\boldsymbol{r}$ des Betrages des Radiusvektors sowie ihrer zeitlichen Ableitungen entwickelt werden. Die jetzt bekannte Fassung der Gesetze der neuen Quantenmechanik nötigt dazu, die Einführung des Polarwinkels φ zu vermeiden. Da dieser nicht innerhalb endlicher Grenzen verbleibt, kann er nämlich formal nicht in der Weise als Matrix dargestellt werden wie die erstgenannten Koordinaten, die in der klassischen Mechanik Librationen vollführen.

Aus eben diesem Grunde erweist sich folgende bei Coulombschen Kräften anwendbare spezielle Intergrationsmethode der klassischen Mechanik, die bereits von Lenz[1]) herangezogen wurde, als zur Übertragung in die neue Quantenmechanik besonders geeignet. Bedeutet

$$\mathfrak{P} = m_0 [\mathfrak{r}\,\mathfrak{v}] \tag{26}$$

den zeitlich konstanten Drehimpuls des Elektrons um den Kern,

$$\mathfrak{p} = m_0 \mathfrak{v}$$

den Linearimpuls, so läßt sich aus den Bewegungsgleichungen in der klassischen Mechanik direkt zeigen, daß der Vektor

$$\mathfrak{A} = \frac{1}{Z e^2 m_0} [\mathfrak{P}\,\mathfrak{p}] + \frac{\mathfrak{r}}{r} \tag{27}$$

zeitlich konstant ist. Durch skalare Multiplikation mit $\mathfrak{r}$ folgt hieraus

$$(\mathfrak{A}\,\mathfrak{r}) = -\frac{1}{Z e^2 m_0} \mathfrak{P}^2 + r. \tag{28}$$

[1]) ZS. f. Phys., l. c.

Dies ist die Gleichung eines Kegelschnittes, und man kann hieraus sehen, daß $\mathfrak{A}$ die Richtung vom Kern zum Aphel der Ellipse und den Betrag ihrer numerischen Exzentrizität hat. Durch Quadrieren von (27) folgt

$$1 - \mathfrak{A}^2 = -\frac{2E}{m_0 Z^2 e^4}\mathfrak{P}^2, \tag{29}$$

worin E die Energie bedeutet.

In § 4 wird gezeigt, daß auch in der neuen Mechanik analog zu (27) eine zeitlich konstante Vektormatrix $\boldsymbol{\mathfrak{A}}$ eingeführt werden kann, für welche zusammen mit der ebenfalls zeitlich konstanten Vektormatrix $\boldsymbol{\mathfrak{P}}$ des Drehimpulses zu (28) und (29) analoge Relationen gelten. Indem man noch die der neuen Mechanik eigentümlichen Quantenbedingungen (II) und die in § 3 entwickelten Beziehungen heranzieht, gelangt man zu einem System von Matrixgleichungen, in welchen nur mehr die zeitlich konstanten Matrizen $\boldsymbol{\mathfrak{A}}$, $\boldsymbol{\mathfrak{P}}$ und $\boldsymbol{E}$ vorkommen, die Koordinaten (d. h. die Übergangswahrscheinlichkeiten) jedoch eliminiert sind. Die Auflösung dieser letzteren Gleichungen, die mit elementaren Methoden gelingt (§ 5), führt dann zu den bereits hier diskutierten Resultaten.

§ 3. Regeln für das Rechnen mit der Matrix des Radiusvektors. Impulssatz für Zentralkräfte. Wir beginnen damit, die Regeln für das gleichzeitige Rechnen mit den Matrizen $\boldsymbol{x}$, $\boldsymbol{y}$, $\boldsymbol{z}$ der kartesischen Koordinaten, welche die Komponenten der Vektormatrix $\boldsymbol{\mathfrak{r}}$ bilden, und der Matrix $\boldsymbol{r}$ des Betrages des Radiusvektors zu entwickeln. Diese muß offenbar der Relation genügen

$$\boldsymbol{r}^2 = \boldsymbol{x}^2 + \boldsymbol{y}^2 + \boldsymbol{z}^2. \tag{30}$$

Die Quantenregeln (II) enthalten neben der Vertauschbarkeit von $\boldsymbol{x}$ und $\boldsymbol{y}$, sowie der von $\boldsymbol{x}$, $\dot{\boldsymbol{y}}$ und von $\dot{\boldsymbol{x}}$, $\dot{\boldsymbol{y}}$ (und entsprechend bei den übrigen Koordinaten)

$$\boldsymbol{x}\boldsymbol{y} = \boldsymbol{y}\boldsymbol{x}, \ldots; \quad \boldsymbol{x}\dot{\boldsymbol{y}} = \dot{\boldsymbol{y}}\boldsymbol{x}, \ldots; \quad \dot{\boldsymbol{x}}\dot{\boldsymbol{y}} = \dot{\boldsymbol{y}}\dot{\boldsymbol{x}}, \ldots \tag{31a}$$

die Relationen

$$\boldsymbol{p}_x\boldsymbol{x} - \boldsymbol{x}\boldsymbol{p}_x = \frac{h}{2\pi i}\cdot\mathbf{1}, \ldots \tag{31b}$$

($\boldsymbol{p}_x = m\dot{\boldsymbol{x}}$ bedeutet die x-Komponente des linearen Impulses $\boldsymbol{\mathfrak{p}} = m\boldsymbol{\mathfrak{v}}$ einer Vektormatrix mit den Komponenten $\boldsymbol{p}_x$, $\boldsymbol{p}_y$, $\boldsymbol{p}_z$). Hier und im folgenden soll stets ... das Gelten von analogen Gleichungen für die übrigen Koordinaten andeuten, die aus der angeschriebenen Gleichung durch zyklische Vertauschung der Koordinaten hervorgehen.

Diese Regeln können bei Heranziehung der Matrix r durch die folgenden zusätzlichen Relationen erweitert werden. Es ist erstens auch r mit x, y, z vertauschbar, oder als Vektorgleichung geschrieben

$$r\mathfrak{r} = \mathfrak{r}r. \tag{32}$$

Es gilt zweitens für eine beliebige rationale Funktion f von r, x, y, z die Relation

$$p_x f - f p_x = \frac{h}{2\pi i}\frac{\partial f}{\partial x}, \cdots \tag{33}$$

insbesondere für $f = r$:

$$\mathfrak{p}r - r\mathfrak{p} = \frac{h}{2\pi i}\frac{\mathfrak{r}}{r}. \tag{34}$$

Aus (31) und (34) folgt umgekehrt allgemein (33) für jede durch Reihen nach positiven und negativen Potenzen von x, y, z und r darstellbare Funktion, wie man durch vollständige Induktion leicht zeigt; auch ist diese Relation (34) mit (30) im Einklang. Das Bestehen der Relationen (32) und (33) muß deshalb notwendig gefordert werden, damit aus dem Energiesatz

$$\frac{m}{2}\mathfrak{v}^2 + F(x, y, z, r) = E \text{ (Diagonalmatrix)} \tag{35}$$

(wir nehmen hier der Einfachheit halber nur ein einziges Teilchen als vorhanden an) und der Frequenzbedingung, die für jede Größe Φ die Gleichung

$$E\Phi - \Phi E = \frac{h}{2\pi i}\dot{\Phi}.$$

zur Folge hat, die Bewegungsgleichungen

$$\frac{dp_x}{dt} = -\frac{\partial F}{\partial x}, \cdots \tag{36}$$

hervorgehen. Wir postulieren hier also die Existenz einer Matrix r, die den Relationen (30), (32), (34) genügt.

Nun führen wir eine Vektormatrix $\mathfrak{P}$ ein, die dem Drehimpuls des Teilchens um den Ursprung entspricht. Zunächst bemerken wir, daß wir im folgenden unter dem skalaren Produkt zweier Vektormatrizen $\mathfrak{A}$ und $\mathfrak{B}$, wie in der gewöhnlichen Vektorrechnung den Ausdruck

$$(\mathfrak{A}\mathfrak{B}) = \mathfrak{A}_x\mathfrak{B}_x + \mathfrak{A}_y\mathfrak{B}_y + \mathfrak{A}_z\mathfrak{B}_z,$$

unter ihrem Vektorprodukt $[\mathfrak{A}\mathfrak{B}]$ eine neue Vektormatrix mit den Komponenten

$$[\mathfrak{A}\mathfrak{B}]_x = \mathfrak{A}_y\mathfrak{B}_z - \mathfrak{A}_z\mathfrak{B}_y, \cdots \tag{37}$$

verstehen wollen. Es ist hierin im allgemeinen auf die Reihenfolge der Multiplikation von $\mathfrak{A}$ und $\mathfrak{B}$ zu achten: Die Ausdrücke $(\mathfrak{A}\,\mathfrak{B}) - (\mathfrak{B}\,\mathfrak{A})$ und $[\mathfrak{A}\,\mathfrak{B}] + [\mathfrak{B}\,\mathfrak{A}]$ verschwinden hier wegen der Ungültigkeit des kommutativen Multiplikationsgesetzes im allgemeinen nicht, auch ist das Vektorprodukt $[\mathfrak{A}\,\mathfrak{A}]$ einer Matrix $\mathfrak{A}$ mit sich selbst mit den Komponenten

$$[\mathfrak{A}\,\mathfrak{A}]_x = \mathfrak{A}_y\,\mathfrak{A}_z - \mathfrak{A}_z\,\mathfrak{A}_y, \ldots \tag{37'}$$

im allgemeinen von Null verschieden. Bilden wir jedoch speziell das Vektorprodukt $[\mathfrak{r}\,\mathfrak{v}]$, so fällt es mit $-[\mathfrak{v}\,\mathfrak{r}]$ wegen der Vertauschbarkeit von x und $\dot{y}$ zusammen, und wir können die Vektormatrix

$$\mathfrak{P} = m\,[\mathfrak{r}\,\mathfrak{v}] = -m\,[\mathfrak{v}\,\mathfrak{r}] \tag{38}$$

als Drehimpuls des Teilchens definieren.

Diese genügt folgenden Vertauschungsregeln, die eine unmittelbare Folge von (31 a) und (31 b) sind:

$$\left.\begin{aligned} x P_x = P_x x, \ldots; \quad x P_y - P_y x = P_x y - y P_x = -\frac{h}{2\pi i} z, \ldots \\ (\mathfrak{r}\,\mathfrak{P}) = (\mathfrak{P}\,\mathfrak{r}) = 0 \end{aligned}\right\} \tag{39}$$

ebenso

$$\left.\begin{aligned} p_x P_x = P_x p_x, \ldots; \\ p_x P_y - P_y p_x = P_x p_y - p_y P_x = -\frac{h}{2\pi i} p_z, \ldots \\ (\mathfrak{P}\,\mathfrak{p}) = (\mathfrak{p}\,\mathfrak{P}) = 0 \end{aligned}\right\} \tag{40}$$

Hieraus folgt weiter, daß $v^2 = \dot{x}^2 + \dot{y}^2 + \dot{z}^2$ mit P_x, P_y, P_z vertauschbar ist[1]):

$$v^2\,\mathfrak{P} = \mathfrak{P}\,v^2, \tag{41}$$

ferner folgt aus (34) die Vertauschbarkeit von r mit P_x, P_y, P_z:

$$r\,\mathfrak{P} = \mathfrak{P}\,r, \tag{42}$$

es ist daher auch jede Funktion $F(r)$ von r allein mit $\mathfrak{P}$ vertauschbar. Hat man es mit einer Zentralkraft zu tun, wo die potentielle Energie allein von r abhängt:

$$\frac{m}{2}\,\mathfrak{v}^2 + F(r) = E \text{ (Diagonalmatrix)}, \tag{35'}$$

so ist demnach

$$E\,\mathfrak{P} = \mathfrak{P}\,E,$$

das heißt die Vektormatrix $\mathfrak{P}$ ist zeitlich konstant (Drehimpulsintegral).

---

[1]) Man beachte die Identität $a^2 b - b a^2 \equiv a(ab - ba) + (ab - ba)a$.

Für das Vektorprodukt von $\mathfrak{P}$ mit sich selbst [vgl. (37′)] findet man leicht die später heranzuziehende Relation [1])

$$[\mathfrak{P}\,\mathfrak{P}] = -\frac{h}{2\pi i}\mathfrak{P}. \tag{43}$$

Z. B. erhält man gemäß (39) und (40) für die z-Komponente von $[\mathfrak{P}\,\mathfrak{P}]$:

$$\begin{aligned} P_x P_y - P_y P_x &= P_x(z p_x - x p_z) - (z p_x - x p_z) P_x \\ &= (P_x z - z P_x) p_x - x(P_x p_z - p_z P_x) \\ &= \frac{h}{2\pi i}(y p_x - x p_y) = -\frac{h}{2\pi i} P_z. \end{aligned}$$

Wir können nunmehr den radialen Impuls

$$p_r = m \dot{r}$$

berechnen. Dieser ist nämlich gleich

$$\begin{aligned} p_r &= \frac{2\pi i}{h} m(Er - rE) = \frac{2\pi i}{h}\cdot\frac{1}{2}[(p_x^2 + p_y^2 + p_z^2) r - r(p_x^2 + p_y^2 + p_z^2)] \\ &= \frac{2\pi i}{h}\cdot\frac{1}{2}[p_x(p_x r - r p_x) + p_y(p_y r - r p_y) + p_z(p_z r - r p_z) \\ &\quad + (p_x r - r p_x) p_x + (p_y r - r p_y) p_y + (p_z r - r p_z) p_z], \end{aligned}$$

also mit Rücksicht auf (34):

$$p_r = \frac{1}{2}\left[\left(\mathfrak{p}\frac{\mathfrak{r}}{r}\right) + \left(\frac{\mathfrak{r}}{r}\mathfrak{p}\right)\right]. \tag{44}$$

Nun ist nach (33)

$$\left(\mathfrak{p}\frac{\mathfrak{r}}{r}\right) - \left(\frac{\mathfrak{r}}{r}\mathfrak{p}\right) = \frac{h}{2\pi i}\operatorname{div}\frac{\mathfrak{r}}{r} = \frac{h}{2\pi i}\frac{2}{r},$$

so daß man (44) auch schreiben kann

$$p_r = (\mathfrak{p}\,\mathfrak{r})\frac{1}{r} - \frac{h}{2\pi i}\frac{1}{r} = \frac{1}{r}(\mathfrak{r}\,\mathfrak{p}) + \frac{h}{2\pi i}\frac{1}{r}. \tag{44′}$$

Durch Multiplikation mit r ergibt sich daraus erstens die Relation

$$p_r r + r p_r = (\mathfrak{p}\,\mathfrak{r}) + (\mathfrak{r}\,\mathfrak{p}), \tag{45}$$

die auch direkt aus (30) durch Differentiation nach der Zeit hervorgeht, und ferner

$$p_r r - r p_r = (\mathfrak{p}\,\mathfrak{r}) - (\mathfrak{r}\,\mathfrak{p}) - \frac{h}{2\pi i}\cdot 2.$$

Die Bedeutung von $(\mathfrak{p}\,\mathfrak{r}) - (\mathfrak{r}\,\mathfrak{p})$ ist nun

$$(p_x x - x p_x) + (p_y y - y p_y) + (p_z z - z p_z),$$

---

1) Vgl. „Quantenmech. II“, S. 597, Gl. (3). Die dort mit M_x, M_y, M_z bezeichneten Größen entsprechen den negativen Drehimpulskomponenten.

und jeder der eingeklammerten Ausdrücke hat nach (31 b) den Wert

$$\frac{h}{2\pi i}\cdot \mathbf{1}.$$

Daher wird im ganzen

$$p_r r - r p_r = \frac{h}{2\pi i}\cdot \mathbf{1}. \tag{46}$$

Wir berechnen endlich noch für die folgende Anwendung die zeitliche Änderung von $\frac{\mathfrak{r}}{r}$. Man hat z. B. für die x-Komponente mit Rücksicht auf (34)

$$\begin{aligned}
\frac{d}{dt}\frac{x}{r} &= \frac{2\pi i}{h}\left(E\frac{x}{r} - \frac{x}{r}E\right) = \frac{2\pi i}{h}\frac{1}{2m}\left(\mathfrak{p}^2\frac{x}{r} - \frac{x}{r}\mathfrak{p}^2\right)\\
&= \frac{2\pi i}{h}\frac{1}{2m}\left\{\mathfrak{p}\left(\mathfrak{p}\frac{x}{r} - \frac{x}{r}\mathfrak{p}\right) + \left(\mathfrak{p}\frac{x}{r} - \frac{x}{r}\mathfrak{p}\right)\mathfrak{p}\right\}\\
&= \frac{1}{2m}\left\{\left(p_x\frac{y^2+z^2}{r^3} - p_y\frac{xy}{r^3} - p_z\frac{xz}{r^3}\right)\right.\\
&\quad \left. + \left(\frac{y^2+z^2}{r^3}p_x - \frac{xy}{r^3}p_y - \frac{xz}{r^3}p_z\right)\right\}\\
&= \frac{1}{2m}\left\{\left(P_y\frac{z}{r^3} - P_z\frac{y}{r^3}\right) + \left(\frac{z}{r^3}P_y - \frac{y}{r^3}P_z\right)\right\}.
\end{aligned}$$

Also allgemein

$$\frac{d}{dt}\frac{\mathfrak{r}}{r} = \frac{1}{2m}\left\{\left[\mathfrak{P}\frac{\mathfrak{r}}{r^3}\right] - \left[\frac{\mathfrak{r}}{r^3}\mathfrak{P}\right]\right\}. \tag{47}$$

§ 4. Einführung der zeitlich konstanten Vektormatrix $\mathfrak{A}$ bei Coulombscher Kraft. Elimination der Koordinaten. Gehen wir nun zur Betrachtung eines Atoms mit einem Elektron der Masse m_0 und der Ladung $-e$ über, das vom festen Kern der Ladung $+Ze$ mit Coulombschen Kräften angezogen wird. Für die Hamiltonsche Funktion haben wir hier anzusetzen

$$\frac{1}{2m_0}\mathfrak{p}^2 - \frac{Ze^2}{r} = E \text{ (Diagonalmatrix)}, \tag{48}$$

d. h. es ist in (35′) speziell

$$F(r) = -\frac{Ze^2}{r}$$

zu setzen. Die aus dem Energiesatz mittels der Quantenregeln abzuleitenden Bewegungsgleichungen (36) lauten hier analog zur klassischen Mechanik

$$\dot{\mathfrak{p}} = m_0\ddot{\mathfrak{r}} = -\frac{Ze^2}{r^3}\mathfrak{r}. \tag{49}$$

Analog zur klassischen Mechanik [vgl. (27)] folgt daher gemäß (47), daß die durch

$$\mathfrak{A} = \frac{1}{Z e^2 m_0} \frac{1}{2} \left\{ \left[\mathfrak{P}\,\mathfrak{p} \right] - \left[\mathfrak{p}\,\mathfrak{P} \right] \right\} + \frac{\mathfrak{r}}{r} \tag{50}$$

definierte Vektormatrix $\mathfrak{A}$ im speziellen Falle eines Coulombschen Kraftfeldes zeitlich konstant ist. Gemäß (40) kann man auch schreiben

$$\begin{aligned} \mathfrak{A} &= \frac{1}{Z e^2 m_0} \left\{ \left[\mathfrak{P}\,\mathfrak{p} \right] + \frac{h}{2\pi i}\,\mathfrak{p} \right\} + \frac{\mathfrak{r}}{r} \\ &= -\frac{1}{Z e^2 m_0} \left\{ \left[\mathfrak{p}\,\mathfrak{P} \right] + \frac{h}{2\pi i}\,\mathfrak{p} \right\} + \frac{\mathfrak{r}}{r} \cdot \end{aligned} \tag{51'}$$

Die weiteren Rechnungen sind ganz elementar mit Zuhilfenahme der im vorigen Paragraphen zusammengestellten Regeln auszuführen. Zunächst erhält man analog zur Kegelschnittgleichung (28) der klassischen Mechanik die Beziehung

$$\frac{1}{2}[(\mathfrak{A}\,\mathfrak{r}) + (\mathfrak{r}\,\mathfrak{A})] = -\frac{1}{Z e^2 m_0}\left[\mathfrak{P}^2 + \frac{3}{2}\frac{h^2}{4\pi^2}\right] + r, \tag{51}$$

ferner die Vertauschungsregel

$$\frac{1}{2}([\mathfrak{A}\,\mathfrak{r}] + [\mathfrak{r}\,\mathfrak{A}]) = -\frac{h}{2\pi i}\frac{3}{2}\frac{1}{Z e^2 m_0}\mathfrak{P}. \tag{52}$$

Weiter erweisen sich folgende Relationen als gültig, in denen die Koordinaten x, y, z, r gänzlich eliminiert sind und in die allein die zeitlich konstanten Matrizen $\mathfrak{A}$, $\mathfrak{P}$ und E eingehen:

$$[\mathfrak{P}\,\mathfrak{P}] = -\frac{h}{2\pi i}\mathfrak{P}. \tag{I}$$

$$\left.\begin{aligned} &A_x P_x = P_x A_x, \ldots, \\ &A_x P_y - P_y A_x = P_x A_y - A_y P_x = -\frac{h}{2\pi i} A_z, \ldots, \\ &(\mathfrak{A}\,\mathfrak{P}) = (\mathfrak{P}\,\mathfrak{A}) = 0 \end{aligned}\right\} \tag{II}$$

$$[\mathfrak{A}\,\mathfrak{A}] = \frac{h}{2\pi i}\frac{2}{m_0 Z^2 e^4} E\,\mathfrak{P}. \tag{III}$$

$$1 - \mathfrak{A}^2 = -\frac{2}{m_0 Z^2 e^4} E\left(\mathfrak{P}^2 + \frac{h^2}{4\pi^2}\right) \cdot \tag{IV}$$

Die Gleichung (I) ist identisch mit der Gleichung (43) des vorigen Abschnittes, (II) ist analog gebaut wie (39), (IV) ist analog der Gleichung (29) der klassischen Mechanik, jedoch ist das Auftreten des Zusatzgliedes $\frac{h^2}{4\pi^2}$ [ebenso das Zusatzglied $\frac{3}{2}\frac{h^2}{4\pi^2}$ in (51)] für die neue Mechanik charakteristisch.

Aus der Existenz der zeitlich konstanten Vektormatrix $\mathfrak{A}$ ist zu schließen, daß das Atom mit einem einzigen Elektron (ebenso wie die Keplerbewegung in der klassischen Mechanik), auch abgesehen von der räumlichen Orientierung des Atoms, ein entartetes System bildet. Es kann nämlich aus den hergeleiteten Relationen leicht gefolgert werden, daß $\mathfrak{A}\mathfrak{P}^2 - \mathfrak{P}^2\mathfrak{A}$ im allgemeinen nicht verschwinden kann; da andererseits $\mathfrak{A}E - E\mathfrak{A}$ verschwindet, kann zu einem bestimmten Wert der Energie E offenbar nicht nur ein einziger Wert von $\mathfrak{P}^2$ gehören, das System ist also in der Tat entartet.

Bei einem solchen System sind in der neuen Quantenmechanik, wie Born, Jordan und Heisenberg[1]) eingehend erörtert haben, die Amplituden der verschiedenen Partialschwingungen, die zu Übergängen zwischen Zuständen mit vorgegebenen Energiewerten gehören, aus den Gleichungen der Quantenmechanik nicht eindeutig festgelegt; auch brauchen zeitlich konstante Matrizen im allgemeinen keine Diagonalmatrizen zu sein, indem ihre Elemente an solchen Plätzen (n, m) von Null verschieden sein können, denen eine verschwindende Frequenz $\nu_m^n = \frac{1}{h}(E_n - E_m) = 0$ entspricht. In unserem Falle gehört zu jedem Energiewert (zu jedem Wert der Hauptquantenzahl) eine Matrix, welche die zeitlich konstanten Teile einer Größe (z. B. x oder r) enthält und deren Grad (Anzahl der Zeilen oder Kolonnen) dem Gewicht des betreffenden Zustandes gleich ist. Diese Matrix, die aus der ursprünglichen Matrix durch Nullsetzen aller Elemente an Plätzen, die einem von einer Änderung des Energiewertes begleiteten Übergangsprozeß entsprechen, entsteht, nennen wir den zeitlichen Mittelwert der betreffenden Größe und bezeichnen ihn durch einen Querstrich (z. B. $\overline{x}$ oder $\overline{r}$).

Wenn auch bei einem entarteten System die einzelnen zur gleichen Frequenz ν_m^n gehörigen Partialschwingungen einer kinematischen Größe nicht eindeutig festgelegt sind, so gilt dies doch von den Energiewerten und statistischen Gewichten dieser Zustände[2]). Es müßte daher im Prinzip möglich sein, aus den Gleichungen (I) bis (IV) ohne nähere spezialisierende Annahmen über die Art der Lösung die Balmerterme und zugehörigen statistischen Gewichte herzuleiten. Dies ist uns jedoch leider nicht gelungen, und wir werden im folgenden einen abweichenden Weg zur Lösung dieser Gleichungen einschlagen, indem wir (auf ver-

---

[1]) Quantenmechanik II, Kap. 2, § 2.

[2]) Ebenda II, Kap. 2, § 2 und Kap. 3.

schiedene Weisen) zusätzliche Bedingungen einführen, welche die Lösung der Gleichungen (I) bis (IV) eindeutig machen.

Wird nämlich die Entartung durch ein zusätzliches Störfeld mit der Hamiltonschen Funktion H_1 aufgehoben, so muß, wie die Durchführung der Störungsrechnung nach Born, Jordan und Heisenberg[1]) zeigt, der über die ungestörte Bewegung erstreckte zeitliche Mittelwert $\overline{H}_1$ der Störungsfunktion eine Diagonalmatrix sein. In unserem Falle wird dieser Mittelwert im allgemeinen außer von der Energie E der ungestörten Bewegung noch von $\mathfrak{P}$ und $\mathfrak{A}$ abhängen.

Ist das Störfeld speziell eine zusätzliche nichtcoulombsche Zentralkraft, so wird ihr Mittelwert (außer von E) nur von $\mathfrak{P}^2$ abhängen, da hier ja keine Richtung im Raum ausgezeichnet ist. Ferner hängt die Störenergie eines Magnetfeldes in der z-Richtung nur von der Impulskomponente P_z parallel dem Feld ab. Wir erhalten daher durch die Forderung, daß $\mathfrak{P}^2$ und P_z Diagonalmatrizen sein sollen, eine spezielle Lösung der Gleichungen (I) bis (IV), die der relativistischen Feinstruktur und einem zusätzlichen schwachen Magnetfeld angepaßt ist. Dieser Fall wird im folgenden Paragraphen zunächst behandelt.

Ein zweiter Fall von besonderem Interesse ist der des Starkeffekts. Hier kommt es darauf an, daß die den elektrischen Mittelpunkt der Bahn darstellende zeitlich konstante Vektormatrix $\bar{\mathfrak{r}}$ eine Komponente $\bar{z}$ in der Feldrichtung (z-Richtung) besitzt, die eine Diagonalmatrix ist. Nun läßt sich aber zeigen, daß diese Matrix $\bar{\mathfrak{r}}$ mit der Matrix $\mathfrak{A}$ ebenso zusammenhängt wie in der klassischen Theorie, nämlich durch die Beziehung

$$\bar{\mathfrak{r}} = \frac{3}{2}\,\frac{Ze^2}{2|E|}\,\mathfrak{A} \tag{53}$$

(in der klassischen Theorie stellt $\frac{Ze^2}{2|E|}$ die große Halbachse a der Keplerellipse dar). Zunächst gelten nämlich für $\bar{\mathfrak{r}}$ und $\mathfrak{P}$ gemäß (39) und (II) dieselben Vertauschungsregeln wie für $\mathfrak{A}$ und $\mathfrak{P}$. Vergleicht man dann ferner (52) und (III), so ergeben sich für die Differenz $\bar{\mathfrak{r}} - \frac{3}{2}\frac{Ze^2}{2|E|}\mathfrak{A}$ außerdem noch die Beziehungen $\left[\mathfrak{A}, \bar{\mathfrak{r}} - \frac{3}{2}\frac{Ze^2}{2|E|}\mathfrak{A}\right] = 0$. Wie die nähere Diskussion auf Grund der im folgenden Paragraphen abgeleiteten Lösungen der Gl. (I) bis (IV) zeigt, sind dies genügend viele homogene lineare Gleichungen, um auf das Verschwinden von $\bar{\mathfrak{r}} - \frac{3}{2}\frac{Ze^2}{2|E|}\mathfrak{A}$

[1]) Quantenmechanik II, Kap. 2, § 2 und Kap. 3.

schließen zu können. Nehmen wir zum elektrischen Feld in der z-Richtung noch ein paralleles Magnetfeld in derselben Richtung hinzu, so können wir gemäß (53) diesen Fall auch durch die Forderung charakterisieren, daß A_z und P_z Diagonalmatrizen sind.

Endlich behandeln wir im folgenden Paragraphen an letzter Stelle noch den Fall gekreuzter elektrischer und magnetischer Felder, der, wie in § 2 näher erläutert wurde, im Hinblick auf die in der bisherigen Theorie auftretenden Zusatzverbote von singulären Bewegungen von besonderem Interesse ist.

§ 5. Auflösung der Gleichungen (I) bis (IV). Herleitung der Balmerterme.

a) P_z und $\mathfrak{P}^2$ sind Diagonalmatrizen.

Um für diesen zuerst zu behandelnden Fall, der der Aufhebung der Entartung durch ein zusätzliches Zentralfeld und ein schwaches Magnetfeld in der z-Richtung entspricht, zunächst die Gleichungen (I) und (II) zu befriedigen, mache man folgenden Ansatz. Bei einem vorgegebenen Wert von $\mathfrak{P}^2$ seien die möglichen Werte von P_z gegeben durch

$$P_{z\,k,m}^{\;k,m} = m\,\frac{h}{2\pi}, \tag{54}$$

worin m von $-k$ bis $+k$ laufe:

$$-k \leqq m \leqq k. \tag{54'}$$

Ferner seien die Partialschwingungen von $\mathfrak{P}$, die zu einer Änderung von m um eine Einheit gehören, links- und rechtszirkular in der (x, y)-Ebene:

$$P_{y\,k,m\pm 1}^{\;k,m} = \pm i\,P_{x\,k,m\pm 1}^{\;k,m}. \tag{55}$$

Aus (I) folgt sodann

$$\left|P_{x\,k,m\mp 1}^{\;k,m}\right|^2 = \left|P_{y\,k,m\mp 1}^{\;k,m}\right|^2 = \frac{1}{4}\,\frac{h^2}{4\pi^2}\,[k(k+1) - m(m\mp 1)]$$

$$= \frac{1}{4}\,\frac{h^2}{4\pi^2}\,(k\pm m)(k+1\mp m). \tag{56}$$

$$(\mathfrak{P}^2)_{k,m}^{k,m} = \frac{h^2}{4\pi^2}\,k(k+1). \tag{57}$$

Weiter setze man für die Matrix $\mathfrak{A}$ entsprechend den Hönl-Kronigschen Formeln für die Intensität der Zeemankomponenten

$$A_{y\,k',m\pm 1}^{\;k,m} = \pm i\,A_{x\,k',m\pm 1}^{\;k,m} \quad (k' = k+1 \text{ oder } k-1), \tag{58}$$

$$\left| A_{x\,k,\,m\pm 1}^{\;k+1,\,m} \right|^2 = \left| A_{y\,k,\,m\pm 1}^{\;k+1,\,m} \right|^2 = \frac{1}{4} C_k^{k+1} (k \mp m)(k \mp m + 1); \quad (59)$$

ersetzt man hierin m durch $m - 1$ bzw. $m + 1$, so folgt noch

$$\left| A_{x\,k+1,\,m\pm 1}^{\;k,\,m} \right|^2 = \left| A_{y\,k+1,\,m\pm 1}^{\;k,\,m} \right|^2$$

$$= \frac{1}{4} C_{k+1}^{k} (k \pm m + 1)(k \pm m + 2); \quad (59\,\text{a})$$

und schließlich gilt für A_z:

$$\left| A_{z\,k,\,m}^{\;k+1,\,m} \right|^2 = C_k^{k+1} [(k+1)^2 - m^2]. \quad (60)$$

Es bleibt hierin noch offen, ob m (also auch k) halb- oder ganzzahlig ist, ferner sind C_k^{k+1} zunächst noch unbestimmt bleibende Funktionen von k, die niemals negative Werte annehmen können und der Symmetrierelation

$$C_k^{k+1} = C_{k+1}^{k} \quad (61)$$

genügen. Bezüglich der Vorzeichen von $\mathfrak{A}$ relativ zu denen von $\mathfrak{P}$ ist noch zu bemerken, daß, wenn P_x und A_z als positiv reell angenommen werden, A_x positiv oder negativ reell anzusetzen ist, je nachdem man es mit Übergängen zu tun hat, bei denen sich k und m im entgegengesetzten Sinne $\left[\text{wie bei } A_{x\,k+1,\,m-1}^{\;k,\,m} \text{ und } A_{x\,k-1,\,m+1}^{\;k,\,m}\right]$ oder im gleichen Sinne $\left[\text{wie bei } A_{x\,k+1,\,m+1}^{\;k,\,m} \text{ und } A_{x\,k-1,\,m-1}^{\;k,\,m}\right]$ ändern. Bei diesen Ansätzen sind, wie die Durchrechnung zeigt, die Gleichungen (I) und (II) des vorigen Paragraphen erfüllt. Überdies geht aus Überlegungen von Born, Jordan und Heisenberg[1]) hervor, daß auch umgekehrt unter der Voraussetzung, daß $\mathfrak{P}^2$ und P_z Diagonalmatrizen sind, die hier angenommenen Ausdrücke für $\mathfrak{A}$ und $\mathfrak{P}$ eine notwendige Folge von (I) und (II) sind.

Um nunmehr auch die Normierung von m und k sowie die Funktion C_{k+1}^{k} zu ermitteln, ziehen wir die Gleichung (III) des vorigen Paragraphen heran. Es genügt jedoch, allein die z-Komponente

$$\mathsf{A}_x \mathsf{A}_y - \mathsf{A}_y \mathsf{A}_x = \frac{h}{2\pi i} \frac{2}{m_0 Z^2 e^4} E \mathsf{P}_z \quad (62)$$

heranzuziehen. Bildet man nämlich den Ausdruck

$$\mathsf{P}_y(\mathsf{A}_x\mathsf{A}_y - \mathsf{A}_y\mathsf{A}_x) - (\mathsf{A}_x\mathsf{A}_y - \mathsf{A}_y\mathsf{A}_x)\mathsf{P}_y = \frac{h}{2\pi i}\frac{2}{m_0 Z^2 e^4} E(\mathsf{P}_y\mathsf{P}_z - \mathsf{P}_z\mathsf{P}_y),$$

---

[1]) Quantenmechanik II, Kap. 4, § 1. Vgl. auch die Ausführungen über den Zeemaneffekt in Kap. 4, § 2.

so erhält man mit Benutzung von (I) und (II) eine Gleichung, die mit der x-Komponente von (III) übereinstimmt. Ebenso ist auch die y-Komponente von (III) eine Folge der z-Komponente dieser Vektorgleichung und der Gleichungen (I) und (II).

Bilden wir nun das am Platz k, m der Diagonalreihe stehende Element der Gleichung (62), so erhalten wir gemäß (58) und (59) für die linke Seite zunächst

$$(\boldsymbol{A}_x \boldsymbol{A}_y - \boldsymbol{A}_y \boldsymbol{A}_x)_{k,m}^{k,m} = 2\,i \left\{ \left| A_{x\,k+1,\,m-1}^{\;k,\,m} \right|^2 - \left| A_{x\,k+1,\,m+1}^{\;k,\,m} \right|^2 \right.$$
$$\left. + \left| A_{x\,k-1,\,m-1}^{\;k,\,m} \right|^2 - \left| A_{x\,k-1,\,m+1}^{\;k,\,m} \right|^2 \right\}$$
$$= i\,m \left\{ -(2\,k+3)\,C_k^{k+1} + (2\,k-1)\,C_{k-1}^{k} \right\}.$$

Berücksichtigen wir noch, daß $\boldsymbol{E}$ negatives Vorzeichen hat und führen wir die Rydbergkonstante

$$R = \frac{2\,\pi^2\,e^4\,m_0}{h^3} \tag{63}$$

und den Wert (54) von $\boldsymbol{P}_z$ ein, so ergibt demnach (62) die Bedingung

$$m \left\{ -(2\,k+3)\,C_k^{k+1} + (2\,k-1)\,C_{k-1}^{k} \right\} = \frac{|\,E\,|}{R\,h\,Z^2}\,m. \tag{64}$$

Betrachten wir zunächst den kleinsten Wert von k, der bei gegebenem $|E|$ möglich ist. Für diesen fällt offenbar der Beitrag des Überganges $k \longrightarrow k-1$ auf der linken Seite fort, der Koeffizient von m auf der linken Seite wird daher sicher nicht positiv, während der Koeffizient von m auf der rechten Seite positiv ist. Die Gleichung (64) ist daher für den Minimalwert von k nur erfüllbar, wenn $m = 0$ ist. Das bedeutet aber nach (54), daß jener Minimalwert von k selbst verschwindet, da ja sonst m noch anderer Werte fähig ist. Es sind demnach k und m notwendig ganzzahlig und wir können für k die Werte ansetzen

$$k = 0,\ 1,\ 2,\ \ldots\ n^*, \tag{65}$$

wenn die ganze Zahl n^* den Maximalwert von k bedeutet, der bei gegebenem $|E|$ erreicht wird.

Nunmehr folgt aus (64)

$$(2\,k-1)\,C_{k-1}^{k} - (2\,k+3)\,C_k^{k+1} = \frac{|E|}{R\,h\,Z^2} \text{ für } k = 1,\ \ldots\ n^* \tag{64'}$$

und es ist dabei überdies

$$C_{n^*}^{n^*+1} = 0 \tag{64''}$$

zu setzen, da für $k = n^*$ der Beitrag des Überganges $k+1 \longrightarrow k$ (zweites Glied) offenbar fortfällt. Man kann, beginnend mit $k = n^*$ und indem man k schrittweise abnehmen läßt, die Werte von

$$C^{n^*}_{n^*-1},\quad C^{n^*-1}_{n^*-2},\quad \ldots\ C^1_0$$

sukzessive aus (64′) berechnen. Das Ergebnis läßt sich in die Formel zusammenfassen

$$\begin{aligned} C^{k+1}_k &= \frac{|E|}{RhZ^2}\,\frac{n^*(n^*+2)-k(k+2)}{(2k+1)(2k+3)} \\ &= \frac{|E|}{RhZ^2}\,\frac{(n^*-k)(n^*+k+2)}{(2k+1)(2k+3)}, \end{aligned} \tag{66}$$

aus welcher mittels Ersetzen von k durch $k-1$ noch folgt

$$\begin{aligned} C^k_{k-1} &= \frac{|E|}{RhZ^2}\,\frac{n^*(n^*+2)-(k-1)(k+1)}{(2k-1)(2k+1)} \\ &= \frac{|E|}{RhZ^2}\,\frac{(n^*-k+1)(n^*+k+1)}{(2k-1)(2k+1)}. \end{aligned} \tag{66'}$$

Man bestätigt mittels dieser Formeln unmittelbar, daß die Relationen (64′), (64″) erfüllt sind.

Um endlich auch den Energiewert selbst zu ermitteln, benutzen wir noch die letzte Gleichung (IV). Zunächst berechnen wir den Wert von $\mathfrak{A}^2$ an der Stelle k, m der Diagonalreihe. Mit Rücksicht auf (59) und (60) erhält man

$$\begin{aligned} (\mathfrak{A}^2)^{k,m}_{k,m} &= 2\left|A_x{}^{k,m}_{k+1,m+1}\right|^2 + 2\left|A_x{}^{k,m}_{k+1,m-1}\right|^2 + \left|A_z{}^{k,m}_{k+1,m}\right|^2 \\ &\quad + 2\left|A_x{}^{k,m}_{k-1,m+1}\right|^2 + 2\left|A_x{}^{k,m}_{k-1,m-1}\right|^2 + \left|A_z{}^{k,m}_{k-1,m}\right|^2 \\ &= (k+1)(2k+3)\,C^{k+1}_k + k(2k-1)\,C^k_{k-1} \end{aligned}$$

und nach Einsetzen von (66), (66′):

$$(\mathfrak{A}^2)^{k,m}_{k,m} = \frac{|E|}{RhZ^2}\,[n^{*2} + 2n^* - k(k+1)]. \tag{67}$$

Dieser Ausdruck für $\mathfrak{A}^2$ und der Ausdruck (57) für $\mathfrak{P}^2$ sind nun in (IV) einzusetzen. Es ergibt sich

$$1 = \frac{|E|}{RhZ^2}(n^{*2} + 2n^* + 1) = \frac{|E|}{RhZ^2}(n^* + 1)^2,$$

also

$$|E| = \frac{RhZ^2}{(n^*+1)^2} = \frac{RhZ^2}{n^2} \tag{68}$$

($n = n^* + 1$ gesetzt) wie in § 2 angegeben wurde. Hiermit ist gezeigt, daß die Balmerterme aus der neuen Quantenmechanik richtig resultieren und daß dem n-quantigen Zustand in der neuen Theorie das Gewicht n^2 zukommt.

b) A_z und P_z sind Diagonalmatrizen (Starkeffekt). Wirkt in der z-Richtung ein homogenes elektrisches Feld der Stärke F, so ist der zeitliche Mittelwert der Störungsenergie gemäß (53) gegeben durch

$$E_1 = \frac{3}{2} eF\bar{\mathbf{z}} = \frac{3}{2} eF \frac{Ze^2}{2|E|} \mathsf{A}_z. \tag{69}$$

Wir bedürfen daher in diesem Falle einer Lösung der Gleichungen (I) bis (IV), bei der A_z eine Diagonalmatrix ist. Wenn wir außerdem die Bedingung stellen, daß auch P_z eine Diagonalmatrix sein soll, so bedeutet dies physikalisch, daß wir die Entartung der säkularen Störung des Starkeffektes durch ein zusätzliches schwaches, zum elektrischen Felde paralleles Magnetfeld aufgehoben denken.

Wir wollen uns hier damit begnügen, das Resultat anzugeben, ohne die Rechnungen im einzelnen durchzuführen und unter Verzicht auf den Nachweis, daß die angegebene Lösung der Gleichungen (I) bis (IV) die einzige ist, die der gestellten Forderung, daß A_z und P_z Diagonalmatrizen sein sollen, genügt. Die Zustände, die zu einem bestimmten, durch (68) gegebenen Wert der ungestörten Energie gehören, sind zu klassifizieren durch zwei Quantenzahlen s und m, von denen die eine den Wert von A_z (und der Zusatzenergie E_1) gemäß

$$A_{z\,s,m}^{\;s,m} = \frac{s}{n}, \quad E_1 = \frac{3}{2} eFa_1 n s \quad (0 \leqq s \leqq n^*), \tag{70}$$

$$\left(a_1 = \frac{h^2}{4\pi^2 Ze^2 m_0}\right),$$

die andere den Wert von P_z gemäß

$$P_{z\,s,m}^{\;s,m} = m \frac{h}{2\pi} \tag{71}$$

bestimmt. Die Wertemannigfaltigkeit von s und m wurde bereits in § 2 durch die Relation (13*) angegeben. Die Matrizen $\mathsf{P}_x, \mathsf{P}_y, \mathsf{A}_x, \mathsf{A}_y$ haben nur an denjenigen Stellen von Null verschiedene Elemente, die Änderungen von s und m um ± 1 entsprechen. Ihre Werte sind gegeben durch

$$P_{y\,s',m\pm 1}^{\;s,m} = \pm i P_{x\,s',m\pm 1}^{\;s,m}, \quad A_{y\,s',m\pm 1}^{\;s,m} = \pm i A_{x\,s',m\pm 1}^{\;s,m} \tag{72}$$

$$(s' = s+1 \quad \text{oder} \quad s-1).$$

$$A_{x\,s\pm 1,m\pm 1}^{\;s,m} = + \frac{2\pi}{h} \frac{1}{n} P_{x\,s\pm 1,m\pm 1}^{\;s,m},$$

$$A_{x\,s\mp 1,m\pm 1}^{\;s,m} = - \frac{2\pi}{h} \frac{1}{n} P_{x\,s\mp 1,m\pm 1}^{\;s,m} \tag{73}$$

(in den letzten Relationen entweder stets das obere oder stets das untere Vorzeichen zu nehmen),

$$\left.\begin{aligned} \left|P_{x\,s-1,\,m-1}^{\;s,\,m}\right|^2 &= \left|P_{y\,s-1,\,m-1}^{\;s,\,m}\right|^2 \\ &= \frac{1}{16}\,\frac{h^2}{4\pi^2}\,[n^* + 2 - (m+s)]\,[n^* + (m+s)], \\ \left|P_{x\,s+1,\,m-1}^{\;s,\,m}\right|^2 &= \left|P_{y\,s+1,\,m-1}^{\;s,\,m}\right|^2 \\ &= \frac{1}{16}\,\frac{h^2}{4\pi^2}\,[n^* + 2 - (m-s)]\,[n^* + (m-s)]. \end{aligned}\right\} \qquad (74)$$

Man rechnet leicht nach, daß durch den Ansatz (70) bis (74) die Gleichungen (I) bis (IV) in der Tat befriedigt sind.

c) Gekreuzte Felder. Sind die Vektoren $\mathfrak{E}$ und $\mathfrak{H}$ die Stärken des äußeren elektrischen und magnetischen Feldes, so ist der zeitliche Mittelwert der Störungsenergie bei gleichzeitiger Anwesenheit beider Felder gegeben durch

$$E_1 = \frac{3}{2}\,e\,a\,(\mathfrak{E}\,\mathfrak{A}) + \frac{e}{2\,m_0\,c}(\mathfrak{H}\,\mathfrak{P}), \qquad (75)$$

worin die Größe a, welche in der bisherigen Theorie die Halbachse der Keplerellipse darstellte, nunmehr einfach als Abkürzung für

$$a = \frac{Z\,e^2}{2\,|E|} \qquad (76)$$

aufzufassen ist. Führt man die Vektoren o_F und o_H ein, die zu $\mathfrak{E}$ bzw. $\mathfrak{H}$ parallel sind und deren Betrag den säkulären Frequenzen gleich ist, die eines der beiden äußeren homogenen Felder für sich allein hervorrufen würde, so daß gilt [vgl. (9) und (10)]

$$o_H = \frac{e\,\mathfrak{H}}{4\,\pi\,m_0\,c}, \quad o_F = \frac{3}{4\pi}\sqrt{\frac{a}{Z m_0}}\;\mathfrak{E} = \frac{3}{4\pi}\,\frac{e\,\mathfrak{E}}{\sqrt{2\,m_0\,|E|}}, \qquad (77)$$

so kann man (75) auch schreiben

$$E_1 = \sqrt{\frac{Z^2 R h}{|E|}}\,(\mathfrak{A}\,o_F)\,h + 2\,\pi\,(\mathfrak{P}\,o_H). \qquad (75\,a)$$

Nun ist es zweckmäßig[1]), die Vektormatrizen $\mathfrak{J}_1$ und $\mathfrak{J}_2$ einzuführen, die definiert sind durch

$$\left.\begin{aligned} 2\,\mathfrak{J}_1 &= \frac{2\pi}{h}\,\mathfrak{P} + \sqrt{\frac{Z^2 R h}{|E|}}\;\mathfrak{A}, \\ 2\,\mathfrak{J}_2 &= \frac{2\pi}{h}\,\mathfrak{P} - \sqrt{\frac{Z^2 R h}{|E|}}\;\mathfrak{A}, \end{aligned}\right\} \qquad (78)$$

---

[1]) Man vgl. zum folgenden die in § 2, S. 341, Fußnote [2]) und [3]) zitierten Arbeiten von Klein und Lenz.

so daß gilt

$$\frac{2\pi}{h}\mathfrak{P} = \mathfrak{J}_1 + \mathfrak{J}_2, \quad \sqrt{\frac{Z^2 R h}{|E|}}\,\mathfrak{A} = \mathfrak{J}_1 - \mathfrak{J}_2; \tag{78a}$$

ferner die beiden Vektoren

$$\mathfrak{o}_1 = \mathfrak{o}_H + \mathfrak{o}_F, \quad \mathfrak{o}_2 = \mathfrak{o}_H - \mathfrak{o}_F, \tag{79}$$

deren absolute Beträge in § 2 mit ω_1 und ω_2 bezeichnet wurden [vgl. Gl. (18)]. Die Störungsenergie (75 a) schreibt sich dann einfach

$$E_1 = (\mathfrak{J}_1\,\mathfrak{o}_1)\,h + (\mathfrak{J}_2\,\mathfrak{o}_2)\,h. \tag{80}$$

In den Gleichungen (I) bis (IV) führen wir nun ebenfalls statt $\mathfrak{A}$ und $\mathfrak{P}$ gemäß (78 a) die neuen Vektormatrizen $\mathfrak{J}_1$ und $\mathfrak{J}_2$ ein. Man erhält dann nach einfacher Rechnung folgende Relationen:

$$I_{1x} I_{2x} = I_{2x} I_{1x} \ldots, \; I_{1x} I_{2y} = I_{2y} I_{1x}, \; I_{2x} I_{1y} = I_{1y} I_{2x}, \ldots \tag{81}$$

$$[\mathfrak{J}_1\,\mathfrak{J}_1] = i\,\mathfrak{J}_1, \quad [\mathfrak{J}_2\,\mathfrak{J}_2] = i\,\mathfrak{J}_2, \tag{82}$$

$$\mathfrak{J}_1^2 = \mathfrak{J}_2^2 = {}^1/_4\left(\frac{R h Z^2}{|E|} - 1\right) = {}^1/_4\,(n^2 - 1) = \frac{n^*}{2}\left(\frac{n^*}{2} + 1\right). \tag{83}$$

Die Relationen (81) besagen, daß jede Komponente von $\mathfrak{J}_1$ mit jeder beliebigen Komponente von $\mathfrak{J}_2$ vertauschbar ist, die Relationen (82) sind ganz analog gebaut wie die Gleichungen (I); in (83) wurden am Schlusse die Energiewerte (68) herangezogen.

Gemäß (80) ist der Fall der gekreuzten Felder dadurch charakterisiert, daß $(\mathfrak{J}_1\,\mathfrak{o}_1)$ und $(\mathfrak{J}_2\,\mathfrak{o}_2)$, oder, was dasselbe ist, die Komponenten von $\mathfrak{J}_1$ und $\mathfrak{J}_2$ parallel zu $\mathfrak{o}_1$ und $\mathfrak{o}_2$, die wir mit $\mathfrak{J}_{1\,\|}$ und $\mathfrak{J}_{2\,\|}$ bezeichnen wollen, Diagonalmatrizen sind. Die Auflösung der Gleichungen (82) erfolgt in diesem Falle völlig analog der Auflösung der Gleichungen (I), wenn in diesen P^2 und P_z als Diagonalmatrix angenommen werden. An Stelle von k tritt hier $\frac{n^*}{2}$, an Stelle von m bei Anschluß an die Bezeichnungen von § 2, Gleichung (15) und (16), die Zahlen $\frac{n^*}{2} - n_1$ und $\frac{n^*}{2} - n_2$, die von $-\frac{n^*}{2}$ bis $+\frac{n^*}{2}$ laufen können. Man hat demnach

$$\left.\begin{aligned} (\mathfrak{J}_{1\,\|})_{n_1}^{n_1} &= \frac{n^*}{2} - n_1, \; (\mathfrak{J}_{2\,\|})_{n_2}^{n_2} = \frac{n^*}{2} - n_2, \; 0 \leqq n_1 \leqq n^*, \; 0 \leqq n_2 \leqq n^*, \\ E_1 &= \left(\frac{n^*}{2} - n_1\right)\omega_1 h + \left(\frac{n^*}{2} - n_2\right)\omega_2 h \\ &\qquad (\omega_1 = |\mathfrak{o}_1|, \quad \omega_2 = |\mathfrak{o}_2|). \end{aligned}\right\} \tag{84}$$

Die Projektionen $\mathfrak{J}_{1\perp}$ bzw. $\mathfrak{J}_{2\perp}$ in die Ebenen senkrecht zur Richtung von $\mathfrak{o}_1$ bzw. $\mathfrak{o}_2$ führen zirkulare Schwingungen aus und werden daher durch Matrizen dargestellt, die zu den durch (56) beschriebenen Matrizen für P_x und P_y analog sind $\left(\text{man ersetze } k \text{ durch } \frac{n^*}{2} \text{ und } m \text{ durch } \frac{n^*}{2} - n_1 \text{ bzw. } \frac{n^*}{2} - n_2;\ \mathfrak{J}_\perp^2 \text{ entspricht der Summe von } P_x^2 \text{ und } P_y^2\right)$:

$$\left.\begin{aligned} \left|\mathfrak{J}_{1\perp}{}^{n_1}_{n_1+1}\right|^2 &= {}^1/_2\,(n_1+1)\,(n^*-n_1), \\ \left|\mathfrak{J}_{2\perp}{}^{n_2}_{n_2+1}\right|^2 &= {}^1/_2\,(n_2+1)\,(n^*-n_2). \end{aligned}\right\} \tag{85}$$

Hierdurch sind die Gleichungen (81) und (82) erfüllt, und da analog zu (57)

$$(\mathfrak{J}_1^2)^{n_1}_{n_1} = (\mathfrak{J}_2^2)^{n_2}_{n_2} = \frac{n^*}{2}\left(\frac{n^*}{2}+1\right)$$

gilt, ist auch (83) mit den Energiewerten (68) befriedigt.

Hiermit sind alle in § 2 angeführten Resultate aus der neuen Mechanik hergeleitet.

§ 6. Zur Frage der Beziehung des Wasserstoffspektrums zu den Spektren der Alkalien.

Bereits in § 2 wurde erwähnt, daß die zur Deutung der anomalen Zeemaneffekte noch erforderlichen Modifikationen der Grundlagen der neuen Quantenmechanik sich möglicherweise bereits bei Atomen mit einem einzigen Elektron bemerkbar machen könnten; insbesondere ist das Ergebnis, daß der Normalzustand eines solchen Atoms unmagnetisch sein sollte, wohl noch nicht als endgültig zu betrachten. Ein spezieller Vorschlag, um dem anomalen Zeemaneffekt Rechnung zu tragen, ist kürzlich von Goudsmit und Uhlenbeck[1]) vorgebracht worden. Gemäß diesem wird das Elektron nicht mehr als eine Punktladung angesehen, sondern es wird ihm eine ausgezeichnete Achse, Drehimpuls und (doppelt anomaler) Magnetismus zugesprochen. Ob diese Annahme ausreicht, um im Verein mit der neuen Quantenmechanik alle Erfahrungsresultate zu erklären, dürfte sich erst entscheiden lassen, wenn auch die Berechnung der relativistischen Feinstruktur auf Grund der neuen Mechanik durchgeführt ist. Diese blieb vorläufig noch außer Betracht, da die hierzu erforderliche Berechnung des zeitlichen Mittelwertes $\overline{\frac{1}{r^2}}$ uns noch nicht gelungen ist.

---

[1]) Naturwissensch. **13**, 953, 1925.

Unabhängig von speziellen Modellvorstellungen ist es jedoch naheliegend, zu fragen, ob nicht das Wasserstoffspektrum (einschließlich Feinstruktur und Beeinflussung durch äußere Felder) aufgefaßt werden kann als Grenzfall der Alkalispektren bzw. Röntgenspektren bei verschwindender Zentralkraft des Atomrestes auf das Leuchtelektron bzw. bei verschwindenden Abschirmungszahlen [so daß Niveaus, die ein Abschirmungsdublett bilden, zusammenfallen] [1]). In diesem Falle würde sich die Feinstruktur der Balmerlinien von der aus der bisherigen Theorie folgenden zwar nicht durch die Lage der Energieniveaus und der Linienkomponenten, wohl aber durch ihre Intensitäten unterscheiden, indem nunmehr an Stelle der Auswahlregel $\Delta k = \pm 1$ die Auswahlregel $\Delta j = 0, \pm 1$ treten würde, die das Auftreten von Komponenten zuläßt, die nach der bisherigen Theorie ausgeschlossen waren. Goudsmit und Uhlenbeck [1]) konnten zeigen, daß die Beobachtungsresultate eine solche Abänderung der Auswahlregel als wahrscheinlich erscheinen lassen. Zugleich weisen sie aber darauf hin, daß der Durchführung der Analogie zwischen Wasserstoff- und Alkalispektrum die Schwierigkeit entgegensteht, daß der Zeemaneffekt in (relativ zur Feinstruktur) schwachen Magnetfeldern bei den Spektren von Atomen mit einem einzigen Elektron nach den vorliegenden Beobachtungen keineswegs alkaliähnlich zu sein scheint.

Obwohl somit die Frage, wieweit sich die in Rede stehende Beziehung zwischen Wasserstoff- und Alkalispektren durchführen läßt, noch nicht als geklärt angesehen werden kann, dürfte es doch gerechtfertigt sein, sich wenigstens in allen Fällen, wo von der relativistischen (bzw. Dublett-) Feinstruktur abgesehen werden kann, von dieser Analogie leiten zu lassen. Dies führt dazu, in Magnetfeldern, in denen die Zeemanaufspaltung groß ist gegenüber dem Abstand der Feinstrukturkomponenten, die magnetischen Energieniveaus in den Spektren der Atome mit einem einzigen Elektron ihrer Anzahl und Lage nach als mit den Paschen-Backtermen der Alkalien übereinstimmend anzunehmen. Es müßten dann in einem äußeren Kraftfeld dem Wasserstoffatom doppelt so viele Zustände zukommen, als im vorigen Paragraphen aus der jetzigen Grundlage der neuen Quantenmechanik abgeleitet wurde (d. h. $2n^2$-Zustände an Stelle von n^2). In einem äußeren Magnetfeld müßten zu jedem Werte der Quantenzahl m (die zwischen $-n^*$ und $+n^*$ liegt), die beiden mag-

[1]) S. Goudsmit und G. E. Uhlenbeck, Physica **5**, 266, 1925. Ähnliche Überlegungen wurden mir bereits vor längerer Zeit von Herrn A. Landé brieflich mitgeteilt.

netischen Energiewerte $(m \pm 1)\ o_H h$ (o_H = Larmorfrequenz) gehören, ebenso müßte in gekreuzten Feldern jeder durch n^*, n_1, n_2 charakterisierte Zustand sich noch in zwei Zustände aufspalten, deren Energiewerte sich von den durch (84) dargestellten um $\pm o_H h$ unterscheiden. Nach dem Korrespondenzprinzip treten dann hier nur solche Übergänge auf, bei denen das Vorzeichen des Zusatzgliedes $\pm o_H h$ unverändert bleibt.

Eine Möglichkeit, um zwischen der im vorigen Paragraphen abgeleiteten Termmannigfaltigkeit, bei welcher der Normalzustand des H-Atoms unmagnetisch ist, und der hier in Analogie zu den Paschen-Backtermen der Alkalien ins Auge gefaßten, bei der dem Normalzustand des H-Atoms im Magnetfeld die Energiewerte $\pm o_H h$ zukommen, zu unterscheiden, würden Ablenkungsversuche mit H-Atomstrahlen im inhomogenen Magnetfeld nach Stern und Gerlach darbieten.

361

Reprint 7

3. *Quantisierung als Eigenwertproblem;*
von E. Schrödinger.

(Erste Mitteilung.)

§ 1. In dieser Mitteilung möchte ich zunächst an dem einfachsten Fall des (nichtrelativistischen und ungestörten) Wasserstoffatoms zeigen, daß die übliche Quantisierungsvorschrift sich durch eine andere Forderung ersetzen läßt, in der kein Wort von „ganzen Zahlen" mehr vorkommt. Vielmehr ergibt sich die Ganzzahligkeit auf dieselbe natürliche Art, wie etwa die Ganzzahligkeit der *Knotenzahl* einer schwingenden Saite. Die neue Auffassung ist verallgemeinerungsfähig und rührt, wie ich glaube, sehr tief an das wahre Wesen der Quantenvorschriften.

Die übliche Form der letzteren knüpft an die Hamiltonsche partielle Differentialgleichung an:

$$H\left(q, \frac{\partial S}{\partial q}\right) = E \,. \tag{1}$$

Es wird von dieser Gleichung eine Lösung gesucht, welche sich darstellt als *Summe* von Funktionen je einer einzigen der unabhängigen Variablen q.

Wir führen nun für S eine neue unbekannte ψ ein derart, daß ψ als ein *Produkt* von eingriffigen Funktionen der einzelnen Koordinaten erscheinen würde. D. h. wir setzen

$$S = K \lg \psi \,. \tag{2}$$

Die Konstante K muß aus dimensionellen Gründen eingeführt werden, sie hat die Dimension einer *Wirkung*. Damit erhält man

$$H\left(q, \frac{K}{\psi}\frac{\partial \psi}{\partial q}\right) = E \,. \tag{1'}$$

Wir suchen nun *nicht* eine Lösung der Gleichung (1'), sondern wir stellen folgende Forderung. Gleichung (1') läßt sich bei Vernachlässigung der Massenveränderlichkeit stets, bei Berücksichtigung derselben wenigstens dann, wenn es sich um das *Ein*elektronenproblem handelt, auf die Gestalt bringen: quadratische

Reprinted from Annalen der Physik, Series 4, Vol. 79, No. 4, pp. 361–376 (1926).

Form von ψ und seinen ersten Ableitungen $= 0$. Wir suchen solche reelle im ganzen Konfigurationenraum eindeutige endliche und zweimal stetig differenzierbare Funktionen ψ, welche das über den ganzen Konfigurationenraum erstreckte Integral der eben genannten quadratischen Form[1]) zu einem *Extremum* machen. *Durch dieses Variationsproblem ersetzen wir die Quantenbedingungen.*

Wir werden für H zunächst die Hamiltonsche Funktion der Keplerbewegung nehmen und zeigen, daß die aufgestellte Forderung für *alle positiven,* aber nur für eine *diskrete Schar von negativen* E-Werten erfüllbar ist. D. h. das genannte Variationsproblem hat ein diskretes und ein kontinuierliches Eigenwertspektrum. Das diskrete Spektrum entspricht den Balmerschen Termen, das kontinuierliche den Energien der Hyperbelbahnen. Damit numerische Übereinstimmung bestehe, muß K den Wert $h/2\pi$ erhalten.

Da für die Aufstellung der Variationsgleichungen die Koordinatenwahl belanglos ist, wählen wir rechtwinkelige kartesische. Dann lautet (1′) in unserem Fall (e, m sind Ladung und Masse des Elektrons):

$$(1'') \qquad \left(\frac{\partial\psi}{\partial x}\right)^2 + \left(\frac{\partial\psi}{\partial y}\right)^2 + \left(\frac{\partial\psi}{\partial z}\right)^2 - \frac{2m}{K^2}\left(E + \frac{e^2}{r}\right)\psi^2 = 0 \; .$$

$$r = \sqrt{x^2 + y^2 + z^2} \; .$$

Und unser Variationsproblem lautet

$$(3) \qquad \begin{cases} \delta J = \delta \iiint dx\, dy\, dz \left[\left(\frac{\partial\psi}{\partial x}\right)^2 + \left(\frac{\partial\psi}{\partial y}\right)^2 + \left(\frac{\partial\psi}{\partial z}\right)^2 - \right. \\ \qquad \left. - \frac{2m}{K^2}\left(E + \frac{e^2}{r}\right)\psi^2\right] = 0 \; , \end{cases}$$

das Integral erstreckt über den ganzen Raum. Man findet daraus in gewohnter Weise

$$(4) \qquad \begin{cases} \frac{1}{2}\delta J = \int df\, \delta\psi \frac{\partial\psi}{\partial n} - \iiint dx\, dy\, dz\, \delta\psi \left[\Delta\psi + \right. \\ \qquad \left. + \frac{2m}{K^2}\left(E + \frac{e^2}{r}\right)\psi\right] = 0 \; . \end{cases}$$

Es muß also erstens

$$(5) \qquad \Delta\psi + \frac{2m}{K^2}\left(E + \frac{e^2}{r}\right)\psi = 0$$

---

1) Es entgeht mir nicht, daß diese Formulierung nicht ganz eindeutig ist.

und zweitens muß das über die unendlich ferne geschlossene Oberfläche zu erstreckende Integral

$$\int df\, \delta\psi \frac{\partial\psi}{\partial n} = 0 \quad . \tag{6}$$

(Es wird sich herausstellen, daß wir wegen dieser letzteren Forderung unser Variationsproblem noch durch eine Forderung über das Verhalten von $\delta\psi$ im Unendlichen zu ergänzen haben, damit auch das oben behauptete *kontinuierliche* Eigenwertspektrum wirklich existiere. Doch davon später.)

Die Lösung von (5) läßt sich (*zum Beispiel*) in räumlichen Polarkoordinaten r, ϑ, φ bewerkstelligen, indem man ψ als *Produkt* je einer Funktion von r, von ϑ, von φ ansetzt. Die Methode ist sattsam bekannt. Für die Abhängigkeit von den Polarwinkeln ergibt sich eine *Kugelflächenfunktion* für die Abhängigkeit von r — die Funktion wollen wir χ nennen — erhält man leicht die Differentialgleichung:

$$\frac{d^2\chi}{dr^2} + \frac{2}{r}\frac{d\chi}{dr} + \left(\frac{2mE}{K^2} + \frac{2me^2}{K^2 r} - \frac{n(n+1)}{r^2}\right)\chi = 0 \quad . \tag{7}$$
$$n = 0,\ 1,\ 2,\ 3 \ldots .$$

Die Beschränkung von n auf ganze Zahlen ist bekanntlich *notwendig*, damit die Abhängigkeit von den Polarwinkeln *eindeutig* werde. — Wir benötigen Lösungen von (7), die für alle nichtnegativen reellen r-Werte endlich bleiben. Nun hat[1]) die Gleichung (7) in der komplexen r-Ebene *zwei* Singularitäten, bei $r = 0$ und $r = \infty$, von denen die zweite eine „Stelle der Unbestimmtheit" (wesentlich singuläre Stelle) *aller* Integrale ist, die erste hingegen nicht (für kein Integral). Diese beiden Singularitäten bilden gerade die *Randpunkte unseres reellen Intervalls*. In einem solchen Falle weiß man nun, daß die Forderung des *Endlichbleibens* in den Randpunkten für die Funktion χ einer *Randbedingung* gleichkommt. Die Gleichung hat *im allgemeinen* überhaupt kein Integral, das in *beiden* Randpunkten endlich bleibt, sondern ein solches Integral existiert

1) Für die Anleitung zur Behandlung der Gleichung (7) bin ich Hermann Weyl zu größtem Dank verpflichtet. Ich verweise für die im folgenden nicht bewiesenen Behauptungen auf L. Schlesinger, Differentialgleichungen (Sammlung Schubert Nr. 13, Göschen 1900, besonders Kap. 3 und 5.)

nur für gewisse ausgezeichnete Werte der in der Gleichung auftretenden Konstanten. Diese ausgezeichneten Werte gilt es zu bestimmen.

Der eben hervorgehobene Sachverhalt ist der *springende Punkt* in der ganzen Untersuchung.

Wir betrachten zunächst die singuläre Stelle $r = 0$. Die sogenannte *determinierende Fundamentalgleichung*, welche das Verhalten der Integrale an dieser Stelle bestimmt, ist

$$\varrho(\varrho - 1) + 2\varrho - n(n+1) = 0 \tag{8}$$

mit den Wurzeln

$$\varrho_1 = n, \quad \varrho_2 = -(n+1) . \tag{8'}$$

Die beiden kanonischen Integrale an dieser Stelle gehören also zu den Exponenten n und $-(n+1)$. Von ihnen ist, da n nicht negativ ist, nur das erste für uns brauchbar. Es wird, da es zu dem *größeren* Exponenten gehört, durch eine gewöhnliche Potenzreihe dargestellt, die mit r^n beginnt. (Das andere Integral, das uns nicht interessiert, kann, wegen der ganzzahligen Differenz zwischen den Exponenten, einen Logarithmus enthalten.) Da der nächste singuläre Punkt erst im Unendlichen liegt, konvergiert die genannte Potenzreihe beständig und stellt eine *Ganze Transzendente* dar. Wir stellen also fest:

Die gesuchte Lösung ist eine (bis auf einen belanglosen konstanten Faktor) eindeutig bestimmte Ganze Transzendente, die bei $r = 0$ *zum Exponenten* n *gehört.*

Es handelt sich jetzt darum, das Verhalten dieser Funktion im *Unendlichen* der positiven reellen Achse zu untersuchen. Dazu vereinfachen wir die Gleichung (7) durch die Substitution

$$\chi = r^{\alpha} U , \tag{9}$$

worin α so gewählt wird, daß das Glied mit $1/r^2$ fortfällt. Dazu muß α einen der beiden Werte n, $-(n+1)$ erhalten, wie man leicht nachrechnet. Gleichung (7) nimmt dann die Form an:

$$\frac{d^2 U}{d r^2} + \frac{2(\alpha + 1)}{r} \frac{d U}{d r} + \frac{2m}{K^2}\left(E + \frac{e^2}{r}\right) U = 0 . \tag{7'}$$

Ihre Integrale gehören bei $r = 0$ zu den Exponenten 0 und $-2\alpha - 1$. Für den ersten α-Wert, $\alpha = n$, ist das *erste*, für den zweiten α-Wert, $\alpha = -(n+1)$, ist das *zweite* dieser Integrale eine Ganze Transzendente und führt nach (9) auf die *gesuchte* Lösung, die ja eindeutig ist. Wir verlieren also nichts, wenn wir uns auf *einen* der beiden α-Werte beschränken. Wir wählen

$$\alpha = n\,. \tag{10}$$

Unsere Lösung U gehört dann also bei $r = 0$ zum Exponenten 0. Gleichung (7′) bezeichnen die Mathematiker als Laplacesche Gleichung. Der allgemeine Typus ist

$$U'' + \left(\delta_0 + \frac{\delta_1}{r}\right) U' + \left(\varepsilon_0 + \frac{\varepsilon_1}{r}\right) U = 0\,. \tag{7''}$$

Bei uns haben die Konstanten die Werte

$$\delta_0 = 0\,,\quad \delta_1 = 2(\alpha+1)\,,\quad \varepsilon_0 = \frac{2mE}{K^2}\,,\quad \varepsilon_1 = \frac{2me^2}{K^2}\,. \tag{11}$$

Dieser Gleichungstypus ist aus *dem* Grunde verhältnismäßig einfach zu behandeln, weil die sogenannte Laplacesche Transformation, die im allgemeinen *wieder* eine Gleichung *zweiter* Ordnung ergibt, *hier* auf die *erste* Ordnung führt, die durch Quadraturen lösbar ist. Dies gestattet eine Darstellung der Lösungen von (7″) selbst durch Integrale im Komplexen. Ich führe hier nur das Endergebnis an.[1]) Das Integral

$$U = \int_L e^{zr}(z-c_1)^{\alpha_1 - 1}(z-c_2)^{\alpha_2 - 1}\,dz \tag{12}$$

ist eine Lösung von (7″) für einen Integrationsweg L, für den

$$\int_L \frac{d}{dz}\left[e^{zr}(z-c_1)^{\alpha_1}(z-c_2)^{\alpha_2}\right] dz = 0\,. \tag{13}$$

Die Konstanten c_1, c_2, α_1, α_2 haben folgende Werte. c_1 und c_2 sind die Wurzeln der quadratischen Gleichung

$$z^2 + \delta_0 z + \varepsilon_0 = 0 \tag{14}$$

und

$$\alpha_1 = \frac{\varepsilon_1 + \delta_1 c_1}{c_1 - c_2}\,,\quad \alpha_2 = \frac{\varepsilon_1 + \delta_1 c_2}{c_2 - c_1}\,. \tag{14'}$$

---

1) Vgl. L. Schlesinger, a. a. O. Die Theorie verdankt man H. Poincaré und J. Horn.

Im Falle der Gleichung (7′) wird also nach (11) und (10)

$$(14'')\quad \begin{cases} c_1 = +\sqrt{\dfrac{-2mE}{K^2}}\,, \quad c_2 = -\sqrt{\dfrac{-2mE}{K^2}}\;; \\ \alpha_1 = \dfrac{me^2}{K\underset{+}{\sqrt{-2mE}}} + n + 1\,, \quad \alpha_2 = -\dfrac{me^2}{K\underset{+}{\sqrt{-2mE}}} + n + 1. \end{cases}$$

Die Integraldarstellung (12) gestattet nicht nur, das asymptotische Verhalten der Gesamtheit von Lösungen, wenn r in bestimmter Weise ins Unendliche geht, zu überblicken, sondern auch, dieses Verhalten für eine *bestimmte* Lösung anzugeben, was immer viel schwieriger ist.

Wir wollen nun zunächst den Fall, daß α_1 und α_2 reelle ganze Zahlen sind, *ausschließen*. Der Fall tritt, wenn er eintritt, stets für beide Größen gleichzeitig ein und zwar dann und nur dann, wenn

$$(15)\qquad \frac{me^2}{K\underset{+}{\sqrt{-2mE}}} = \text{reelle ganze Zahl}\;.$$

Wir nehmen also jetzt an, daß (15) nicht erfüllt ist.

Das Verhalten der Gesamtheit von Lösungen für eine bestimmte Art des Unendlichwerdens von r — wir wollen stets denken für realpositives Unendlichwerden — wird alsdann[1]) charakterisiert durch das Verhalten der beiden linear unabhängigen Lösungen, die durch folgende *zwei Spezialisierungen* des Integrationsweges L erhalten werden, und die wir U_1 und U_2 nennen wollen. *Beide* Male komme z aus dem Unendlichen und gehe auf demselben Wege dorthin zurück, und zwar in solcher Richtung, daß

$$(16)\qquad \lim_{z=\infty} e^{zr} = 0\;,$$

d. h. der Realteil von zr soll negativ unendlich werden. Hierdurch wird der Bedingung (13) genügt. *Dazwischen* werde im *einen* Falle (Lösung U_1) die Stelle c_1, im *anderen* Falle (Lösung U_2) die Stelle c_2 je einmal umlaufen.

Diese beiden Lösungen werden nun für sehr große realpositive r-Werte *asymptotisch* (im Sinne Poincarés) dargestellt durch

1) Wenn (15) erfüllt ist, wird mindestens der eine von den beiden im Text beschriebenen Integrationswegen unbrauchbar, da er ein verschwindendes Ergebnis liefert.

$$(17) \quad \begin{cases} U_1 \sim e^{c_1 r} r^{-\alpha_1} (-1)^{\alpha_1} (e^{2\pi i \alpha_1} - 1) \Gamma(\alpha_1)(c_1 - c_2)^{\alpha_2 - 1}, \\ U_2 \sim e^{c_2 r} r^{-\alpha_2} (-1)^{\alpha_2} (e^{2\pi i \alpha_2} - 1) \Gamma(\alpha_2)(c_2 - c_1)^{\alpha_1 - 1}, \end{cases}$$

wobei wir uns hier mit dem ersten Glied der nach ganzen negativen Potenzen von r fortschreitenden asymptotischen Reihen begnügen.

Wir haben nun die beiden Fälle $E \gtrless 0$ zu unterscheiden. Sei zunächst

1. $\underline{E > 0.}$ Wir bemerken erstens, daß hierdurch das Nichtzutreffen von (15) eo ipso gewährleistet ist, weil diese Größe reinimaginär wird. Ferner werden nach (14″) auch c_1 und c_2 reinimaginär. Die Exponentialfunktionen in (17) sind also, da r reell ist, endlichbleibende periodische Funktionen. Die Werte von α_1 und α_2 nach (14″) zeigen, daß U_1 und U_2 *beide* wie r^{-n-1} gegen Null gehen. *Dasselbe muß also von unserer ganzen transzendenten Lösung U gelten*, deren Verhalten wir suchen, *wie immer sie sich aus U_1 und U_2 linear zusammensetzen möge.* Ferner lehrt (9) mit Beachtung von (10), daß die Funktion χ, d. i. die ganze transzendente Lösung der *ursprünglich vorliegenden* Gleichung (7), immer noch wie $1/r$ gegen Null geht, da sie aus U durch Multiplikation mit r^n entsteht. Wir können also aussprechen:

Die Eulersche Differenzialgleichung (5) unseres Variationsproblems hat für jedes positive E Lösungen, die im ganzen Raum eindeutig endlich und stetig sind und im Unendlichen unter beständigen Oszillationen wie $1/r$ gegen Null gehen. — Von der Oberflächenbedingung (6) wird noch zu sprechen sein.

2. $\underline{E < 0.}$ In diesem Fall ist die Möglichkeit (15) nicht *eo ipso* ausgeschlossen, doch halten wir vorläufig an ihrem verabredeten Ausschluß fest. Dann wächst nach (14″) und (17) U_1 für $r = \infty$ über alle Grenzen, U_2 dagegen verschwindet *exponentiell.* Unsere Ganze Transzendente U (und dasselbe gilt von χ) wird also dann und nur dann endlich bleiben, wenn U mit U_2 bis auf einen numerischen Faktor identisch ist. *Das ist aber nicht der Fall.* Man erkennt das so: wählt man in (12) für den Integrationsweg L einen *geschlossenen* Umlauf um *beide* Punkte c_1 und c_2, welcher Umlauf wegen der Ganzzahligkeit der *Summe* $\alpha_1 + \alpha_2$ dann *wirklich* auf der Riemannschen Fläche des Integranden *geschlossen* ist, mithin eo ipso der

Bedingung (13) genügt, so läßt sich leicht zeigen, daß das Integral (12) alsdann *unsere Ganze Transzendente* U darstellt. Es läßt sich nämlich in eine Reihe nach positiven Potenzen von r entwickeln, die jedenfalls für hinreichend kleine r konvergiert, daher der Differentialgleichung (7′) genügt, daher mit derjenigen für U zusammenfallen muß. Also: U wird durch (12) dargestellt, wenn L ein geschlossener Umlauf um beide Punkte, c_1 und c_2, ist. Dieser geschlossene Umlauf läßt sich aber so verzerren, daß er aus den beiden früher betrachteten Integrationswegen, die zu U_1 und U_2 gehören, *additiv kombiniert* erscheint und zwar *mit nicht verschwindenden Faktoren*, etwa 1 und $e^{2\pi i \alpha_1}$. Daher kann U nicht mit U_2 übereinstimmen, sondern muß auch U_1 enthalten. W. z. b. w.

Unsere Ganze Transzendente U, die unter den Lösungen von (7′) allein für die Problemlösung in Betracht kommt, bleibt also unter den gemachten Voraussetzungen für große r *nicht* endlich. — Vorbehaltlich der *Vollständigkeits*untersuchung, d. h. des Nachweises, daß unser Verfahren *alle* linear unabhängigen Problemlösungen finden läßt, dürfen wir also aussprechen:

Für negative E, welche der Bedingung (15) nicht genügen, hat unser Variationsproblem keine Lösung.

Wir haben jetzt nur noch diejenige diskrete Schar von negativen E-Werten zu untersuchen, welche *der Bedingung (15) genügen.* α_1 und α_2 sind dann also beide ganzzahlig. Von den zwei Integrationswegen, welche uns früher das Fundamentalsystem U_1, U_2 geliefert haben, muß sicherlich der erste abgeändert werden, um Nichtverschwindendes zu liefern. Denn $\alpha_1 - 1$ ist sicherlich positiv, die Stelle c_1 ist also jetzt weder ein Verzweigungspunkt noch ein Pol des Integranden, sondern eine gewöhnliche Nullstelle. Es kann auch die Stelle c_2 regulär werden, wenn nämlich auch $\alpha_2 - 1$ nicht negativ ist. In *jedem* Falle aber lassen sich zwei passende Integrationswege leicht angeben und die Integration auf ihnen sogar in geschlossener Form durch bekannte Funktionen ausführen, so daß das Verhalten der Lösungen vollkommen überblickt wird.

Sei nämlich

$$\frac{m e^2}{K\sqrt{-2mE}} = l\,; \quad l = 1,\ 2,\ 3,\ 4\,\ldots. \tag{15'}$$

Dann ist nach (14″)

$$(14''')\qquad \alpha_1 - 1 = l + n\,,\quad \alpha_2 - 1 = -l + n\,.$$

Man hat nun die beiden Fälle zu unterscheiden $l \leqslant n$ und $l > n$. Sei zunächst

a) $l \leqslant n$. Dann verlieren c_2 *und* c_1 jedweden singulären Charakter, gewinnen aber dafür die Eignung, als Anfangs- oder Endpunkte des Integrationsweges zu fungieren, behufs Erfüllung der Bedingung (13). Ein dritter hiefür geeigneter Punkt ist das Negativreellunendliche. Jeder Weg zwischen zweien dieser drei Punkte liefert eine Lösung und von diesen drei Lösungen sind je zwei linear unabhängig, wie man leicht bestätigt, indem man die Integrale in geschlossener Form ausrechnet. Im besonderen wird die *ganze transzendente Lösung* durch den Integrationsweg von c_1 nach c_2 geliefert. Denn daß *dieses* Integral für $r = 0$ regulär bleibt, erkennt man sofort, ohne es auszurechnen. Ich betone das, weil die wirkliche Ausrechnung eher geeignet ist, diesen Sachverhalt zu verschleiern. Dagegen zeigt *sie*, daß das Integral für positiv unendlich große r über alle Grenzen wächst. *Endlich* bleibt für große r eines von den beiden *anderen* Integralen, das aber dafür für $r = 0$ unendlich wird.

Wir erhalten also im Falle $l \leqslant n$ *keine* Problemlösung.

b) $l > n$. Dann ist nach (14‴) c_1 eine Nullstelle, c_2 ein Pol mindestens erster Ordnung des Integranden. Zwei unabhängige Integrale werden dann geliefert: das eine durch den Weg, der von $z = -\infty$, vorsichtshalber mit Vermeidung des Pols, zur Nullstelle führt; das andere durch das *Residuum* im Pol. *Letzteres* ist die Ganze Transzendente. Wir wollen seinen ausgerechneten Wert angeben, aber gleich mit r^n multipliziert, wodurch wir nach (9) und (10) die Lösung χ der ursprünglich vorliegenden Gleichung (7) erhalten. (Die belanglose multiplikative Konstante ist willkürlich adjustiert.) Man findet:

$$(18)\;\chi = f\left(r\,\frac{\sqrt{-2mE}}{K}\right);\quad f(x) = x^n\, e^{-x} \sum_{k=0}^{l-n-1} \frac{(-2x)^k}{k!} \binom{l+n}{l-n-1-k}.$$

Man erkennt, daß dies wirklich eine brauchbare Lösung ist, da sie für alle reellen nichtnegativen r endlich bleibt. Durch ihr exponentielles Verschwinden im Unendlichen wird überdies die Oberflächenbedingung (6) verbürgt. Wir fassen die Resultate für negative E zusammen:

Bei negativem E hat unser Variationsproblem dann und nur dann Lösungen, wenn E der Bedingung (15) genügt. Der ganzen Zahl n, welche die Ordnung der in der Lösung auftretenden Kugelflächenfunktion angibt, dürfen dann immer nur Werte kleiner als l erteilt werden (wovon stets mindestens einer zur Verfügung steht). Der von r abhängige Teil der Lösung wird durch (18) gegeben.

Durch Abzählung der Konstanten in den Kugelflächenfunktionen (bekanntlich $2n + 1$) findet man ferner:

Die gefundene Lösung enthält für eine zulässige Kombination (n, l) genau $2n + 1$ willkürliche Konstanten; für einen vorgegebenen l-Wert also l^2 willkürliche Konstanten.

Wir haben damit die eingangs aufgestellten Behauptungen über das Eigenwertspektrum unseres Variationsproblems in den Hauptzügen bestätigt, immerhin bestehen noch Lücken.

Erstens der Nachweis der Vollständigkeit des *gesamten* nachgewiesenen Systems von Eigenfunktionen. Damit will ich mich in dieser Note nicht befassen. Nach anderweitigen Erfahrungen darf man vermuten, daß uns keine Eigenwerte entgangen sind.

Zweitens ist jetzt daran zu erinnern, daß die für positives E nachgewiesenen Eigenfunktionen nicht ohne weiteres das Variationsproblem in der Form, in der es anfangs gestellt wurde, lösen, weil sie im Unendlichen nur wie $1/r$, $\frac{\partial \psi}{\partial r}$ auf einer großen Kugel also nur wie $1/r^2$ gegen Null geht. Das Oberflächenintegral (6) bleibt daher gerade noch von der Ordnung des $\delta\psi$ im Unendlichen. Wünscht man also das kontinuierliche Spektrum wirklich mit zu erhalten, so muß man dem *Problem* noch eine Bedingung hinzufügen: etwa daß $\delta\psi$ im Unendlichen verschwinden, oder wenigstens, daß es einem konstanten Wert zustreben soll, unabhängig von der Richtung, in der man ins räumlich Unendliche geht; in letzterem Fall bringen die Kugelflächenfunktionen das Oberflächenintegral zum Verschwinden.

§ 2. Die Bedingung (15) ergibt

$$-E_l = \frac{m e^4}{2 K^2 l^2}\,. \tag{19}$$

Es ergeben sich also die wohlbekannten Bohrschen Energieniveaus, die den Balmertermen entsprechen, wenn man der Konstante K, die wir in (2) aus dimensionellen Gründen einführen mußten, den Wert erteilt

$$K = \frac{h}{2\pi}\,. \tag{20}$$

Dann wird ja

$$-E_l = \frac{2\pi^2 m e^4}{h^2 l^2}\,. \tag{19'}$$

Unser l ist die Hauptquantenzahl. $n+1$ hat Analogie mit der Azimutalquantenzahl, die weitere Aufspaltung dieser Zahl bei der näheren Bestimmung der Kugelflächenfunktionen kann mit der Aufspaltung des Azimutalquants in ein „äquatoriales“ und ein „polares“ Quant in Analogie gesetzt werden. Diese Zahlen bestimmen *hier* das System der Knotenlinien auf der Kugel. Auch die „radiale Quantenzahl,“ $l-n-1$ bestimmt genau die Zahl der „Knotenkugeln“, denn man kann sich leicht überzeugen, daß die Funktion $f(x)$ in (18) genau $l-n-1$ positive reelle Wurzeln hat. — Die positiven E-Werte entsprechen dem Kontinuum der hyperbolischen Bahnen, denen man in gewissem Sinn die radiale Quantenzahl ∞ zuschreiben kann. Dem entspricht, daß, wie wir gesehen haben, die betreffenden Lösungsfunktionen unter *beständigen* Oszillationen ins Unendliche hinaus laufen.

Von Interesse ist noch, daß der Bereich, innerhalb dessen die Funktionen (18) merklich von Null verschieden sind und innerhalb dessen sich ihre Oszillationen abspielen, jedenfalls von der *allgemeinen Größenordnung* der großen Achse der zugeordneten Ellipse ist. Der Faktor, mit dem multipliziert der Radiusvektor als Argument der konstantenfreien Funktion f auftritt, ist — selbstverständlich — das Reziproke einer Länge, und diese Länge ist

$$\frac{K}{\sqrt{-2mE}} = \frac{K^2 l}{m e^2} = \frac{h^2 l}{4\pi^2 m e^2} = \frac{a_l}{l}\,, \tag{21}$$

wo a_l die Halbachse der l-ten Ellipsenbahn. $\left(\text{Die Gleichungen folgen aus (19) zusammen mit der bekannten Beziehung } E_l = -\frac{e^2}{2a_l}\right)$.

Die Größe (21) gibt die Größenordnung des Wurzelbereiches für kleinzahliges l und n; denn dann darf angenommen werden, daß die Wurzeln von $f(x)$ von der Größenordnung Eins sind. Das ist natürlich nicht mehr der Fall, wenn die Koeffizienten des Polynoms große Zahlen sind. Ich möchte auf die genauere Abschätzung der Wurzeln jetzt nicht eingehen, glaube aber, daß obige Behauptung sich dabei ziemlich genau bestätigen wird.

§ 3. Es liegt natürlich sehr nahe, die Funktion ψ auf *einen Schwingungsvorgang* im Atom zu beziehen, dem die den Elektronenbahnen heute vielfach bezweifelte Realität in höherem Maße zukommt als ihnen. Ich hatte auch ursprünglich die Absicht, die neue Fassung der Quantenvorschrift in dieser mehr anschaulichen Art zu begründen, habe aber dann die obige neutral mathematische Form vorgezogen, weil sie das Wesentliche klarer zutage treten läßt. Als das Wesentliche erscheint mir, daß in der Quantenvorschrift nicht mehr die geheimnisvolle „Ganzzahligkeitsforderung“ auftritt, sondern diese ist sozusagen einen Schritt weiter zurückverfolgt: sie hat ihren Grund in der Endlichkeit und Eindeutigkeit einer gewissen Raumfunktion.

Ich möchte auch jetzt noch nicht näher auf die Erörterung der Vorstellungsmöglichkeiten über diesen Schwingungsvorgang eingehen, bevor etwas kompliziertere Fälle in der neuen Fassung mit Erfolg durchgerechnet sind. Es ist nicht ausgemacht, daß dieselbe in ihren Ergebnissen ein bloßer Abklatsch der üblichen Quantentheorie sein wird. Z. B. führt das relativistische Keplerproblem, wenn man es genau nach der eingangs gegebenen Vorschrift durchrechnet, merkwürdigerweise auf *halbzahlige Teil*quanten (Radial- und Azimutquant).

Immerhin seien zu der Schwingungsvorstellung hier noch einige Bemerkungen erlaubt. Vor allem möchte ich nicht unerwähnt lassen, daß ich die Anregung zu diesen Überlegungen in erster Linie den geistvollen Thèses des Hrn. Louis de Broglie[1]) verdanke und dem Nachdenken über die räumliche Verteilung jener „Phasenwellen“, von denen er gezeigt hat, daß ihrer stets eine *ganze Zahl*, entlang der Bahn gemessen, auf jede

1) L. de Broglie, Ann. de Physique (10) 3. S. 22. 1925 (Thèses, Paris 1924)

Periode oder Quasiperiode des Elektrons entfallen. Der Hauptunterschied ist, daß de Broglie an fortschreitende Wellen denkt, während wir, wenn wir unseren Formeln die Schwingungsvorstellung unterlegen, auf stehende Eigenschwingungen geführt werden. Ich habe kürzlich[1]) gezeigt, daß man die Einsteinsche Gastheorie auf die Betrachtung solcher stehender Eigenschwingungen, für welche man das Dispersionsgesetz der de Broglieschen Phasenwellen ansetzt, gründen kann. Die obigen Betrachtungen für das Atom hätten sich als Verallgemeinerung jener Überlegungen am Gasmodell darstellen lassen.

Faßt man die einzelnen Funktionen (18), multipliziert mit einer Kugelflächenfunktion der Ordnung n, als die Beschreibung von Eigenschwingungsvorgängen auf, dann muß die Größe E etwas mit der *Frequenz* des betreffenden Vorganges zu tun haben. Nun ist man gewöhnt, daß bei Schwingungsproblemen der „Parameter" (gewöhnlich λ genannt) dem *Quadrat* der Frequenz proportional ist. Aber erstens würde ein solcher Ansatz im vorliegenden Fall gerade für die *negativen* E-Werte zu *imaginären* Frequenzen führen, zweitens sagt dem Quantentheoretiker sein Gefühl, daß die Energie der Frequenz selbst und nicht ihrem Quadrat proportional sein muß.

Der Widerspruch löst sich folgendermaßen. Für den „Parameter" E der Variationsgleichung (5) ist ja vorläufig *kein natürliches Nullniveau* festgelegt, besonders da die unbekannte Funktion ψ außer mit E noch mit einer Funktion von r multipliziert erscheint, die, unter entsprechender Änderung des Nullniveaus von E, um eine Konstante abgeändert werden kann. Folglich ist die „Erwartung des Schwingungstheoretikers" dahin zu berichtigen, daß nicht E selbst — das, was wir bisher so nannten und auch weiter so nennen wollen — sondern E vermehrt um eine gewisse Konstante dem Quadrat der Frequenz proportional erwartet wird. Sei diese Konstante nun *sehr groß* im Vergleich zu den Beträgen aller vorkommenden negativen E-Werte [die ja durch (15) beschränkt sind]. Dann werden erstens die Frequenzen *reell,* zweitens aber werden unsere E-Werte, da sie nur relativ kleinen Frequenz*unterschieden* entsprechen, tatsächlich sehr angenähert diesen Frequenzunter-

1) Erscheint demnächst in der Physik. Zeitschr.

schieden proportional. Das hinwiederum ist alles, was das „natürliche Gefühl" des Quantentheoretikers verlangen kann, solange das Nullniveau der *Energie* nicht festgelegt ist.

Die Auffassung, daß die Frequenz des Schwingungsvorganges etwa durch

$$\nu = C' \sqrt{C + E} = C' \sqrt{C} + \frac{C'}{2\sqrt{C}} E + \ldots \tag{22}$$

gegeben sei, wo C eine gegen alle E sehr große Konstante, hat aber noch einen anderen sehr schätzenswerten Vorzug. *Sie vermittelt ein Verständnis für die Bohrsche Frequenzbedingung.* Nach der letzteren sind doch die *Emissionsfrequenzen* den *E-Differenzen* proportional, also nach (22) auch den Differenzen der Eigenfrequenzen ν jener hypothetischen Schwingungsvorgänge. Und zwar sind die Eigenfrequenzen alle sehr groß gegen die Emissionsfrequenzen, stimmen unter sich nahe überein. Die Emissionsfrequenzen erscheinen demnach als tiefe „Differenztöne" der mit viel höherer Frequenz erfolgenden Eigenschwingungen selbst. Daß beim Hinüberwandern der Energie aus der einen in die andere Normalschwingung *irgendetwas* — ich meine die Lichtwelle — in Erscheinung tritt, dem als *Frequenz* jene Frequenz*differenz* zukommt, ist sehr verständlich; man braucht sich nur vorzustellen, daß die Lichtwelle ursächlich verknüpft ist mit den während des Überganges an jeder Raumstelle notwendig auftretenden *Schwebungen* und daß die Frequenz des Lichtes bestimmt wird durch die Häufigkeit, mit der das Intensitätsmaximum des Schwebungsvorganges pro Sekunde wiederkehrt.

Es mag Bedenken erregen, daß diese Schlüsse sich auf die Beziehung (22) in ihrer *näherungsweisen* Gestalt (nach Entwicklung der Quadratwurzel) gründen, wodurch die Bohrsche Frequenzbedingung selbst scheinbar den Charakter einer Näherungsformel erhält. Das ist aber nur scheinbar und wird völlig vermieden, wenn man die *relativistische* Theorie entwickelt, durch welche überhaupt erst ein tieferes Verständnis vermittelt wird. Die große additive Konstante C hängt natürlich aufs innigste zusammen mit der Ruhenergie $m c^2$ des Elektrons. Auch das scheinbar *nochmalige* und *unabhängige* Auftreten der Konstante h [die doch schon durch (20) eingeführt war] in der Frequenzbedingung wird durch die relativistische Theorie aufgeklärt bzw. vermieden. Aber leider begegnet ihre einwandfreie Durchführung vorläufig noch gewissen, oben berührten Schwierigkeiten.

Es ist kaum nötig, hervorzuheben, um wie vieles sympathischer die Vorstellung sein würde, daß bei einem Quantenübergang die Energie aus einer Schwingungsform in eine andere übergeht, als die Vorstellung von den springenden Elektronen. Die Änderung der Schwingungsform kann sich stetig in Raum und Zeit vollziehen, sie kann gern solange dauern, als erfahrungsgemäß (Kanalstrahlversuche von W. Wien) der Emissionsprozeß dauert; und gleichwohl werden, wenn während dieses Übergangs das Atom für verhältnismäßig kurze Zeit einem elektrischen Feld ausgesetzt wird, das die Eigenfrequenzen verstimmt, die Schwebungsfrequenzen sogleich mitverstimmt werden, und zwar gerade nur solange als das Feld einwirkt. Diese experimentell festgestellte Tatsache bereitet bekanntlich dem Verständnis bisher die größten Schwierigkeiten, man vergleiche etwa die Diskussion in dem bekannten Lösungsversuch von Bohr-Kramers-Slater.

Im übrigen darf man in der Freude über das menschliche Näherrücken all dieser Dinge doch nicht vergessen, daß die Vorstellung, das Atom schwinge, wenn es nicht strahlt, jeweilig in Form *einer* Eigenschwingung, daß, sage ich, diese Vorstellung, *wenn* sie festgehalten werden muß, sich doch immer noch sehr stark von dem *natürlichen* Bild eines schwingenden Systems entfernt. Denn ein makroskopisches System verhält sich ja bekanntlich nicht so, sondern liefert im allgemeinen ein Potpourri seiner Eigenschwingungen. Man darf aber seine Ansicht in diesem Punkt nicht voreilig festlegen. Auch ein Potpourri von Eigenschwingungen am einzelnen Atom würde nichts verschlagen, sofern dabei ja auch keine anderen Schwebungsfrequenzen auftreten als diejenigen, zu deren Emission das Atom erfahrungsgemäß *unter Umständen* befähigt ist. Auch die gleichzeitige wirkliche Aussendung vieler von diesen Spektrallinien durch dasselbe Atom widerspricht keiner Erfahrung. Man könnte sich also gut denken, daß bloß im Normalzustand (und näherungsweise in gewissen „metastabilen" Zuständen) das Atom mit *einer* Eigenfrequenz schwingt und eben aus diesem Grunde *nicht* strahlt, weil nämlich keine Schwebungen auftreten.

Die *Anregung* bestünde in einer gleichzeitigen Erregung einer oder mehrerer anderer Eigenfrequenzen, wodurch dann Schwebungen entstehen, welche die Lichtemission hervorrufen.

Unter allen Umständen möchte ich glauben, daß die zur *gleichen* Frequenz gehörigen Eigenfunktionen im allgemeinen alle gleichzeitig angeregt sind. Mehrfachheit der Eigenwerte entspricht nämlich in der Sprache der bisherigen Theorie der *Entartung*. Der Reduktion der Quantisierung entarteter Systeme dürfte die willkürliche Aufteilung der Energie auf die zu *einem* Eigenwert gehörigen Eigenfunktionen entsprechen.

Zusatz bei der Korrektur am 28. II. 1926.

Für den Fall der klassischen Mechanik konservativer Systeme läßt sich die Variationsaufgabe schöner, als eingangs geschehen, ohne ausdrückliche Beziehung auf die Hamiltonsche partielle Differentialgleichung folgendermaßen formulieren. Sei $T(q, p)$ die kinetische Energie als Funktion der Koordinaten und Impulse, V die potentielle Energie, $d\tau$ das Volumelement des Konfigurationenraums „rationell gemessen“, d. h. nicht einfach das Produkt $dq_1, dq_2 \ldots dq_n$, sondern noch dividiert durch die Quadratwurzel aus der Diskriminante der quadratischen Form $T(q, p)$. (Vgl. Gibbs, Statistische Mechanik.) Dann soll ψ das „Hamiltonsche Integral“

$$\int d\tau \left\{ K^2 T\left(q, \frac{\partial \psi}{\partial q}\right) + \psi^2 V \right\} \tag{23}$$

stationär machen unter der *normierenden Nebenbedingung*

$$\int \psi^2 d\tau = 1 . \tag{24}$$

Die *Eigenwerte* dieses Variationsproblems sind bekanntlich *die stationären Werte* des Integrals (23) und liefern nach unserer These *die Quantenniveaus der Energie*.

Zu (14″) sei noch bemerkt, daß man in der Größe α_2 im wesentlichen den bekannten Sommerfeldschen Ausdruck $-\frac{B}{\sqrt{A}} + \sqrt{C}$ vor sich hat (vgl. „Atombau“, 4. Aufl., S. 775).

Zürich, Physikalisches Institut der Universität.

(Eingegangen 27. Januar 1926.)

Druck von Metzger & Wittig in Leipzig.

610

Reprint 8

The Quantum Theory of the Electron.

By P. A. M. Dirac, St. John's College, Cambridge.

(Communicated by R. H. Fowler, F.R.S.—Received January 2, 1928.)

The new quantum mechanics, when applied to the problem of the structure of the atom with point-charge electrons, does not give results in agreement with experiment. The discrepancies consist of " duplexity " phenomena, the observed number of stationary states for an electron in an atom being twice the number given by the theory. To meet the difficulty, Goudsmit and Uhlenbeck have introduced the idea of an electron with a spin angular momentum of half a quantum and a magnetic moment of one Bohr magneton. This model for the electron has been fitted into the new mechanics by Pauli,* and Darwin,† working with an equivalent theory, has shown that it gives results in agreement with experiment for hydrogen-like spectra to the first order of accuracy.

The question remains as to why Nature should have chosen this particular model for the electron instead of being satisfied with the point-charge. One would like to find some incompleteness in the previous methods of applying quantum mechanics to the point-charge electron such that, when removed, the whole of the duplexity phenomena follow without arbitrary assumptions. In the present paper it is shown that this is the case, the incompleteness of the previous theories lying in their disagreement with relativity, or, alternatetively, with the general transformation theory of quantum mechanics. It appears that the simplest Hamiltonian for a point-charge electron satisfying the requirements of both relativity and the general transformation theory leads to an explanation of all duplexity phenomena without further assumption. All the same there is a great deal of truth in the spinning electron model, at least as a first approximation. The most important failure of the model seems to be that the magnitude of the resultant orbital angular momentum of an electron moving in an orbit in a central field of force is not a constant, as the model leads one to expect.

* Pauli, ' Z. f. Physik,' vol. 43, p. 601 (1927).

† Darwin, ' Roy. Soc. Proc.,' A, vol. 116, p. 227 (1927).

Reprinted from Proceedings of the Royal Society of London, Series A, Vol. 117, No. 778, pp. 610–624 (1928).

§ 1. *Previous Relativity Treatments.*

The relativity Hamiltonian according to the classical theory for a point electron moving in an arbitrary electro-magnetic field with scalar potential A_0 and vector potential $\mathbf{A}$ is

$$F \equiv \left(\frac{W}{c} + \frac{e}{c}A_0\right)^2 + \left(\mathbf{p} + \frac{e}{c}\mathbf{A}\right)^2 + m^2c^2,$$

where $\mathbf{p}$ is the momentum vector. It has been suggested by Gordon* that the operator of the wave equation of the quantum theory should be obtained from this F by the same procedure as in non-relativity theory, namely, by putting

$$W = ih\frac{\partial}{\partial t},$$

$$p_r = -ih\frac{\partial}{\partial x_r}, \qquad r = 1, 2, 3,$$

in it. This gives the wave equation

$$F\psi \equiv \left[\left(ih\frac{\partial}{c\,\partial t} + \frac{e}{c}A_0\right)^2 + \Sigma_r\left(-ih\frac{\partial}{\partial x_r} + \frac{e}{c}A_r\right)^2 + m^2c^2\right]\psi = 0, \qquad (1)$$

the wave function ψ being a function of x_1, x_2, x_3, t. This gives rise to two difficulties.

The first is in connection with the physical interpretation of ψ. Gordon, and also independently Klein,† from considerations of the conservation theorems, make the assumption that if ψ_m, ψ_n are two solutions

$$\rho_{mn} = -\frac{e}{2mc^2}\left\{ih\left(\psi_m\frac{\partial\bar{\psi}_n}{\partial t} - \bar{\psi}_n\frac{\partial\psi_m}{\partial t}\right) + 2eA_0\psi_m\bar{\psi}_n\right\}$$

and

$$\mathbf{I}_{mn} = -\frac{e}{2m}\left\{-ih\left(\psi_m \operatorname{grad}\bar{\psi}_n - \bar{\psi}_n \operatorname{grad}\psi_m\right) + 2\frac{e}{c}\mathbf{A}_m\psi_m\bar{\psi}_n\right\}$$

are to be interpreted as the charge and current associated with the transition $m \to n$. This appears to be satisfactory so far as emission and absorption of radiation are concerned, but is not so general as the interpretation of the non-relativity quantum mechanics, which has been developed‡ sufficiently to enable one to answer the question: What is the probability of any dynamical variable

* Gordon, 'Z. f. Physik,' vol. 40, p. 117 (1926).

† Klein, 'Z. f. Physik,' vol. 41, p. 407 (1927).

‡ Jordan, 'Z. f. Physik,' vol. 40, p. 809 (1927); Dirac, 'Roy. Soc. Proc.,' A, vol. 113, p. 621 (1927).

at any specified time having a value lying between any specified limits, when the system is represented by a given wave function ψ_n? The Gordon-Klein interpretation can answer such questions if they refer to the position of the electron (by the use of ρ_{nn}), but not if they refer to its momentum, or angular momentum or any other dynamical variable. We should expect the interpretation of the relativity theory to be just as general as that of the non-relativity theory.

The general interpretation of non-relativity quantum mechanics is based on the transformation theory, and is made possible by the wave equation being of the form

$$(\mathrm{H} - \mathrm{W})\,\psi = 0, \tag{2}$$

i.e., being linear in W or $\partial/\partial t$, so that the wave function at any time determines the wave function at any later time. The wave equation of the relativity theory must also be linear in W if the general interpretation is to be possible.

The second difficulty in Gordon's interpretation arises from the fact that if one takes the conjugate imaginary of equation (1), one gets

$$\left[\left(-\frac{\mathrm{W}}{c} + \frac{e}{c}\mathrm{A}_0\right)^2 + \left(-\mathbf{p} + \frac{e}{c}\mathbf{A}\right)^2 + m^2c^2\right]\psi = 0,$$

which is the same as one would get if one put $-e$ for e. The wave equation (1) thus refers equally well to an electron with charge e as to one with charge $-e$. If one considers for definiteness the limiting case of large quantum numbers one would find that some of the solutions of the wave equation are wave packets moving in the way a particle of charge $-e$ would move on the classical theory, while others are wave packets moving in the way a particle of charge e would move classically. For this second class of solutions W has a negative value. One gets over the difficulty on the classical theory by arbitrarily excluding those solutions that have a negative W. One cannot do this on the quantum theory, since in general a perturbation will cause transitions from states with W positive to states with W negative. Such a transition would appear experimentally as the electron suddenly changing its charge from $-e$ to e, a phenomenon which has not been observed. The true relativity wave equation should thus be such that its solutions split up into two non-combining sets, referring respectively to the charge $-e$ and the charge e.

In the present paper we shall be concerned only with the removal of the first of these two difficulties. The resulting theory is therefore still only an approximation, but it appears to be good enough to account for all the duplexity phenomena without arbitrary assumptions.

§ 2. *The Hamiltonian for No Field.*

Our problem is to obtain a wave equation of the form (2) which shall be invariant under a Lorentz transformation and shall be equivalent to (1) in the limit of large quantum numbers. We shall consider first the case of no field, when equation (1) reduces to

$$(-p_0^2 + \mathbf{p}^2 + m^2c^2)\,\psi = 0 \tag{3}$$

if one puts

$$p_0 = \frac{W}{c} = ih\frac{\partial}{c\,\partial t}.$$

The symmetry between p_0 and p_1, p_2, p_3 required by relativity shows that, since the Hamiltonian we want is linear in p_0, it must also be linear in p_1, p_2 and p_3. Our wave equation is therefore of the form

$$(p_0 + \alpha_1 p_1 + \alpha_2 p_2 + \alpha_3 p_3 + \beta)\,\psi = 0, \tag{4}$$

where for the present all that is known about the dynamical variables or operators α_1, α_2, α_3, β is that they are independent of p_0, p_1, p_2, p_3, *i.e.*, that they commute with t, x_1, x_2, x_3. Since we are considering the case of a particle moving in empty space, so that all points in space are equivalent, we should expect the Hamiltonian not to involve t, x_1, x_2, x_3. This means that α_1, α_2, α_3, β are independent of t, x_1, x_2, x_3, *i.e.*, that they commute with p_0, p_1, p_2, p_3. We are therefore obliged to have other dynamical variables besides the co-ordinates and momenta of the electron, in order that α_1, α_2, α_3, β may be functions of them. The wave function ψ must then involve more variables than merely x_1, x_2, x_3, t.

Equation (4) leads to

$$\begin{aligned} 0 &= (-p_0 + \alpha_1 p_1 + \alpha_2 p_2 + \alpha_3 p_3 + \beta)(p_0 + \alpha_1 p_1 + \alpha_2 p_2 + \alpha_3 p_3 + \beta)\,\psi \\ &= [-p_0^2 + \Sigma\,\alpha_1^2 p_1^2 + \Sigma\,(\alpha_1\alpha_2 + \alpha_2\alpha_1)\,p_1 p_2 + \beta^2 + \Sigma\,(\alpha_1\beta + \beta\alpha_1)\,p_1]\,\psi, \end{aligned} \tag{5}$$

where the Σ refers to cyclic permutation of the suffixes 1, 2, 3. This agrees with (3) if

$$\left.\begin{aligned} &\alpha_r^2 = 1, \qquad &\alpha_r\alpha_s + \alpha_s\alpha_r = 0 \quad (r \neq s) \\ &\beta^2 = m^2c^2, \qquad &\alpha_r\beta + \beta\alpha_r = 0 \end{aligned}\right\} \quad r, s = 1, 2, 3.$$

If we put $\beta = \alpha_4 mc$, these conditions become

$$\alpha_\mu^2 = 1 \qquad \alpha_\mu\alpha_\nu + \alpha_\nu\alpha_\mu = 0 \;(\mu \neq \nu) \qquad \mu, \nu = 1, 2, 3, 4. \tag{6}$$

We can suppose the α_μ's to be expressed as matrices in some matrix scheme, the matrix elements of α_μ being, say, $\alpha_\mu(\zeta'\,\zeta'')$. The wave function ψ must now be a function of ζ as well as x_1, x_2, x_3, t. The result of α_μ multiplied into ψ will be a function $(\alpha_\mu \psi)$ of x_1, x_2, x_3, t, ζ defined by

$$(\alpha_\mu \psi)(x, t, \zeta) = \Sigma_{\zeta'} \alpha_\mu (\zeta \zeta') \psi (x, t, \zeta').$$

We must now find four matrices α_μ to satisfy the conditions (6). We make use of the matrices

$$\sigma_1 = \begin{pmatrix} 0 & 1 \\ 1 & 0 \end{pmatrix} \qquad \sigma_2 = \begin{pmatrix} 0 & -i \\ i & 0 \end{pmatrix} \qquad \sigma_3 = \begin{pmatrix} 1 & 0 \\ 0 & -1 \end{pmatrix}$$

which Pauli introduced* to describe the three components of spin angular momentum. These matrices have just the properties

$$\sigma_r^2 = 1 \qquad \sigma_r \sigma_s + \sigma_s \sigma_r = 0; \qquad (r \neq s), \tag{7}$$

that we require for our α's. We cannot, however, just take the σ's to be three of our α's, because then it would not be possible to find the fourth. We must extend the σ's in a diagonal manner to bring in two more rows and columns, so that we can introduce three more matrices ρ_1, ρ_2, ρ_3 of the same form as σ_1, σ_2, σ_3, but referring to different rows and columns, thus :—

$$\sigma_1 = \begin{Bmatrix} 0 & 1 & 0 & 0 \\ 1 & 0 & 0 & 0 \\ 0 & 0 & 0 & 1 \\ 0 & 0 & 1 & 0 \end{Bmatrix} \quad \sigma_2 = \begin{Bmatrix} 0 & -i & 0 & 0 \\ i & 0 & 0 & 0 \\ 0 & 0 & 0 & -i \\ 0 & 0 & i & 0 \end{Bmatrix} \quad \sigma_3 = \begin{Bmatrix} 1 & 0 & 0 & 0 \\ 0 & -1 & 0 & 0 \\ 0 & 0 & 1 & 0 \\ 0 & 0 & 0 & -1 \end{Bmatrix},$$

$$\rho_1 = \begin{Bmatrix} 0 & 0 & 1 & 0 \\ 0 & 0 & 0 & 1 \\ 1 & 0 & 0 & 0 \\ 0 & 1 & 0 & 0 \end{Bmatrix} \quad \rho_2 = \begin{Bmatrix} 0 & 0 & -i & 0 \\ 0 & 0 & 0 & -i \\ i & 0 & 0 & 0 \\ 0 & i & 0 & 0 \end{Bmatrix} \quad \rho_3 = \begin{Bmatrix} 1 & 0 & 0 & 0 \\ 0 & 1 & 0 & 0 \\ 0 & 0 & -1 & 0 \\ 0 & 0 & 0 & -1 \end{Bmatrix}.$$

The ρ's are obtained from the σ's by interchanging the second and third rows, and the second and third columns. We now have, in addition to equations (7)

$$\text{and also} \quad \left.\begin{aligned} &\rho_r^2 = 1 \qquad \rho_r \rho_s + \rho_s \rho_r = 0 \qquad (r \neq s), \\ &\rho_r \sigma_t = \sigma_t \rho_r. \end{aligned}\right\}. \tag{7'}$$

* Pauli, *loc. cit.*

If we now take

$$\alpha_1 = \rho_1\sigma_1, \qquad \alpha_2 = \rho_1\sigma_2, \qquad \alpha_3 = \rho_1\sigma_3, \qquad \alpha_4 = \rho_3,$$

all the conditions (6) are satisfied, *e.g.*,

$$\alpha_1^2 = \rho_1\sigma_1\rho_1\sigma_1 = \rho_1^2\sigma_1^2 = 1$$

$$\alpha_1\alpha_2 = \rho_1\sigma_1\rho_1\sigma_2 = \rho_1^2\sigma_1\sigma_2 = -\rho_1^2\sigma_2\sigma_1 = -\alpha_2\alpha_1.$$

The following equations are to be noted for later reference

$$\left.\begin{array}{l} \rho_1\rho_2 = i\rho_3 = -\rho_2\rho_1 \\ \sigma_1\sigma_2 = i\sigma_3 = -\sigma_2\sigma_1 \end{array}\right\}, \tag{8}$$

together with the equations obtained by cyclic permutation of the suffixes.

The wave equation (4) now takes the form

$$[p_0 + \rho_1(\boldsymbol{\sigma}, \mathbf{p}) + \rho_3 mc]\,\psi = 0, \tag{9}$$

where $\boldsymbol{\sigma}$ denotes the vector $(\sigma_1, \sigma_2, \sigma_3)$.

§ 3. *Proof of Invariance under a Lorentz Transformation.*

Multiply equation (9) by ρ_3 on the left-hand side. It becomes, with the help of (8),

$$[\rho_3 p_0 + i\rho_2(\sigma_1 p_1 + \sigma_2 p_2 + \sigma_3 p_3) + mc]\,\psi = 0.$$

Putting

$$p_0 = ip_4,$$

$$\rho_3 = \gamma_4, \qquad \rho_2\sigma_r = \gamma_r, \qquad r = 1, 2, 3, \tag{10}$$

we have

$$[i\Sigma\gamma_\mu p_\mu + mc]\,\psi = 0, \qquad \mu = 1, 2, 3, 4. \tag{11}$$

The p_μ transform under a Lorentz transformation according to the law

$$p_\mu' = \Sigma_\nu a_{\mu\nu} p_\nu,$$

where the coefficients $a_{\mu\nu}$ are c-numbers satisfying

$$\Sigma_\mu a_{\mu\nu} a_{\mu\tau} = \delta_{\nu\tau}, \qquad \Sigma_\tau a_{\mu\tau} a_{\nu\tau} = \delta_{\mu\nu}.$$

The wave equation therefore transforms into

$$[i\Sigma\gamma_\mu' p_\mu' + mc]\,\psi = 0, \tag{12}$$

where

$$\gamma_\mu' = \Sigma_\nu a_{\mu\nu}\gamma_\nu.$$

Now the γ_μ, like the α_μ, satisfy

$$\gamma_\mu^2 = 1, \qquad \gamma_\mu\gamma_\nu + \gamma_\nu\gamma_\mu = 0, \qquad (\mu \neq \nu).$$

These relations can be summed up in the single equation

$$\gamma_\mu \gamma_\nu + \gamma_\nu \gamma_\mu = 2\delta_{\mu\nu}.$$

We have

$$\begin{aligned}\gamma_\mu' \gamma_\nu' + \gamma_\nu' \gamma_\mu' &= \Sigma_{\tau\lambda}\, a_{\mu\tau}\, a_{\nu\lambda}\, (\gamma_\tau \gamma_\lambda + \gamma_\lambda \gamma_\tau) \\ &= 2\Sigma_{\tau\lambda}\, a_{\mu\tau}\, a_{\nu\lambda}\, \delta_{\tau\lambda} \\ &= 2\Sigma_\tau\, a_{\mu\tau}\, a_{\nu\tau} = 2\delta_{\mu\nu}.\end{aligned}$$

Thus the γ_μ' satisfy the same relations as the γ_μ. Thus we can put, analogously to (10)

$$\gamma_4' = \rho_3' \qquad \gamma_r' = \rho_2' \sigma_r'$$

where the ρ''s and σ''s are easily verified to satisfy the relations corresponding to (7), (7′) and (8), if ρ_2' and ρ_1' are defined by $\rho_2' = -i\gamma_1'\gamma_2'\gamma_3'$, $\rho_1' = -i\rho_2'\rho_3'$.

We shall now show that, by a canonical transformation, the ρ''s and σ''s may be brought into the form of the ρ's and σ's. From the equation $\rho_3'^2 = 1$, it follows that the only possible characteristic values for ρ_3' are ± 1. If one applies to ρ_3' a canonical transformation with the transformation function ρ_1', the result is

$$\rho_1' \rho_3' (\rho_1')^{-1} = -\rho_3' \rho_1' (\rho_1')^{-1} = -\rho_3'.$$

Since characteristic values are not changed by a canonical transformation, ρ_3' must have the same characteristic values as $-\rho_3'$. Hence the characteristic values of ρ_3' are $+1$ twice and -1 twice. The same argument applies to each of the other ρ''s, and to each of the σ''s.

Since ρ_3' and σ_3' commute, they can be brought simultaneously to the diagonal form by a canonical transformation. They will then have for their diagonal elements each $+1$ twice and -1 twice. Thus, by suitably rearranging the rows and columns, they can be brought into the form ρ_3 and σ_3 respectively. (The possibility $\rho_3' = \pm \sigma_3'$ is excluded by the existence of matrices that commute with one but not with the other.)

Any matrix containing four rows and columns can be expressed as

$$c + \Sigma_r c_r \sigma_r + \Sigma_r c_r' \rho_r + \Sigma_{rs} c_{rs} \rho_r \sigma_s \tag{13}$$

where the sixteen coefficients c, c_r, c_r', c_{rs} are c-numbers. By expressing σ_1' in this way, we see, from the fact that it commutes with $\rho_3' = \rho_3$ and anticommutes* with $\sigma_3' = \sigma_3$, that it must be of the form

$$\sigma_1' = c_1 \sigma_1 + c_2 \sigma_2 + c_{31} \rho_3 \sigma_1 + c_{32} \rho_3 \sigma_2,$$

* We say that a anticommutes with b when $ab = -ba$.

i.e., of the form

$$\sigma_1' = \begin{Bmatrix} 0 & a_{12} & 0 & 0 \\ a_{21} & 0 & 0 & 0 \\ 0 & 0 & 0 & a_{34} \\ 0 & 0 & a_{43} & 0 \end{Bmatrix}$$

The condition $\sigma_1'^2 = 1$ shows that $a_{12}a_{21} = 1$, $a_{34}a_{43} = 1$. If we now apply the canonical transformation: first row to be multiplied by $(a_{21}/a_{12})^{\frac{1}{2}}$ and third row to be multiplied by $(a_{43}/a_{34})^{\frac{1}{2}}$, and first and third columns to be divided by the same expressions, σ_1' will be brought into the form of σ_1, and the diagonal matrices σ_3' and ρ_3' will not be changed.

If we now express ρ_1' in the form (13) and use the conditions that it commutes with $\sigma_1' = \sigma_1$ and $\sigma_3' = \sigma_3$ and anticommutes with $\rho_3' = \rho_3$, we see that it must be of the form

$$\rho_1' = c_1'\rho_1 + c_2'\rho_2.$$

The condition $\rho_1'^2 = 1$ shows that $c_1'^2 + c_2'^2 = 1$, or $c_1' = \cos\theta$, $c_2' = \sin\theta$. Hence ρ_1' is of the form

$$\rho_1' = \begin{Bmatrix} 0 & 0 & e^{-i\theta} & 0 \\ 0 & 0 & 0 & e^{-i\theta} \\ e^{i\theta} & 0 & 0 & 0 \\ 0 & e^{i\theta} & 0 & 0 \end{Bmatrix}$$

If we now apply the canonical transformation: first and second rows to be multiplied by $e^{i\theta}$ and first and second columns to be divided by the same expression, ρ_1' will be brought into the form ρ_1, and σ_1, σ_3, ρ_3 will not be altered. ρ_2' and σ_2' must now be of the form ρ_2 and σ_2, on account of the relations $i\rho_2' = \rho_3'\rho_1'$, $i\sigma_2' = \sigma_3'\sigma_1'$.

Thus by a succession of canonical transformations, which can be combined to form a single canonical transformation, the ρ''s and σ''s can be brought into the form of the ρ's and σ's. The new wave equation (12) can in this way be brought back into the form of the original wave equation (11) or (9), so that the results that follow from this original wave equation must be independent of the frame of reference used.

§ 4. *The Hamiltonian for an Arbitrary Field.*

To obtain the Hamiltonian for an electron in an electromagnetic field with scalar potential A_0 and vector potential $\mathbf{A}$, we adopt the usual procedure of substituting $p_0 + e/c \,.\, A_0$ for p_0 and $\mathbf{p} + e/c \,.\, \mathbf{A}$ for $\mathbf{p}$ in the Hamiltonian for no field. From equation (9) we thus obtain

$$\left[p_0 + \frac{e}{c}A_0 + \rho_1\left(\boldsymbol{\sigma}, \mathbf{p} + \frac{e}{c}\mathbf{A}\right) + \rho_3 mc\right]\psi = 0. \tag{14}$$

This wave equation appears to be sufficient to account for all the duplexity phenomena. On account of the matrices ρ and σ containing four rows and columns, it will have four times as many solutions as the non-relativity wave equation, and twice as many as the previous relativity wave equation (1). Since half the solutions must be rejected as referring to the charge $+e$ on the electron, the correct number will be left to account for duplexity phenomena. The proof given in the preceding section of invariance under a Lorentz transformation applies equally well to the more general wave equation (14).

We can obtain a rough idea of how (14) differs from the previous relativity wave equation (1) by multiplying it up analogously to (5). This gives, if we write e' for e/c

$$\begin{aligned} 0 &= [-(p_0 + e'A_0) + \rho_1(\boldsymbol{\sigma}, \mathbf{p} + e'\mathbf{A}) + \rho_3 mc] \\ &\qquad \times [(p_0 + e'A_0) + \rho_1(\boldsymbol{\sigma}, \mathbf{p} + e'\mathbf{A}) + \rho_3 mc]\,\psi \\ &= [-(p_0 + e'A_0)^2 + (\boldsymbol{\sigma}, \mathbf{p} + e'\mathbf{A})^2 + m^2c^2 \\ &\qquad + \rho_1\{(\boldsymbol{\sigma}, \mathbf{p} + e'\mathbf{A})(p_0 + e'A_0) - (p_0 + e'A_0)(\boldsymbol{\sigma}, \mathbf{p} + e'\mathbf{A})\}]\,\psi. \end{aligned} \tag{15}$$

We now use the general formula, that if $\mathbf{B}$ and $\mathbf{C}$ are any two vectors that commute with $\boldsymbol{\sigma}$

$$\begin{aligned} (\boldsymbol{\sigma}, \mathbf{B})(\boldsymbol{\sigma}, \mathbf{C}) &= \Sigma\, \sigma_1^2 B_1C_1 + \Sigma\,(\sigma_1\sigma_2 B_1C_2 + \sigma_2\sigma_1 B_2C_1) \\ &= (\mathbf{B}, \mathbf{C}) + i\,\Sigma\,\sigma_3(B_1C_2 - B_2C_1) \\ &= (\mathbf{B}, \mathbf{C}) + i\,(\boldsymbol{\sigma}, \mathbf{B} \times \mathbf{C}). \end{aligned} \tag{16}$$

Taking $\mathbf{B} = \mathbf{C} = \mathbf{p} + e'\mathbf{A}$, we find

$$\begin{aligned} (\boldsymbol{\sigma}, \mathbf{p} + e'\mathbf{A})^2 &= (\mathbf{p} + e'\mathbf{A})^2 + i\,\Sigma\,\sigma_3 \\ &\qquad [(p_1 + e'A_1)(p_2 + e'A_2) - (p_2 + e'A_2)(p_1 + e'A_1)] \\ &= (\mathbf{p} + e'\mathbf{A})^2 + he'(\boldsymbol{\sigma}, \text{curl } \mathbf{A}). \end{aligned}$$

Thus (15) becomes

$$0 = \Big[-(p_0 + e'\mathrm{A}_0)^2 + (\mathbf{p} + e'\mathbf{A})^2 + m^2c^2 + e'h\,(\boldsymbol{\sigma}, \operatorname{curl} \mathbf{A}) - ie'h\rho_1\Big(\boldsymbol{\sigma}, \operatorname{grad} \mathrm{A}_0 + \frac{1}{c}\frac{\partial \mathbf{A}}{\partial t}\Big)\Big]\psi$$

$$= [-(p_0 + e'\mathrm{A}_0)^2 + (\mathbf{p} + e'\mathbf{A})^2 + m^2c^2 + e'h\,(\boldsymbol{\sigma}, \mathbf{H}) + ie'h\rho_1\,(\boldsymbol{\sigma}, \mathbf{E})]\,\psi,$$

where **E** and **H** are the electric and magnetic vectors of the field.

This differs from (1) by the two extra terms

$$\frac{eh}{c}(\boldsymbol{\sigma}, \mathbf{H}) + \frac{ieh}{c}\rho_1\,(\boldsymbol{\sigma}, \mathbf{E})$$

in F. These two terms, when divided by the factor $2m$, can be regarded as the additional potential energy of the electron due to its new degree of freedom. The electron will therefore behave as though it has a magnetic moment $eh/2mc\,.\,\boldsymbol{\sigma}$ and an electric moment $ieh/2mc\,.\,\rho_1\boldsymbol{\sigma}$. This magnetic moment is just that assumed in the spinning electron model. The electric moment, being a pure imaginary, we should not expect to appear in the model. It is doubtful whether the electric moment has any physical meaning, since the Hamiltonian in (14) that we started from is real, and the imaginary part only appeared when we multiplied it up in an artificial way in order to make it resemble the Hamiltonian of previous theories.

§ 5. *The Angular Momentum Integrals for Motion in a Central Field.*

We shall consider in greater detail the motion of an electron in a central field of force. We put $\mathbf{A} = 0$ and $e'\mathrm{A}_0 = \mathrm{V}(r)$, an arbitrary function of the radius r, so that the Hamiltonian in (14) becomes

$$\mathrm{F} \equiv p_0 + \mathrm{V} + \rho_1(\boldsymbol{\sigma}, \mathbf{p}) + \rho_3 mc.$$

We shall determine the periodic solutions of the wave equation $\mathrm{F}\,\psi = 0$, which means that p_0 is to be counted as a parameter instead of an operator; it is, in fact, just $1/c$ times the energy level.

We shall first find the angular momentum integrals of the motion. The orbital angular momentum **m** is defined by

$$\mathbf{m} = \mathbf{x} \times \mathbf{p},$$

and satisfies the following " Vertauschungs " relations

$$\left.\begin{aligned} m_1x_1 - x_1m_1 &= 0, \qquad & m_1x_2 - x_2m_1 &= ihx_3 \\ m_1p_1 - p_1m_1 &= 0, \qquad & m_1p_2 - p_2m_1 &= ihp_3 \\ \mathbf{m} \times \mathbf{m} &= ih\mathbf{m}, \qquad & \mathbf{m}^2m_1 - m_1\mathbf{m}^2 &= 0, \end{aligned}\right\}, \tag{17}$$

together with similar relations obtained by permuting the suffixes. Also **m** commutes with r, and with p_r, the momentum canonically conjugate to r.

We have

$$\begin{aligned} m_1 F - F m_1 &= \rho_1 \{ m_1 (\boldsymbol{\sigma}, \mathbf{p}) - (\boldsymbol{\sigma}, \mathbf{p}) m_1 \} \\ &= \rho_1 (\boldsymbol{\sigma}, m_1 \mathbf{p} - \mathbf{p} m_1) \\ &= ih\rho_1 (\sigma_2 p_3 - \sigma_3 p_2), \end{aligned}$$

and so

$$\mathbf{m}F - F\mathbf{m} = ih\rho_1 \, \boldsymbol{\sigma} \times \mathbf{p}. \tag{18}$$

Thus **m** is not a constant of the motion. We have further

$$\begin{aligned} \sigma_1 F - F \sigma_1 &= \rho_1 \{ \sigma_1 (\boldsymbol{\sigma}, \mathbf{p}) - (\boldsymbol{\sigma}, \mathbf{p}) \sigma_1 \} \\ &= \rho_1 (\sigma_1 \boldsymbol{\sigma} - \boldsymbol{\sigma} \sigma_1, \mathbf{p}) \\ &= 2i\rho_1 (\sigma_3 p_2 - \sigma_2 p_3), \end{aligned}$$

with the help of (8), and so

$$\boldsymbol{\sigma} F - F \boldsymbol{\sigma} = -2i\rho_1 \, \boldsymbol{\sigma} \times \mathbf{p}.$$

Hence

$$(\mathbf{m} + \tfrac{1}{2}h\boldsymbol{\sigma}) F - F (\mathbf{m} + \tfrac{1}{2}h\boldsymbol{\sigma}) = 0.$$

Thus $\mathbf{m} + \frac{1}{2}h\boldsymbol{\sigma}$ (= **M** say) is a constant of the motion. We can interpret this result by saying that the electron has a spin angular momentum of $\frac{1}{2}h\boldsymbol{\sigma}$, which, added to the orbital angular momentum **m**, gives the total angular momentum **M**, which is a constant of the motion.

The Vertauschungs relations (17) all hold when M's are written for the m's. In particular

$$\mathbf{M} \times \mathbf{M} = ih\mathbf{M} \quad \text{and} \quad \mathbf{M}^2 M_3 = M_3 \mathbf{M}^2.$$

M_3 will be an action variable of the system. Since the characteristic values of m_3 must be integral multiples of h in order that the wave function may be single-valued, the characteristic values of M_3 must be half odd integral multiples of h. If we put

$$\mathbf{M}^2 = (j^2 - \tfrac{1}{4}) h^2, \tag{19}$$

j will be another quantum number, and the characteristic values of M_3 will extend from $(j - \frac{1}{2}) h$ to $(-j + \frac{1}{2}) h$.* Thus j takes integral values.

One easily verifies from (18) that $\mathbf{m}^2$ does not commute with F, and is thus not a constant of the motion. This makes a difference between the present theory and the previous spinning electron theory, in which $\mathbf{m}^2$ is constant, and defines the azimuthal quantum number k by a relation similar to (19). We shall find that our j plays the same part as the k of the previous theory.

* See 'Roy. Soc. Proc.,' A, vol. 111, p. 281 (1926).

§ 6. *The Energy Levels for Motion in a Central Field.*

We shall now obtain the wave equation as a differential equation in r, with the variables that specify the orientation of the whole system removed. We can do this by the use only of elementary non-commutative algebra in the following way.

In formula (16) take $\mathbf{B} = \mathbf{C} = \mathbf{m}$. This gives

$$(\boldsymbol{\sigma}, \mathbf{m})^2 = \mathbf{m}^2 + i\,(\boldsymbol{\sigma}, \mathbf{m} \times \mathbf{m}) \tag{20}$$
$$= (\mathbf{m} + \tfrac{1}{2}h\,\boldsymbol{\sigma})^2 - h\,(\boldsymbol{\sigma}, \mathbf{m}) - \tfrac{1}{4}h^2\,\boldsymbol{\sigma}^2 - h\,(\boldsymbol{\sigma}, \mathbf{m})$$
$$= \mathbf{M}^2 - 2h\,(\boldsymbol{\sigma}, \mathbf{m}) - \tfrac{3}{4}h^2.$$

Hence

$$\{(\boldsymbol{\sigma}, \mathbf{m}) + h\}^2 = \mathbf{M}^2 + \tfrac{1}{4}h^2 = j^2h^2.$$

Up to the present we have defined j only through j^2, so that we could now, if we liked, take jh equal to $(\boldsymbol{\sigma}, \mathbf{m}) + h$. This would not be convenient since we want j to be a constant of the motion while $(\boldsymbol{\sigma}, \mathbf{m}) + h$ is not, although its square is. We have, in fact, by another application of (16),

$$(\boldsymbol{\sigma}, \mathbf{m})\,(\boldsymbol{\sigma}, \mathbf{p}) = i\,(\boldsymbol{\sigma}, \mathbf{m} \times \mathbf{p})$$

since $(\mathbf{m}, \mathbf{p}) = 0$, and similarly

$$(\boldsymbol{\sigma}, \mathbf{p})\,(\boldsymbol{\sigma}, \mathbf{m}) = i\,(\boldsymbol{\sigma}, \mathbf{p} \times \mathbf{m}),$$

so that

$$(\boldsymbol{\sigma}, \mathbf{m})\,(\boldsymbol{\sigma}, \mathbf{p}) + (\boldsymbol{\sigma}, \mathbf{p})\,(\boldsymbol{\sigma}, \mathbf{m}) = i\Sigma\sigma_1\,(m_2p_3 - m_3p_2 + p_2m_3 - p_3m_2)$$
$$= i\Sigma\sigma_1\,.\,2ihp_1 = -2h\,(\boldsymbol{\sigma}, \mathbf{p}),$$

or

$$\{(\boldsymbol{\sigma}, \mathbf{m}) + h\}\,(\boldsymbol{\sigma}, \mathbf{p}) + (\boldsymbol{\sigma}, \mathbf{p})\,\{(\boldsymbol{\sigma}, \mathbf{m}) + h\} = 0.$$

Thus $(\boldsymbol{\sigma}, \mathbf{m}) + h$ anticommutes with one of the terms in F, namely, $\rho_1\,(\boldsymbol{\sigma}, \mathbf{p})$, and commutes with the other three. Hence $\rho_3\,\{(\boldsymbol{\sigma}, \mathbf{m}) + h\}$ commutes with all four, and is therefore a constant of the motion. But the square of $\rho_3\{(\boldsymbol{\sigma}, \mathbf{m}) + h\}$ must also equal j^2h^2. We therefore take

$$jh = \rho_3\,\{(\boldsymbol{\sigma}, \mathbf{m}) + h\}. \tag{21}$$

We have, by a further application of (16)

$$(\boldsymbol{\sigma}, \mathbf{x})\,(\boldsymbol{\sigma}, \mathbf{p}) = (\mathbf{x}, \mathbf{p}) + i\,(\boldsymbol{\sigma}, \mathbf{m}).$$

Now a permissible definition of p_r is

$$(\mathbf{x}, \mathbf{p}) = rp_r + ih,$$

and from (21)

$$(\boldsymbol{\sigma}, \mathbf{m}) = \rho_3 jh - h.$$

Hence

$$(\boldsymbol{\sigma}, \mathbf{x})\,(\boldsymbol{\sigma}, \mathbf{p}) = rp_r + i\rho_3 jh. \tag{22}$$

Introduce the quantity ε defined by

$$r\varepsilon = \rho_1 (\boldsymbol{\sigma}, \mathbf{x}). \tag{23}$$

Since r commutes with ρ_1 and with $(\boldsymbol{\sigma}, \mathbf{x})$, it must commute with ε. We thus have

$$r^2\varepsilon^2 = [\rho_1 (\boldsymbol{\sigma}, \mathbf{x})]^2 = (\boldsymbol{\sigma}, \mathbf{x})^2 = \mathbf{x}^2 = r^2$$

or

$$\varepsilon^2 = 1.$$

Since there is symmetry between $\mathbf{x}$ and $\mathbf{p}$ so far as angular momentum is concerned, $\rho_1 (\boldsymbol{\sigma}, \mathbf{x})$, like $\rho_1 (\boldsymbol{\sigma}, \mathbf{p})$, must commute with $\mathbf{M}$ and j. Hence ε commutes with $\mathbf{M}$ and j. Further, ε must commute with p_r, since we have

$$(\boldsymbol{\sigma}, \mathbf{x}) (\mathbf{x}, \mathbf{p}) - (\mathbf{x}, \mathbf{p}) (\boldsymbol{\sigma}, \mathbf{x}) = ih (\boldsymbol{\sigma}, \mathbf{x}),$$

which gives

$$r\varepsilon (rp_r + ih) - (rp_r + ih) r\varepsilon = ihr\varepsilon,$$

which reduces to

$$\varepsilon p_r - p_r\varepsilon = 0.$$

From (22) and (23) we now have

$$r\varepsilon\rho_1 (\boldsymbol{\sigma}, \mathbf{p}) = rp_r + i\rho_3 jh$$

or

$$\rho_1 (\boldsymbol{\sigma}, \mathbf{p}) = \varepsilon p_r + i\varepsilon\rho_3 jh/r.$$

Thus

$$\mathrm{F} = p_0 + \mathrm{V} + \varepsilon p_r + i\varepsilon\rho_3 jh/r + \rho_3 mc. \tag{24}$$

Equation (23) shows that ε anticommutes with ρ_3. We can therefore by a canonical transformation (involving perhaps the x's and p's as well as the σ's and ρ's) bring ε into the form of the ρ_2 of § 2 without changing ρ_3, and without changing any of the other variables occurring on the right-hand side of (24), since these other variables all commute with ε. $i\varepsilon\rho_3$ will now be of the form $i\rho_2\rho_3 = -\rho_1$, so that the wave equation takes the form

$$\mathrm{F}\psi \equiv [p_0 + \mathrm{V} + \rho_2 p_r - \rho_1 jh/r + \rho_3 mc]\,\psi = 0.$$

If we write this equation out in full, calling the components of ψ referring to the first and third rows (or columns) of the matrices ψ_α and ψ_β respectively, we get

$$(\mathrm{F}\psi)_\alpha \equiv (p_0 + \mathrm{V})\,\psi_\alpha - h\frac{\partial}{\partial r}\psi_\beta - \frac{jh}{r}\psi_\beta + mc\psi_\alpha = 0,$$

$$(\mathrm{F}\psi)_\beta \equiv (p_0 + \mathrm{V})\,\psi_\beta + h\frac{\partial}{\partial r}\psi_\alpha - \frac{jh}{r}\psi_\alpha - mc\psi_\beta = 0.$$

The second and fourth components give just a repetition of these two equations. We shall now eliminate ψ_α. If we write $h\text{B}$ for $p_0 + \text{V} + mc$, the first equation becomes

$$\left(\frac{\partial}{\partial r} + \frac{j}{r}\right)\psi_\beta = \text{B}\psi_\alpha,$$

which gives on differentiating

$$\begin{aligned}\frac{\partial^2}{\partial r^2}\psi_\beta + \frac{j}{r}\frac{\partial}{\partial r}\psi_\beta - \frac{j}{r^2}\psi_\beta &= \text{B}\frac{\partial}{\partial r}\psi_\alpha + \frac{\partial \text{B}}{\partial r}\psi_\alpha \\ &= \frac{\text{B}}{h}\left[-(p_0 + \text{V} - mc)\,\psi_\beta + \frac{jh}{r}\psi_\alpha\right] + \frac{1}{h}\frac{\partial \text{V}}{\partial r}\psi_\alpha \\ &= -\frac{(p_0+\text{V})^2 - m^2c^2}{h^2}\psi_\beta + \left(\frac{j}{r} + \frac{1}{\text{B}h}\frac{\partial \text{V}}{\partial r}\right)\left(\frac{\partial}{\partial r} + \frac{j}{r}\right)\psi_\beta.\end{aligned}$$

This reduces to

$$\frac{\partial^2}{\partial r^2}\psi_\beta + \left[\frac{(p_0+\text{V})^2 - m^2c^2}{h^2} - \frac{j(j+1)}{r^2}\right]\psi_\beta - \frac{1}{\text{B}h}\frac{\partial \text{V}}{\partial r}\left(\frac{\partial}{\partial r} + \frac{j}{r}\right)\psi_\beta = 0. \quad (25)$$

The values of the parameter p_0 for which this equation has a solution finite at $r = 0$ and $r = \infty$ are $1/c$ times the energy levels of the system. To compare this equation with those of previous theories, we put $\psi_\beta = r\chi$, so that

$$\frac{\partial^2}{\partial r^2}\chi + \frac{2}{r}\frac{\partial}{\partial r}\chi + \left[\frac{(p_0+\text{V})^2 - m^2c^2}{h^2} - \frac{j(j+1)}{r^2}\right]\chi - \frac{1}{\text{B}h}\frac{\partial \text{V}}{\partial r}\left(\frac{\partial}{\partial r} + \frac{j+1}{r}\right)\chi = 0. \quad (26)$$

If one neglects the last term, which is small on account of B being large, this equation becomes the same as the ordinary Schroedinger equation for the system, with relativity correction included. Since j has, from its definition, both positive and negative integral characteristic values, our equation will give twice as many energy levels when the last term is not neglected.

We shall now compare the last term of (26), which is of the same order of magnitude as the relativity correction, with the spin correction given by Darwin and Pauli. To do this we must eliminate the $\partial\chi/\partial r$ term by a further transformation of the wave function. We put

$$\chi = \text{B}^{-\frac{1}{2}}\chi_1,$$

which gives

$$\begin{aligned}\frac{\partial^2}{\partial r^2}\chi_1 + \frac{2}{r}\frac{\partial}{\partial r}\chi_1 + \left[\frac{(p_0+\text{V})^2 - m^2c^2}{h^2} - \frac{j(j+1)}{r^2}\right]\chi_1 \\ + \left[\frac{1}{\text{B}h}\frac{j}{r}\frac{\partial \text{V}}{\partial r} - \frac{1}{2}\frac{1}{\text{B}h}\frac{\partial^2 \text{V}}{\partial r^2} + \frac{1}{4}\frac{1}{\text{B}^2h^2}\left(\frac{\partial \text{V}}{\partial r}\right)^2\right]\chi_1 = 0. \quad (27)\end{aligned}$$

The correction is now, to the first order of accuracy

$$\frac{1}{\mathrm{B}h}\left(\frac{j}{r}\frac{\partial \mathrm{V}}{\partial r} - \frac{1}{2}\frac{\partial^2 \mathrm{V}}{\partial r^2}\right),$$

where $\mathrm{B}h = 2mc$ (provided p_0 is positive). For the hydrogen atom we must put $\mathrm{V} = e^2/cr$. The first order correction now becomes

$$-\frac{e^2}{2mc^2r^3}(j+1). \tag{28}$$

If we write $-j$ for $j+1$ in (27), we do not alter the terms representing the unperturbed system, so

$$\frac{e^2}{2mc^2r^3}\,j \tag{28'}$$

will give a second possible correction for the same unperturbed term.

In the theory of Pauli and Darwin, the corresponding correcting term is

$$\frac{e^2}{2mhc^2r^3}(\boldsymbol{\sigma}, \mathbf{m})$$

when the Thomas factor $\frac{1}{2}$ is included. We must remember that in the Pauli-Darwin theory, the resultant orbital angular momentum k plays the part of our j. We must define k by

$$\mathbf{m}^2 = k\,(k+1)\,h^2$$

instead of by the exact analogue of (19), in order that it may have integral characteristic values, like j. We have from (20)

$$(\boldsymbol{\sigma}, \mathbf{m})^2 = k\,(k+1)\,h^2 - h\,(\boldsymbol{\sigma}, \mathbf{m})$$

or

$$\{(\boldsymbol{\sigma}, \mathbf{m}) + \tfrac{1}{2}h\}^2 = (k+\tfrac{1}{2})^2h^2,$$

hence

$$(\boldsymbol{\sigma}, \mathbf{m}) = kh \text{ or } -(k+1)\,h.$$

The correction thus becomes

$$\frac{e^2}{2mc^2r^3}\,k \quad \text{or} \quad -\frac{e^2}{2mc^2r^3}(k+1),$$

which agrees with (28) and (28′). The present theory will thus, in the first approximation, lead to the same energy levels as those obtained by Darwin, which are in agreement with experiment.

351

Reprint 9

The Quantum Theory of the Electron. Part II.

By P. A. M. Dirac, St. John's College, Cambridge.

(Communicated by R. H. Fowler, F.R.S.—Received February 2, 1928.)

In a previous paper by the author* it is shown that the general theory of quantum mechanics together with relativity require the wave equation for an electron moving in an arbitrary electromagnetic field of potentials, A_0, A_1, A_2, A_3 to be of the form

$$F\psi \equiv \left[p_0 + \frac{e}{c} A_0 + \alpha_1 \left(p_1 + \frac{e}{c} A_1\right) + \alpha_2 \left(p_2 + \frac{e}{c} A_2\right) + \alpha_3 \left(p_3 + \frac{e}{c} A_3\right) + \alpha_4 mc\right] \psi = 0. \quad (1)$$

The α's are new dynamical variables which it is necessary to introduce in order to satisfy the conditions of the problem. They may be regarded as describing some internal motion of the electron, which for most purposes may be taken to be the spin of the electron postulated in previous theories. We shall call them the spin variables.

The α's must satisfy the conditions

$$\alpha_\mu^2 = 1, \qquad \alpha_\mu \alpha_\nu + \alpha_\nu \alpha_\mu = 0. \qquad (\mu \neq \nu.)$$

They may conveniently be expressed in terms of six variables ρ_1, ρ_2, ρ_3, σ_1, σ_2, σ_3 that satisfy

$$\left.\begin{array}{ll} & \rho_r^2 = 1, \qquad \sigma_r^2 = 1, \qquad \rho_r\sigma_s = \sigma_s\rho_r, \qquad (r, s = 1, 2, 3) \\ \text{and} & \\ & \rho_1\rho_2 = i\rho_3 = -\rho_2\rho_1, \qquad \sigma_1\sigma_2 = i\sigma_3 = -\sigma_2\sigma_1 \end{array}\right\}, \quad (2)$$

together with the relations obtained from these by cyclic permutation of the suffixes, by means of the equations

$$\alpha_1 = \rho_1\sigma_1, \qquad \alpha_2 = \rho_1\sigma_2, \qquad \alpha_3 = \rho_1\sigma_3, \qquad \alpha_4 = \rho_3.$$

The variables σ_1, σ_2, σ_3 now form the three components of a vector, which corresponds (apart from a constant factor) to the spin angular momentum vector that appears in Pauli's theory of the spinning electron. The ρ's and σ's vary with the time, like other dynamical variables. Their equations of motion, written in the Poisson Bracket notation [ ], are

$$\dot{\rho}_r = c\,[\rho_r, F], \qquad \dot{\sigma}_r = c\,[\sigma_r, F].$$

* 'Roy. Soc. Proc.,' A, vol. 117, p. 610 (1928). This is referred to later by *loc. cit.*

Reprinted from Proceedings of the Royal Society of London, Series A, Vol. 118, No. 779, pp. 351–361 (1928).

It should be observed that these equations of motion are consistent with the conditions (2), so that if the conditions are satisfied initially they always remain satisfied. For example, we have

$$ih/c \,.\, \dot{\sigma}_1 = \sigma_1 \mathrm{F} - \mathrm{F}\sigma_1 = 2i\rho_1\sigma_3\,(p_2 + e/c \,.\, \mathrm{A}_2) - 2i\rho_1\sigma_2\,(p_3 + e/c \,.\, \mathrm{A}_3).$$

Thus $\dot{\sigma}_1$ anticommutes with σ_1, so that

$$d\sigma_1{}^2/dt = \dot{\sigma}_1\sigma_1 + \sigma_1\dot{\sigma}_1 = 0.$$

The ρ's and σ's, and therefore also any function of them, can be represented by matrices with four rows and columns. A possible representation, in which ρ_3 and σ_3 are diagonal matrices, is given in (*loc. cit.*) § 2. Such a representation can apply only to a single instant of time, since the ρ's and σ's vary with the time. To get a scheme of representation which holds for all times, so that the equations of motion are valid in it, we should have to have only constants of the motion as diagonal matrices. It is, however, quite correct for the purpose of solving the wave equation (1) to take a matrix representation for the ρ's and σ's which holds only for a single instant of time (as was done in *loc. cit.*), since the wave function is then the transformation function connecting the ρ's, σ's and x's at this particular time with a set of variables that are constants of the motion, as is required for the general interpretation of quantum mechanics.

Before we proceed with the theory of atoms with single electrons that was begun in *loc. cit.*, the proof will be given of the conservation theorem, which states that the change in the probability of the electron being in a given volume during a given time is equal to the probability of its having crossed the boundary. This proof is supplementary to the work of *loc. cit.* § 3, and is necessary before one can infer that the theory will give consistent results that are invariant under a Lorentz transformation.

§ 1. *The Conservation Theorem.*

We shall first make a slight generalisation of the usual interpretation of wave mechanics to apply to cases when the Hamiltonian is not Hermitian. Let the wave equation, written in certain variables q, be

$$(\mathrm{H} - \mathrm{W})\,\psi = 0. \tag{i}$$

Consider also the equation

$$(\tilde{\mathrm{H}} - \tilde{\mathrm{W}})\,\phi = 0$$

or

$$(\tilde{\mathrm{H}} + \mathrm{W})\,\phi = 0, \tag{ii}$$

where the symbol $\tilde{a}$ denotes the matrix obtained from the matrix a by transposing rows and columns. If ψ_m, ϕ_n are suitably normalised solutions of (i) and (ii) respectively, referring to the states m and n, we take $\phi_n \psi_m$ to be the corresponding matrix element of the probability of the q's having specified values. If H is Hermitian, $\tilde{\text{H}}$ is the conjugate imaginary of H (obtained by writing $-i$ for i) and the solutions of (ii) are just the conjugate imaginaries to the solutions of (i), so that in this case our probability $\phi_n \psi_m$ becomes the usual one $\bar{\psi}_n \psi_m$. In the general case it is necessary to use the transposed Hamiltonian instead of the conjugate imaginary Hamiltonian in (ii) in order to secure that if ϕ_n, ψ_m are initially orthogonal or mutually normalised $\left(i.e., \int \phi_n \psi_m \, dq = 1\right)$, they always remain orthogonal or mutually normalised respectively.

Our wave equation for an electron in an electromagnetic field is

$$[p_0 + e'\text{A}_0 + \rho_1 (\boldsymbol{\sigma}, \mathbf{p} + e'\mathbf{A}) + \rho_3 mc] \psi = 0 \tag{3}$$

where $e' = e/c$. The Hamiltonian here will be Hermitian if a matrix scheme for the spin variables is chosen in which they are Hermitian. However, if one now applies a Lorentz transformation to this wave equation and divides out by the coefficient of the new p_0, the resulting new Hamiltonian will not, in general, be Hermitian, although, as shown in *loc. cit.*, § 3, it may be brought back to its original Hermitian form by a canonical transformation of the matrix scheme for the spin variables. In the following work we require to have the same matrix representation of the spin variables for all frames of reference, so we cannot assume our Hamiltonian is Hermitian, and must use the above generalised interpretation.

The equation obtained by transposing rows and columns in the operator of (3) is

$$[-p_0 + e'\text{A}_0 + \tilde{\rho}_1 (\tilde{\boldsymbol{\sigma}}, -\mathbf{p} + e'\mathbf{A}) + \tilde{\rho}_3 mc] \phi = 0. \tag{4}$$

The probability per unit volume of the electron being in the neighbourhood of any point is given, according to the above assumption, by $\phi\psi$, where this product must now be understood to mean the sum of the products of each of the four components of ϕ (referring respectively to the four rows or columns of the matrices ρ, σ) into the corresponding component of ψ. We have to prove that this probability is the time component of a 4-vector, and that the divergence of this 4-vector vanishes.

From (3)

$$[\rho_3(p_0+e'A_0)+\rho_1\rho_3(\sigma,\ \mathbf{p}+e'\mathbf{A})+mc]\,\rho_3\psi=0$$

or

$$[\gamma_0(p_0+e'A_0)+\Sigma_{r=1,2,3}\gamma_r(p_r+e'A_r)+mc]\chi=0, \tag{5}$$

where

$$\gamma_0=\rho_3, \qquad \gamma_r=\rho_1\rho_3\sigma_r, \qquad \chi=\rho_3\psi.$$

Equation (5) is symmetrical between the four dimensions of space and time, and shows that γ_0, $-\gamma_1$, $-\gamma_2$, $-\gamma_3$ are the contravariant components of a 4-vector. If we multiply (4) by $\tilde{\rho}_3$ on the left-hand side, we get

$$[\tilde{\gamma}_0(-p_0+e'A_0)+\Sigma_r\tilde{\gamma}_r(-p_r+e'A_r)+mc]\,\phi=0, \tag{6}$$

since

$$\tilde{\gamma}_0=\rho_3. \qquad \tilde{\gamma}_r=\tilde{\sigma}_r\tilde{\rho}_3\tilde{\rho}_1=\tilde{\rho}_3\tilde{\sigma}_1\tilde{\rho}_r.$$

The operator in this equation is just the transposed operator of (5). The probability per unit volume of the electron being in any place is now given by

$$\phi\psi=\phi\rho_3\chi=\phi\gamma_0\chi, \tag{7}$$

where $\phi\alpha\chi$ denotes the sum of the products of each component of ϕ into the corresponding component of $\alpha\chi$, α being any function of the spin variables, represented by a matrix with four rows and columns. [Note that quite generally $\phi\alpha\chi=\chi\tilde{\alpha}\phi$.] Expression (7) is the time component of a 4-vector, whose spacial components, namely,

$$-\phi\gamma_1\chi, \qquad -\phi\gamma_2\chi, \qquad -\phi\gamma_3\chi,$$

must give $1/c$ times the probability per unit time of the electron crossing unit area perpendicular to each of the three axes respectively.

We must now show that the divergence of this 4-vector vanishes, *i.e.*, that

$$\frac{1}{c}\frac{\partial}{\partial t}(\phi\gamma_0\chi)-\Sigma_r\frac{\partial}{\partial x_r}(\phi\gamma_r\chi)=0. \tag{8}$$

Multiplying (5) by ϕ and (6) by χ and subtracting, we get

$$\phi\,[\gamma_0p_0+\Sigma_r\gamma_rp_r]\,\chi+\chi\,[\tilde{\gamma}_0p_0+\Sigma_r\tilde{\gamma}_rp_r]\,\phi=0,$$

which gives

$$\phi\left[\gamma_0\frac{\partial}{c\partial t}-\Sigma_r\gamma_r\frac{\partial}{\partial x_r}\right]\chi+\chi\left[\tilde{\gamma}_0\frac{\partial}{c\partial t}-\Sigma_r\tilde{\gamma}_r\frac{\partial}{\partial x_r}\right]\phi=0,$$

or

$$\phi\left[\gamma_0\frac{\partial}{c\partial t}-\Sigma_r\gamma_r\frac{\partial}{\partial x_r}\right]\chi+\frac{1}{c}\frac{\partial\phi}{\partial t}\gamma_0\chi-\Sigma_r\frac{\partial\phi}{\partial x_r}\gamma_r\chi=0.$$

This gives immediately the conservation equation (8) as the γ's are here constant matrices.

§ 2. *The Selection Principle.*

In *loc. cit.* the quantum number j was introduced, which determines the magnitude of the resultant angular momentum for an electron moving in a central field of force. j can take both positive and negative integral values. Again, the magnetic quantum number $u = M_3/h$, say, that determines the component of the total angular momentum in some specified direction, was shown to take half odd integral values from $-|j| + \frac{1}{2}$ to $|j| - \frac{1}{2}$. The state $j = 0$ is thus excluded, and the weight of any state j is $2|j|$. The equation obtained to determine the energy levels, *i.e.*, equation (25) or (26), involves j only through the combination $j(j+1)$ except in the last term, which represents the spin correction. Thus two values of j which give the same value for $j(j+1)$ form a spin doublet, so that $j = j'$ and $j = -(j'+1)$ form a spin doublet when $j' > 0$. The connection between j-values and the usual notation for alkali spectra is therefore given by the following scheme :—

$$\begin{array}{cccc} j = -1 & \underbrace{1 \quad -2} & \underbrace{2 \quad -3} & \underbrace{3 \quad -4} \quad \ldots \\ \mathrm{S} & \mathrm{P} & \mathrm{D} & \mathrm{F} \end{array}$$

There is no azimuthal quantum number k in the present theory, an orbit for an electron in an atom being defined by three quantum numbers n, j, u only. One might on this account expect the selection rules, the relative intensities of the lines of a multiplet, etc., in the usual derivation of which k plays an important part, to be different in the present theory, but it will be found that they do just happen to be the same.

We shall first determine the selection rule for j. We use the following two theorems :—

(i) If a dynamical variable X anticommutes with j, its matrix elements all refer to transitions of the type $j \rightarrow -j$.

(ii) If a dynamical variable Y satisfies

$$[[\mathrm{Y}, jh], jh] = -\mathrm{Y}, \tag{9}$$

its matrix elements all refer to transitions of the type $j \rightarrow j \pm 1$.

To prove (i) we observe that the condition $j\mathrm{X} + \mathrm{X}j = 0$ gives

$$j' \,.\, \mathrm{X}(j'j'') + \mathrm{X}(j'j'') \,.\, j'' = 0$$

or

$$(j' + j'') \,.\, \mathrm{X}(j'j'') = 0.$$

Hence $\mathrm{X}(j'j'') = 0$ unless $j'' = -j'$.

A proof of (ii) involving angle variables has been given in a previous paper.* A simple proof analogous to the foregoing proof of (i) is as follows. Equation (9) gives

$$\mathrm{Y}j^2 - 2j\,\mathrm{Y}j + j^2\mathrm{Y} = \mathrm{Y}$$

or

$$\mathrm{Y}\,(j'\,j'')\,.\,j''^2 - 2j'\,.\,\mathrm{Y}\,(j'j'')\,.\,j'' + j'^2\,.\,\mathrm{Y}\,(j'j'') = \mathrm{Y}\,(j'j'').$$

Hence $\mathrm{Y}\,(j'\,j'') = 0$ except when

$$j''^2 - 2j'j'' + j'^2 = 1,$$

i.e., when

$$j'' = j' \pm 1.$$

We shall now evaluate $[\,[x_3, jh], jh]$. The definition of j is

$$jh = \rho_3\{(\boldsymbol{\sigma}, \mathbf{m}) + h\}.$$

Hence

$$\begin{aligned}[x_3, jh] &= \rho_3\,\{\sigma_1\,[x_3, m_1] + \sigma_2\,[x_3, m_2]\}\\ &= \rho_3\,(\sigma_1 x_2 - \sigma_2 x_1),\end{aligned} \tag{10}$$

so that

$$[\,[x_3, jh], jh] = [\sigma_1\,x_2 - \sigma_2 x_1,\,(\boldsymbol{\sigma}, \mathbf{m})].$$

Now

$$ih\,[\sigma_1,\,(\boldsymbol{\sigma}, \mathbf{m})] = \sigma_1\,(\boldsymbol{\sigma}, \mathbf{m}) - (\boldsymbol{\sigma}, \mathbf{m})\,\sigma_1 = 2i\,(\sigma_3 m_2 - \sigma_2 m_3)$$

or

$$\tfrac{1}{2}h\,[\sigma_1,\,(\boldsymbol{\sigma}, \mathbf{m})] = \sigma_3 m_2 - \sigma_2 m_3,$$

and similarly

$$\tfrac{1}{2}h\,[\sigma_2,\,(\boldsymbol{\sigma}, \mathbf{m})] = \sigma_1 m_3 - \sigma_3 m_1.$$

Hence

$$\begin{aligned}\tfrac{1}{2}h\,[\,[x_3, jh], jh] &= (\sigma_3 m_2 - \sigma_2 m_3)\,x_2 + \tfrac{1}{2}h\,\sigma_1\,(\sigma_3 x_1 - \sigma_1 x_3)\\ &\qquad - (\sigma_1 m_3 - \sigma_3 m_1)\,x_1 - \tfrac{1}{2}h\,\sigma_2\,(\sigma_2 x_3 - \sigma_3 x_2)\\ &= \sigma_3\,(\mathbf{m}, \mathbf{x}) - m_3(\boldsymbol{\sigma}, \mathbf{x}) + \tfrac{1}{2}h\,\{-\sigma_3\,(\boldsymbol{\sigma}, \mathbf{x}) - x_3\}\\ &= -\,\mathrm{M}_3\,(\boldsymbol{\sigma}, \mathbf{x}) - \tfrac{1}{2}hx_3,\end{aligned}$$

so that

$$[\,[x_3, jh], jh] = -\,2u\,(\boldsymbol{\sigma}, \mathbf{x}) - x_3.$$

Thus x_3 does not quite satisfy the condition that Y satisfies in (9), owing to the extra term $-\,2u\,(\boldsymbol{\sigma}, \mathbf{x})$ This extra term, however, anticommutes with j. If we now form the expression $x_3 - cu\,(\boldsymbol{\sigma}, \mathbf{x})$, where c is some quantity that commutes with j, we can choose c so as to make this expression satisfy completely the condition that Y satisfies in (9). We have, in fact,

$$\begin{aligned}[\,[x_3 - cu\,(\boldsymbol{\sigma}, \mathbf{x}), jh], jh] &= -\,2u\,(\boldsymbol{\sigma}, \mathbf{x}) - x_3 + cu\,.\,4j^2\,(\boldsymbol{\sigma}, \mathbf{x})\\ &= -\,\{x_3 - cu\,(\boldsymbol{\sigma}, \mathbf{x})\}\end{aligned}$$

* 'Roy. Soc. Proc.,' A, vol. 111, p. 281 (1926), § 3.

if c is chosen such that

$$-2 + 4j^2 c = c,$$

i.e., if

$$c = 1/2\,(j^2 - \tfrac{1}{4}).$$

Hence x_3 can be expressed as the sum of two terms, namely,

$$\frac{u}{2\,(j^2 - \frac{1}{4})}\,(\boldsymbol{\sigma}, \mathbf{x}) \qquad \text{and} \qquad x_3 - \frac{u}{2\,(j^2 - \frac{1}{4})}\,(\boldsymbol{\sigma}, \mathbf{x}),$$

of which the first anticommutes with j, and therefore contains only matrix elements referring to transitions of the type $j \to -j$, while the second satisfies the condition that Y satisfies in (9), and therefore contains only matrix elements referring to transitions of the type $j \to j \pm 1$. A similar result holds for x_1 and x_2. Hence the selection rule for j is

$$j \to -j \qquad \text{or} \qquad j \to j \pm 1.$$

Thus from states with $j = 2$ transitions can take place to states with $j = 1$, -2 or 3. Comparing this selection rule with the above scheme connecting j-values with the S, P, D notation, we see that it is exactly equivalent to the two selection rules for j and k of the usual theory, and is therefore in agreement with experiment.

§ 3. *The Relative Intensities of the Lines of a Multiplet.*

The relative intensities of the various components into which a line is split up in a weak magnetic field must be the same on the present theory as on previous theories, as they depend only on the Vertauschungs relations connecting the co-ordinates x_r with the components of total angular momentum M_r, which are taken over unchanged into the present theory. It will therefore be sufficient, for determining the relative intensities of the lines of a multiplet, to consider only one Zeeman component of each line, say, the component for which $\Delta u = 0$, *i.e.*, the component that comes from x_3.

We shall determine the matrix elements of x_3, when expressed as a matrix in a scheme in which r, j, u and ρ_3 are diagonal. x_3 is diagonal in (*i.e.*, commutes with) all of these variables except j. The part of x_3 referring to transitions $j \to -j$ we found to be

$$\frac{u}{2(j^2 - \frac{1}{4})}\,(\boldsymbol{\sigma}, \mathbf{x}) = \frac{u}{2(j^2 - \frac{1}{4})}\,\varepsilon\,\rho_1\,r, \tag{11}$$

using the ε introduced in *loc. cit.* § 6. $\varepsilon\rho_1$ anticommutes with j, so that it can contain only matrix elements of the type $\varepsilon\rho_1(j, -j)$, and from the condition $(\varepsilon\rho_1)^2 = 1$ we must have

$$|\varepsilon\rho_1(j, -j)| = 1.$$

Hence

$$|x_3(j, -j)| = \frac{u}{2(j^2 - \frac{1}{4})} r\, |\varepsilon\rho_1(j, -j)| = \frac{u}{2(j^2 - \frac{1}{4})} r. \tag{12}$$

Again, we have from (10)

$$\{x_3 - i[x_3, jh]\}\{x_3 + i[x_3, jh]\} = \{x_3 - i\rho_3(\sigma_1 x_2 - \sigma_2 x_1)\}\{x_3 + i\rho_3(\sigma_1 x_2 - \sigma_2 x_1)\}$$
$$= x_3^2 + (\sigma_1 x_2 - \sigma_2 x_1)^2 = r^2,$$

which gives

$$\{(j+1)x_3 - x_3 j\}\{x_3(j+1) - jx_3\} = r^2.$$

If we equate the (j, j) matrix elements of each side of this equation, we get on the left-hand side the sum of three terms, namely, the $(j, -j)$ matrix element of the first { } bracket times the $(-j, j)$ element of the second, the $(j, j+1)$ element of the first times the $(j+1, j)$ element of the second, and the $(j, j-1)$ element of the first times the $(j-1, j)$ element of the second. The second of these three terms vanishes, leaving

$$(2j+1)^2|x_3(j, -j)|^2 + 4|x_3(j, j-1)|^2 = r^2.$$

Hence

$$|x_3(j, j-1)|^2 = \tfrac{1}{4}r^2\left\{1 - \frac{u^2}{(j-\frac{1}{2})^2}\right\} = \tfrac{1}{4}r^2\,\frac{(j+u-\frac{1}{2})(j-u-\frac{1}{2})}{(j-\frac{1}{2})^2}. \tag{13}$$

Writing $-j$ for j, we get

$$|x_3(-j, -j-1)|^2 = \tfrac{1}{4}r^2\,\frac{(j+u+\frac{1}{2})(j-u+\frac{1}{2})}{(j+\frac{1}{2})^2}. \tag{14}$$

The three matrix elements of x_3 given in (12), (13) and (14) are associated with the three components of the multiplet formed by the combination of two doublets. The ratios of these matrix elements will, to a first approximation, remain unchanged when one makes a transformation from the matrix scheme in which r, j, u, ρ_3 are diagonal to a scheme in which the Hamiltonian is diagonal, and will therefore give the relative intensities of the Zeeman components $\Delta u = 0$ of the lines in a combination doublet. These ratios are in agreement with those of previous theories based on the spinning electron model.

§ 4. *The Zeeman Effect.*

If there is a uniform magnetic field of intensity H in the direction of the x_3 axis, we can take the magnetic potentials to be

$$A_1 = -\tfrac{1}{2}Hx_2, \qquad A_2 = \tfrac{1}{2}Hx_1 \qquad A_3 = 0.$$

The additional terms appearing in the Hamiltonian F will now be

$$\Delta F = \rho_1 e' (\boldsymbol{\sigma}, A) = -\tfrac{1}{2} He' \rho_1 (\sigma_1 x_2 - \sigma_2 x_1).$$

From (10) it follows that $\rho_3 (\sigma_1 x_2 - \sigma_2 x_1)$ or $(\sigma_1 x_2 - \sigma_2 x_1)$, like x_3, contains only matrix elements of the type $(j, -j)$ or $(j, j \pm 1)$. Now ρ_1 anticommutes with j, and therefore contains only matrix elements of the type $(j, -j)$. Hence ΔF contains only matrix elements of the type (j, j) or $(j, -j \pm 1)$.

In *loc. cit.*, § 6, it was found [see equation (24)] that the Hamiltonian could be expressed as

$$F \equiv p_0 + V + \varepsilon p_r + i\varepsilon \rho_3 jh/r + \rho_3 mc. \tag{15}$$

It follows from (10) that $(\sigma_1 x_2 - \sigma_2 x_1)$ anticommutes with $(\boldsymbol{\sigma}, \mathbf{x})$, and therefore also with ε. Thus if we put

$$\Delta F = ih\varepsilon \rho_3 \eta r,$$

so that

$$\eta = \tfrac{1}{2} He/ch \,.\, \varepsilon \rho_2 (\sigma_1 x_2 - \sigma_2 x_1)/r,$$

η commutes with ε. Further, η commutes with ρ_3, r and p_r, so that it commutes with all the variables occurring in (15) except j. If we now express η as a matrix in j, we shall have obtained an expression for ΔF in terms of the variables occurring in (15). We have from (10) and (13)

$$| \rho_3 (\sigma_1 x_2 - \sigma_2 x_1) (j, j-1) |^2 = | x_3 (j, j-1) |^2 = \tfrac{1}{4} r^2 \frac{(j + u - \frac{1}{2})(j - u - \frac{1}{2})}{(j - \frac{1}{2})^2},$$

and similarly

$$\rho_3 (\sigma_1 x_2 - \sigma_2 x_1) (j, j+1) |^2 = | x_3 (j, j+1) |^2 = \tfrac{1}{4} r^2 \frac{(j + u + \frac{1}{2})(j - u + \frac{1}{2})}{(j + \frac{1}{2})^2}.$$

We have seen that the matrix elements of $\varepsilon \rho_1$, all of which are of the type $(j, -j)$, must be of modulus unity. Hence

$$\left.\begin{aligned} |\eta (j, -j-1)|^2 &= \left(\frac{He}{2chr}\right)^2 |i\varepsilon \rho_1 (j, -j)|^2 |\rho_3 (\sigma_1 x_2 - \sigma_2 x_1)(-j, -j-1)|^2 \\ &= \left(\frac{He}{4ch}\right)^2 \frac{(j + u + \frac{1}{2})(j - u + \frac{1}{2})}{(j + \frac{1}{2})^2} \\ \text{and similarly} \qquad & \\ | \eta (j, -j+1)|^2 &= \left(\frac{He}{4ch}\right)^2 \frac{(j + u - \frac{1}{2})(j - u - \frac{1}{2})}{(j - \frac{1}{2})^2} \end{aligned}\right\} . \tag{16}$$

Again, from (10) and (11)

$$\rho_3 (\sigma_1 x_2 - \sigma_2 x_1)(-j, j) = -2ij \,.\, x_3 (-j, j) = -\frac{u}{(j^2 - \frac{1}{4})} irj \,.\, (\varepsilon \rho_1)(-j, j),$$

so that

$$\eta (j, j) = \frac{He}{2ch} \frac{uj}{j^2 - \frac{1}{4}}. \tag{17}$$

If we now write down in full, as in *loc. cit.*, the wave equation corresponding to (15), and include the extra term ΔF, we shall have

$$[(F + \Delta F)\,\psi]_\alpha = (p_0 + V)\,\psi_\alpha - h\frac{\partial}{\partial r}\psi_\beta - \left(\frac{j}{r} + \eta r\right) h\psi_\beta + mc\psi_\alpha = 0,$$

$$[(F + \Delta F)\,\psi]_\beta = (p_0 + V)\,\psi_\beta + h\frac{\partial}{\partial r}\psi_\alpha - \left(\frac{j}{r} + \eta r\right) h\psi_\alpha - mc\psi_\beta = 0,$$

where η is now an operator, operating on ψ_α and ψ_β, that commutes with everything except j. On eliminating ψ_α this gives, corresponding to (25) of *loc. cit.*,

$$\frac{\partial^2}{\partial r^2}\psi_\beta + \left[\frac{(p_0 + V)^2 - m^2c^2}{h^2} - \frac{j(j+1)}{r^2} + \eta - \eta j - j\eta - \eta^2 r^2\right]\psi_\beta$$
$$- \frac{1}{p_0 + V + mc}\frac{\partial V}{\partial r}\left[\frac{\partial}{\partial r} + \frac{j}{r} + \eta r\right]\psi_\beta = 0.$$

We can neglect the $\eta^2 r^2$ term, which is proportional to the square of the field strength, and also the ηr term in the last bracket, which is of the order of magnitude of field strength times spin correction. The only first order effect of the field is the insertion of the terms $\eta - \eta j - j\eta$ in the first bracket. This bracket may now be written as

$$\left[\frac{2mE}{h^2} + \frac{E^2}{c^2h^2} + \frac{2\cdot(E + mc^2)}{ch^2}V + \frac{V^2}{h^2} - \frac{j(j+1)}{r^2} + \eta - \eta j - j\eta\right], \quad (18)$$

where E is the energy level, equal to $p_0 c - mc^2$.

If the field is weak compared with the doublet separation, we can obtain a first approximation to the change in the energy levels by neglecting the non-diagonal matrix elements of ΔF or of η. The extra terms $\eta - \eta j - j\eta$ in (18) are now a constant instead of an operator, namely, the constant

$$-(2j - 1)\,\eta(j, j) = -\frac{He}{ch}\frac{uj}{j + \frac{1}{2}}$$

from (17). The energy levels will be reduced by $h^2/2m$ times this constant, if we neglect the fact that the characteristic E occurs in (18) in other places besides the term $2mE/h^2$, which means neglecting the interaction of the magnetic field with the relativity variation of mass with velocity. The increase in the energy levels caused by the magnetic field is thus

$$\frac{He}{2mc}\frac{j}{j + \frac{1}{2}}uh = \omega g u h$$

where ω is the Larmor frequency $\mathrm{H}e/2mc$, and g, the Landê splitting factor, has the value

$$g = j/(j+\tfrac{1}{2}).$$

For the succession of j-values, $-1, 1, -2, 2, -3 \ldots g$ has the values, $2, \frac{2}{3}, \frac{4}{3}, \frac{4}{5}, \frac{6}{5} \ldots$, in agreement with Landê's formula for alkali spectra.

We now take the case of a magnetic field that is strong compared with the doublet separation, but weak compared with the separations of terms of different series. This requires that the matrix elements of η of the type $\eta\,(j, -j-1)$ with $j > 0$ shall be taken into account, although those of the type $\eta\,(j, -j+1)$ can still be neglected. The reduction in the energy levels will now be approximately $h^2/2m$ times one or other of the characteristic values of the extra terms $\eta - \eta j - j\eta$ in (18). These characteristic values are the roots ξ of the equation

$$\begin{vmatrix} (\eta - \eta j - j\eta)\,(j, j) - \xi & (\eta - \eta j - j\eta)\,(j, -j-1) \\ (\eta - \eta j - j\eta)\,(-j-1, j) & (\eta - \eta j - j\eta)\,(-j-1, -j-1) - \xi \end{vmatrix} = 0.$$

or

$$\begin{vmatrix} -(2j-1)\,.\,\eta\,(j, j) - \xi & 2\eta\,(j, -j-1) \\ 2\eta\,(-j-1, j) & (2j+3)\,.\,\eta\,(-j-1, -j-1) - \xi \end{vmatrix} = 0.$$

This gives, with the help of (16) and (17)

$$\xi^2 + \frac{\mathrm{H}e}{ch}\left[\frac{uj}{j+\frac{1}{2}} + \frac{u\,(j+1)}{j+\frac{1}{2}}\right]\xi + \left(\frac{\mathrm{H}e}{ch}\right)^2\left[\frac{u^2 j\,(j+1)}{(j+\frac{1}{2})^2} - \frac{(j+\frac{1}{2})^2 - u^2}{4\,(j+\frac{1}{2})^2}\right] = 0,$$

which reduces to

$$\xi^2 + \frac{\mathrm{H}e}{ch} 2u\xi + \left(\frac{\mathrm{H}e}{ch}\right)^2 (u^2 - \tfrac{1}{4}) = 0.$$

Hence

$$\xi = -\frac{\mathrm{H}e}{ch}(u \pm \tfrac{1}{2}).$$

The increase in the energy levels due to the magnetic field is therefore

$$-\frac{h^2}{2m}\xi = \frac{h^2}{2m}\frac{\mathrm{H}e}{ch}(u \pm \tfrac{1}{2}) = \omega\,(u \pm \tfrac{1}{2})\,h,$$

in agreement with the previous spinning electron theory of the Paschen-Back effect.

One might expect that with still stronger magnetic fields the matrix elements $(j, -j+1)$ of η would come into play, and would cause interference between the Zeeman patterns of terms whose quantum numbers k in the usual notation differ by 2. The matrix elements $(j, -j+1)$ of $\eta - \eta j - j\eta$, however, vanish for arbitrary η, so that no effect of this nature occurs.

Reprint 10

Hydrogen in Electrodynamics. VI
The General Solution

Hans Sallhofer
Forschungs- und Versuchsanstalt der Austria Metall AG, Braunau, Austria

Z. Naturforsch. **45a**, 1361–1366 (1990); received November 12, 1990

After a discussion of the one-component Schrödinger (1926) and the four-component Dirac (1928) representation of hydrogen it is shown that the six-component electrodynamic picture turns out to be considerably simpler and clearer. The computational effort is reduced to a fraction.

1. The Transversality of Hydrogen

Again we start with covariant source-free electrodynamics [1] (22) with its complex components

$$\left.\begin{cases}\operatorname{rot} E+\frac{\mu}{c}\dot{H}=0\\ \operatorname{rot} H-\frac{\varepsilon}{c}\dot{E}=0\\ \operatorname{div}\varepsilon E=0\\ \operatorname{div}\mu H=0\end{cases}\right\}_{\operatorname{div}E=\operatorname{div}H=0} \equiv \left.\begin{cases}\operatorname{rot} E+\frac{\mu}{c}\dot{H}=0\\ \operatorname{rot} H-\frac{\varepsilon}{c}\dot{E}=0\end{cases}\right\}_{\substack{\operatorname{div}E=\operatorname{div}H=0\\ E\perp \operatorname{gerad}\varepsilon\\ H\perp\operatorname{grad}\mu}} \tag{1}$$

where we have ensured the absence of sources by the requirement

$$\operatorname{div} E=\operatorname{div} H=0\,. \tag{2}$$

From (2) follows

$$E\cdot\operatorname{grad}\varepsilon=0 \quad\text{and}\quad H\cdot\operatorname{grad}\mu=0 \tag{3}$$

or

$$E\perp\operatorname{grad}\varepsilon \quad\text{and}\quad H\perp\operatorname{grad}\mu\,. \tag{4}$$

For the case of hydrogen we have to insert the hydrogen-interfraction

$$(\varepsilon_{\rm int}^{\rm hyd},\mu_{\rm int}^{\rm hyd})=\left(1-\frac{\Phi-m_0c^2}{\hbar\omega},\;1-\frac{\Phi+m_0c^2}{\hbar\omega}\right) \tag{5}$$
$$=\left(1-\frac{-e_0^2/r-m_0c^2}{\hbar\omega},\;1-\frac{-e_0^2/r+m_0c^2}{\hbar\omega}\right)$$

according to [2] (2). grad $\varepsilon_{\rm int}^{\rm hyd}$ and $\mu_{\rm int}^{\rm hyd}$ then both point in the direction of the radius vector and, according to (4), the field vectors are therefore tangent to spheres around the origin. We call this essential feature, which immediately provides us with a basic and comprehensive picture of the field structure of hydrogen, "central transversality".

2. The Fermion Ansatz

In [4] (21) we have seen that all components of the hydrogen field have to vanish always and everywhere – with nonvanishing energy – because of the half-integer sign ambiguity of angular momentum and spin. This is in best agreement with the experience that the hydrogen atom doesn't reveal itself to the outside as an electromagnetic phenomenon and is thus perceptible only mechanically. For the general solution we therefore require

$$Q_k=0 \quad (Q=E,H;\; k=1,\ldots,3) \tag{6}$$

with

$$U=Q_k\overset{*}{Q}_k\neq 0\,. \tag{7}$$

Furthermore we especially put

$$Q_2^L=\pm i\,Q_1^L \quad (L=\text{I, II}), \tag{8}$$

noticing that this requirement is contained in the ansatz (6), as long as both values of the right-hand side of (8) are taken into account simultaneously. – Since only half-integer spin or momentum functions on the right-hand side of (8) can be double-valued, (8) in the end requires solutions with half-integer azimuthal functions. We therefore call (6)–(8) the fermion ansatz of electrodynamics. The purely electrodynamic derivation of the of the Maxwell-Dirac isomorphism from [3] (7) to [3] (20) shows that the solutions of (1) can be represented by means of the Dirac functions Ψ_k^L (L = I,

Reprint requests to Dr. H. Sallhofer, Bahnhofstrasse 36, A-5280 Braunau, Austria.

II; $k=1,\ldots,4$) as follows:

$$\begin{pmatrix} iE_3^{\mathrm{I}} & i(E_1^{\mathrm{II}}-iE_2^{\mathrm{II}}) \\ i(E_1^{\mathrm{I}}+iE_2^{\mathrm{I}}) & -iE_3^{\mathrm{II}} \\ H_3^{\mathrm{I}} & H_1^{\mathrm{II}}-iH_2^{\mathrm{II}} \\ (H_1^{\mathrm{I}}+iH_2^{\mathrm{I}}) & -H_2^{\mathrm{II}} \end{pmatrix} = \begin{pmatrix} \Psi_1^{\mathrm{I}} & \Psi_1^{\mathrm{II}} \\ \Psi_2^{\mathrm{I}} & \Psi_2^{\mathrm{II}} \\ \Psi_3^{\mathrm{I}} & \Psi_2^{\mathrm{II}} \\ \Psi_4^{\mathrm{I}} & \Psi_4^{\mathrm{II}} \end{pmatrix} \tag{9}$$

(Maxwell) (Dirac)

Inserting into the Maxwell side of (9) successively both alternatives of (8) and adding the results we get

$$\begin{pmatrix} iE_3^{\mathrm{I}} & iE_1^{\mathrm{II}} \\ iE_1^{\mathrm{I}} & -iE_3^{\mathrm{II}} \\ H_3^{\mathrm{I}} & H_1^{\mathrm{II}} \\ H_1^{\mathrm{I}} & -H_3^{\mathrm{II}} \end{pmatrix} = \frac{1}{2} \begin{pmatrix} \Psi_1^{\mathrm{I}} & \Psi_1^{\mathrm{II}} \\ \Psi_2^{\mathrm{I}} & \Psi_2^{\mathrm{II}} \\ \Psi_3^{\mathrm{I}} & \Psi_3^{\mathrm{II}} \\ \Psi_4^{\mathrm{I}} & \Psi_4^{\mathrm{II}} \end{pmatrix}. \tag{10}$$

This equation relates the electromagnetic field components unambiguously to the Dirac functions, and thus enables us to calculate without much effort the electromagnetic field components according to the scheme

$$Q_k = Q_k^{\mathrm{I}} + Q_k^{\mathrm{II}}, \tag{11}$$

once we know the Dirac functions.

3. The General Hydrogen Solution

Let us now insert into the Dirac side of (10) those Dirac functions which we have obtained in [5] (47)–(50). With

$$\Psi_k^L = \overset{-}{\Psi}{}_k^L + \overset{+}{\Psi}{}_k^L \quad (L=\mathrm{I},\mathrm{II};\ k=1,\ldots,4) \tag{12}$$

they are characterized by the D'Alembert structure and, to begin with, permit the following decomposition

$$\begin{pmatrix} iE_3^{\mathrm{I}} & iE_1^{\mathrm{II}} \\ iE_1^{\mathrm{I}} & -iE_3^{\mathrm{II}} \\ H_3^{\mathrm{I}} & H_1^{\mathrm{II}} \\ H_1^{\mathrm{I}} & -H_3^{\mathrm{II}} \end{pmatrix} + \begin{pmatrix} iE_3^{\mathrm{I}} & iE_1^{\mathrm{II}} \\ iE_1^{\mathrm{I}} & -iE_3^{\mathrm{II}} \\ H_3^{\mathrm{I}} & H_1^{\mathrm{II}} \\ H_1^{\mathrm{I}} & -H_3^{\mathrm{II}} \end{pmatrix}$$

$$= \begin{pmatrix} \overset{-}{\Psi}{}_1^{\mathrm{I}} & \overset{+}{\Psi}{}_1^{\mathrm{II}} \\ \overset{-}{\Psi}{}_2^{\mathrm{I}} & \overset{+}{\Psi}{}_2^{\mathrm{II}} \\ \overset{-}{\Psi}{}_3^{\mathrm{I}} & \overset{+}{\Psi}{}_3^{\mathrm{II}} \\ \overset{-}{\Psi}{}_4^{\mathrm{I}} & \overset{+}{\Psi}{}_4^{\mathrm{II}} \end{pmatrix} + \begin{pmatrix} \overset{+}{\Psi}{}_1^{\mathrm{I}} & \overset{-}{\Psi}{}_1^{\mathrm{II}} \\ \overset{+}{\Psi}{}_2^{\mathrm{I}} & \overset{-}{\Psi}{}_2^{\mathrm{II}} \\ \overset{+}{\Psi}{}_3^{\mathrm{I}} & \overset{-}{\Psi}{}_3^{\mathrm{II}} \\ \overset{+}{\Psi}{}_4^{\mathrm{I}} & \overset{-}{\Psi}{}_4^{\mathrm{II}} \end{pmatrix}. \tag{13}$$

or simply

$$\begin{pmatrix} \overset{-}{\Psi}{}_1^{\mathrm{I}} & \overset{+}{\Psi}{}_1^{\mathrm{II}} \\ \overset{-}{\Psi}{}_2^{\mathrm{I}} & \overset{+}{\Psi}{}_2^{\mathrm{II}} \\ \overset{-}{\Psi}{}_3^{\mathrm{I}} & \overset{+}{\Psi}{}_3^{\mathrm{II}} \\ \overset{-}{\Psi}{}_4^{\mathrm{I}} & \overset{+}{\Psi}{}_4^{\mathrm{II}} \end{pmatrix} = \begin{pmatrix} i\check{E}_3^{\mathrm{I}} & i\check{E}_1^{\mathrm{II}} \\ i\check{E}_1^{\mathrm{I}} & -i\check{E}_3^{\mathrm{II}} \\ \check{H}_3^{\mathrm{I}} & \check{H}_1^{\mathrm{II}} \\ \check{H}_1^{\mathrm{I}} & -\check{H}_3^{\mathrm{II}} \end{pmatrix} = \begin{pmatrix} \overset{+}{\Psi}{}_1^{\mathrm{I}} & \overset{-}{\Psi}{}_1^{\mathrm{II}} \\ \overset{+}{\Psi}{}_2^{\mathrm{I}} & \overset{-}{\Psi}{}_2^{\mathrm{II}} \\ \overset{+}{\Psi}{}_3^{\mathrm{I}} & \overset{-}{\Psi}{}_3^{\mathrm{II}} \\ \overset{+}{\Psi}{}_4^{\mathrm{I}} & \overset{-}{\Psi}{}_4^{\mathrm{II}} \end{pmatrix}. \tag{14}$$

Continuing, after these preparations, towards the solution of (1) we have to solve the purely electromagnetic equation [3] (19), according to the derivation between [3] (7) and [3] (20), or, taking into account (5), the purely electromagnetic equation [3] (20), that is

$$\left[\gamma\cdot\nabla + \frac{i}{\hbar c}\begin{pmatrix} \left(\hbar\omega + \frac{e_0^2}{r} + m_0c^2\right)\mathbf{1} & 0 \\ 0 & \left(\hbar\omega + \frac{e_0^2}{r} - m_0c^2\right)\mathbf{1} \end{pmatrix}\right]\Psi = 0, \tag{15}$$

which is identical with the Dirac equation of hydrogen. Since the solution of the Dirac equation of hydrogen is known, we may in the following take over the solutions $\Psi_k^{\mathrm{I,II}}$, as well as any needed details of the method of solution from the Dirac theory.

According to [5] (47)–(50) the radial components R_Q of the Dirac functions, taken as solutions of the two pairs of differential equations [5] (31)–(32) and [5] (36)–(37), appear in the two states "I" and "II" as follows

$$\Psi^{\mathrm{I}} = CR_Q^{\mathrm{I}}\,P^{m^{\mathrm{I}}}\,e^{im^{\mathrm{I}}\phi}\,e^{-i\omega^{\mathrm{I}}t} \qquad 0 \leqq |m^{\mathrm{I}}| \leqq l^{\mathrm{I}}, \tag{16}$$

$$\Psi^{\mathrm{II}} = CR_Q^{\mathrm{II}}\,P^{m^{\mathrm{II}}}\,e^{im^{\mathrm{II}}\phi}\,e^{-i\omega^{\mathrm{II}}t} \qquad 0 \leqq |m^{\mathrm{II}}| \leqq l^{\mathrm{II}}. \tag{17}$$

Here the electric radial component R_E, for instance, has the following form for the states I and II:

$$R_E^{\mathrm{I}} = e^{-\lambda r}\sum_{k=0}^{n_r} a_k^{\mathrm{I}}\,r^{q^{\mathrm{I}}+k}, \quad \lambda = \frac{2\pi}{hc}\sqrt{m_0^2c^4-(h\nu)^2}, \tag{18}$$

$$q^{\mathrm{I}} = -1 + \sqrt{(-l^{\mathrm{I}}-1)^2-\alpha^2}, \quad (l^{\mathrm{I}}=0,1,\ldots), \tag{19}$$

$$a_k^{\mathrm{I}} = \left[\alpha-(q^{\mathrm{I}}+k-l^{\mathrm{I}})\sqrt{\frac{m_0c^2-h\nu}{m_0c^2+h\nu}}\right]c_k^{\mathrm{I}}, \tag{20}$$

$$[(q_k^{\mathrm{I}}+k-l^{\mathrm{I}}+1)(q^{\mathrm{I}}+k+l^{\mathrm{I}}+3)+\alpha^2]\,c_{k+1}^{\mathrm{I}} = \frac{4\pi}{hc}[(q^{\mathrm{I}}+k+1)\sqrt{m_0^2c^4-(h\nu)^2}-\alpha h\nu]\,c_k^{\mathrm{I}}, \tag{21}$$

$$\omega^{\mathrm{I}} = \frac{m_0c^2}{\hbar\sqrt{1+\dfrac{\alpha^2}{(n_r+\sqrt{(-l^{\mathrm{I}}-1)^2-\alpha^2})^2}}} = \omega(-l^{\mathrm{I}}-1), \tag{22}$$

and

$$R_E^{II} = e^{-\lambda r} \sum_{k=0}^{n_r} a_k^{II} r^{q^{II}+k}, \quad \lambda = \frac{2\pi}{hc}\sqrt{m_0^2 c^4 - (h\nu)^2}, \quad (23)$$

$$q^{II} = -1 + \sqrt{(+l^{II})^2 - \alpha^2}, \quad (l^{II} = 1, 2, \ldots), \quad (24)$$

$$a_k^{II} = \left[\alpha - (q^{II} + k + l^{II} + 1)\sqrt{\frac{m_0 c^2 - h\nu}{m_0 c^2 + h\nu}}\right] c_k^{II}, \quad (25)$$

$$[(q_k^{II} + k + l^{II} + 2)(q^{II} + k - l^{II} + 2) + \alpha^2]\, c_{k+1}^{II} = \frac{4\pi}{hc}[(q^{II} + k + 1)\sqrt{m_0^2 c^4 - (h\nu)^2} - \alpha h\nu]\, c_k^{II}, \quad (26)$$

the different bounds of l^{I} and l^{II} are to be taken into account in such a way that the magnetic quantum number m is lowered by 1 in the azimuthal function of the state II.

We are now going to compute the general solution E from (11) and (14). In (14) we have, in anticipation, denoted the prospective solution as $\check{E}$, since, as we shall see, it only represents one half of the general solution, – the other half $\hat{E}$ will be obtained by an appropriate substitution. The general solution, therefore, will be of the form

$$E_k = \check{E}_k + \hat{E}_k. \quad (32)$$

To begin with, let us turn to the left-hand side of the douple equation (14) and write it out explicitly according to [5] (47)–(50):

$$C\begin{pmatrix} -iR_E^{I} P_{l^I+1}^{m+1} e^{-i(m+1)\phi - i\omega^I t} & -i(l^{II}+m) R_E^{II} P_{l^{II}-1}^{m} e^{i(m-1)\phi - i\omega^{II} t} \\ i(l^I - m + 1) R_E^{I} P_{l^I+1}^{m} e^{-im\phi - i\omega^I t} & iR_E^{II} P_{l^{II}-1}^{m+1} e^{im\phi - i\omega^{II} t} \\ R_H^{I} P_{l^I}^{m+1} e^{-i(m+1)\phi - i\omega^I t} & (l^{II} - m) R_H^{II} P_{l^{II}}^{m} e^{i(m-1)\phi - i\omega^{II} t} \\ (l^I + m + 1) R_H^{I} P_{l^I}^{m} e^{-im\phi - i\omega^I t} & R_H^{II} P_{l^{II}}^{m+1} e^{im\phi - i\omega^{II} t} \end{pmatrix} = \begin{pmatrix} i\check{E}_3^{I} & i\check{E}_1^{II} \\ i\check{E}_1^{I} & -i\check{E}_3^{II} \\ \check{H}_3^{I} & \check{H}_1^{II} \\ \check{H}_1^{I} & -\check{H}_3^{II} \end{pmatrix}. \quad (33)$$

$$\omega^{II} = \frac{m_0 c^2}{\hbar\sqrt{1 + \dfrac{\alpha^2}{(n_r + \sqrt{(l^{II})^2 - \alpha^2})^2}}} = \omega(l^{II}). \quad (27)$$

Obviously the relations

$$R_E^{I} = R_E(-l^I - 1), \quad (28)$$

$$R_E^{II} = R_E(l^{II}) \quad (29)$$

hold good. In (19) and (24), l^{I} and l^{II} have different bounds. The first of the quantum numbers corresponds to the usual bound as in (16). On the other hand, l^{II} starts with 1, a fact which has been discussed extensively by von Laue [6]. Since m^{II} has to go through zero for the smallest l^{II}-value ($=1$), it has to be raised in the spherical harmonic of the ansatz (17): (30)

$$\Psi^{II} = CR_Q^{II} P^{|m^{II}+1|} e^{im^{II}\phi} e^{-i\omega^{II} t}, \quad 1 \leq |m^{II}+1| \leq l^{II}.$$

The spherical harmonics should, however, possess the same upper index in the two states I and II in view of their intended separation. We will therefore lower again the upper index of the spherical harmonics in (26), the azimuthal function being taken into account accordingly:

$$\Psi^{II} = CR_Q^{II} P^{m^{II}} e^{i(m^{II}-1)\phi} e^{-i\omega^{II} t}. \quad (31)$$

This is possible since we are free to choose the levels of m^{I} and m^{II} in the scheme (11). On balance, therefore,

Here we have lowered the quantum numbers m of Ψ^{II} by one in view of the remarks below (31) to achieve the adjustment of the l-bounds. In addition we have put all constants equal to C.

For $\check{E}_1$ we now get, according to (33) and (11):

$$\check{E}_1 = \check{E}_1^{I} + \check{E}_1^{II} = C[(l^I - m + 1) R_E^{I} P_{l^I+1}^{m} e^{-im\phi} e^{-i\omega^I t} - (l^{II} + m) R_E^{II} P_{l^{II}-1}^{m} e^{i(m-1)\phi} e^{-i\omega^{II} t}] \quad (34)$$

or, with (28) and (29), as well as (22) and (27),

$$\check{E}_1 = C[(l^I - m + 1) R_E(-l^I - 1) P_{l^I+1}^{m} e^{-im\phi} e^{-i\omega(-l^I-1)t} - (l^{II} + m) R_E(l^{II}) P_{l^{II}-1}^{m} e^{i(m-1)\phi} e^{-i\omega(l^{II})t}]. \quad (35)$$

We are now trying to separate the two state parameters l^{I} and l^{II} by means of the averaging double requirement

$$-l^I - 1 = l^{II} = l = l^I = -l^{II} - 1. \quad (36)$$

Inserting first the left-hand, then the right-hand requirement of (36) into (35), taking into account the well-known relation for the spherical harmonics

$$P_l^m = P_{-l-1}^m, \quad (37)$$

we get the two wave fields

$$-\frac{\check{E}_1}{R_E(l)} = C[(l+m) P_{l-1}^{m} e^{-im\phi} + (l+m) P_{l-1}^{m} e^{i(m-1)\phi}] e^{-i\omega(l)t}, \quad (38)$$

$$\frac{\check{E}_1}{R_E(-l-1)} = C\,[(l-m+1)\,P_{l+1}^{m}\,e^{-im\phi} + (l-m+1)\,P_{l+1}^{m}\,e^{i(m-1)\phi}]\,e^{-i\omega(-l-1)t}, \quad (39)$$

According to (22), or (27) respectively, the wave (38) oscillates with the frequency

$$\omega(l) = \frac{m_0 c^2}{\hbar\sqrt{1+\dfrac{\alpha^2}{(n_r+\sqrt{(l^2-\alpha^2)})^2}}} \quad (40)$$

and the wave (39) of approximately the same amplitude with an almost equal frequency, that is

$$\omega(-l-1) = \frac{m_0 c^2}{\hbar\sqrt{1+\dfrac{\alpha^2}{(n_r+\sqrt{(l+1)^2-\alpha^2})^2}}} = \omega(l+1), \quad (41)$$

According to well-known theorems of wave theory thus a mean hydrogen frequency

$$\omega = \frac{m_0 c^2}{\hbar\sqrt{1+\dfrac{\alpha^2}{(n_r+\sqrt{(l+1/2)^2-\alpha^2})^2}}} = \omega(l+1/2), \quad (42)$$

arises. Equation (42) perfectly agrees with the spectroscopically determined hydrogen freqencies [6].

By accepting the mean frequency (42) for (38) and (39), the two fields can be added with the help of [7] (11), and we finally get

$$\check{E}_1 = 4\,C R_E^R\,P_l^m \cos\theta\,[\pm\cos(m-\tfrac{1}{2})\,\phi]\,(\pm e^{-i\phi/2})\,e^{-i\omega t}, \quad (43)$$

if we introduce

$$R_E^R = \frac{(l+1/2)\,R_E(l)\,R_E(-l-1)}{R_E(l)-R_E(-l-1)} \quad (44)$$

as the "reduced electric radial function". This follows the concept of reduced mass [1] (6) because of the similarity in structure and similar invariance with respect to the interchange ($m^{\mathrm{Re}} \leftrightarrow m^{\mathrm{Im}}$, $l \leftrightarrow -l-1$).

As we have already seen in [4] (21), $\check{E}_1$ has to vanish everywhere and always, in view of the linearity of (1), because of the two azimuthal functions with their sign ambiguities

$$\pm\cos(m-\tfrac{1}{2})\,\phi \quad \text{(momentum function)} \quad \text{and}$$
$$\pm e^{-i\phi/2} \quad \text{(spin function)} \quad (45)$$

in contrast to the energy, which is positive everywhere and always. Keeping this basic fact in mind we are going to write (43), and all other components or partial components, in a simpler form, that is

$$\check{E}_{1-} = 4\,C R_E^R\,P_l^m \cos\theta\,\cos(m-\tfrac{1}{2})\,\phi\,e^{-i\phi/2}\,e^{-i\omega t} = 0. \quad (46)$$

Here the sign after the component index denotes which sense of rotation is indicated by the half-integer spin factor.

Because of the vanishing of all components, or partial components respectively, we do not have to pay any attention to their sign and thus simply write them as basically positive quantities.

If we analogously repeat the arguments leading from (33) to (43) for $\check{E}_3$, taking into account the relation [7] (10) instead of [7] (11), we get for this component

$$\check{E}_{3-} = 4\,C R_E^R\,P_l^m \sin\theta\,\cos(m+\tfrac{1}{2})\,\phi\,e^{-i\phi/2}\,e^{-i\omega t} = 0. \quad (47)$$

Turning now to the right-hand side of (14), and proceeding as on the left-hand side, we get

$$\check{E}_{1+} = i4\,C R_E^R\,P_l^m \sin\theta\,\sin(m+\tfrac{1}{2})\,\phi\,e^{i\phi/2}\,e^{-i\omega t} = 0. \quad (48)$$

$$\check{E}_{3+} = i4\,C R_E^R\,P_l^m \cos\theta\,\sin(m-\tfrac{1}{2})\,\phi\,e^{i\phi/2}\,e^{-i\omega t} = 0. \quad (49)$$

In order to obtain the electric partial component $\hat{E}_k$, which together with $\check{E}_k$ should yield the general components according to (32), we interchange the columns in (9), either on the Maxwell or on the Dirac side. This amounts to an interchange of the columns of the central matrix in the double relation (14) – by keeping the Roman index fixed:

$$\begin{pmatrix} \overset{-}{\Psi}{}_1^{\mathrm{I}} & \overset{+}{\Psi}{}_1^{\mathrm{II}} \\ \overset{-}{\Psi}{}_2^{\mathrm{I}} & \overset{+}{\Psi}{}_2^{\mathrm{II}} \\ \overset{-}{\Psi}{}_3^{\mathrm{I}} & \overset{+}{\Psi}{}_3^{\mathrm{II}} \\ \overset{-}{\Psi}{}_4^{\mathrm{I}} & \overset{+}{\Psi}{}_4^{\mathrm{II}} \end{pmatrix} = \begin{pmatrix} i\hat{E}_1^{\mathrm{I}} & i\hat{E}_3^{\mathrm{II}} \\ -i\hat{E}_3^{\mathrm{I}} & i\hat{E}_1^{\mathrm{II}} \\ \hat{H}_1^{\mathrm{I}} & \hat{H}_3^{\mathrm{II}} \\ -\hat{H}_3^{\mathrm{I}} & \hat{H}_1^{\mathrm{II}} \end{pmatrix} = \begin{pmatrix} \overset{+}{\Psi}{}_1^{\mathrm{I}} & \overset{-}{\Psi}{}_1^{\mathrm{II}} \\ \overset{+}{\Psi}{}_2^{\mathrm{I}} & \overset{-}{\Psi}{}_2^{\mathrm{II}} \\ \overset{+}{\Psi}{}_3^{\mathrm{I}} & \overset{-}{\Psi}{}_3^{\mathrm{II}} \\ \overset{+}{\Psi}{}_4^{\mathrm{I}} & \overset{-}{\Psi}{}_4^{\mathrm{II}} \end{pmatrix}. \quad (50)$$

As an be immediately seen, we get as a by-product of this interchange of columns an interchange of components of the form

$$\hat{E}_1 \leftrightarrow \hat{E}_3 \quad \text{and} \quad \hat{H}_1 \leftrightarrow \hat{H}_3, \quad (51)$$

which has to be taken into account below.

Writing out explicitly the central matrix of the double relation (50) and then, exactly as before with (14), turning successively to the two sides, the respective electric partial components – taking into account (51) – are obtained as:

$$\hat{E}_{1-} = i4\,C R_E^R\,P_l^m \cos\theta\,\sin(m-\tfrac{1}{2})\,\phi\,e^{-i\phi/2}\,e^{-i\omega t} = 0. \quad (52)$$

$$\hat{E}_{3-} = i4CR_E^R P_l^m \sin\theta \sin(m+\tfrac{1}{2})\phi\, e^{-i\phi/2} e^{-i\omega t} = 0. \quad (53)$$

$$\hat{E}_{1+} = 4CR_E^R P_l^m \sin\theta \cos(m+\tfrac{1}{2})\phi\, e^{i\phi/2} e^{-i\omega t} = 0. \quad (54)$$

$$\hat{E}_{3+} = 4CR_E^R P_l^m \cos\theta \cos(m-\tfrac{1}{2})\phi\, e^{i\phi/2} e^{-i\omega t} = 0. \quad (55)$$

By finally putting together the partial components of equal spin from (46)–(49) and (52)–(55) according to (32), and by taking into account (8) and (11) for the 2-components, we get for the left-spinning hydrogen field

$$E_-^{\mathrm{hyd}} = \begin{cases} E_{1-} = 4CR_E^R P_l^{|m|} \cos\theta\, e^{i(m-1/2)\phi} e^{i(-\phi/2-\omega t)} = 0, & (56) \\ E_{2-} = i4CR_E^R P_l^{|m|} \cos\theta\, e^{i(m-1/2)\phi} e^{i(-\phi/2-\omega t)} = 0, & (57) \\ E_{3-} = 4CR_E^R P_l^{|m|} \sin\theta\, e^{i(m+1/2)\phi} e^{i(-\phi/2-\omega t)} = 0, & (58) \end{cases}$$

and an analogous expression for the right-spinning one. Putting together both spin states, we can write

$$E_\pm^{\mathrm{hyd}} = \begin{pmatrix} 4CR_E^R P_l^{|m|} \cos\theta\, e^{i(m-1/2)\phi} e^{i(\pm\phi/2-\omega t)} \\ i\,4CR_E^R P_l^{|m|} \cos\theta\, e^{i(m-1/2)\phi} e^{i(\pm\phi/2-\omega t)} \\ 4CR_E^R P_l^{|m|} \sin\theta\, e^{i(m+1/2)\phi} e^{i(\pm\phi/2-\omega t)} \end{pmatrix} = 0 \quad (59)$$

The magnetic field components we obtain in the same manner as the electric ones by first inserting the relations [5] (44), (45)

$$l_H^{\mathrm{I}} = l_E^{\mathrm{I}} - 1 \quad \text{and} \quad l_H^{\mathrm{II}} = l_E^{\mathrm{II}} + 1\,. \quad (60)$$

They read

$$H_\pm^{\mathrm{hyd}} = \begin{pmatrix} 4CR_H^R P_l^{|m|} \cos\theta\, e^{i(m-1/2)\phi} e^{i(\pm\phi/2-\omega t)} \\ i\,4CR_H^R P_l^{|m|} \cos\theta\, e^{i(m-1/2)\phi} e^{i(\pm\phi/2-\omega t)} \\ 4CR_H^R P_l^{|m|} \sin\theta\, e^{i(m+1/2)\phi} e^{i(\pm\phi/2-\omega t)} \end{pmatrix} = 0 \quad (61)$$

Obviously we are back to Schrödinger again. In essence only the spin function has been added. This represents an improvement which had to appear imperatively, since electrodynamics as a theory with clear concepts necessarily has to bring about a conceptually clear description of spin. The Dirac theory, in comparison, is not able to explicitly express spin in the spinor components of its hydrogen field since it cannot accomplish the spin generating interchange of components achieved by electrodynamics in its equation (10). The Maxwell theory shows spin to be a pure phase phenomenon, i.e., the spin wave revolves, in the coordinate system chosen here, around the 3-axis as a pure phase wave, slowly near the center, far away in the far-field with arbitrary speed. A deeper insight into the perculiarities of the electrodynamic spin wave fields of half-integer spin particles requires profound analyses. –Especially the exciting idea of a substantiated spin field which rotatingly transports not the phases but the field strengths themselves reminds of Feinberg's tachyons, and still is far from a conceivable credibility and from the almost unforeseeable applications in a future interstellar communication technology.

The most essential improvement of the spin and momentum representation obviously lies in the cancelling action of the azimuthal functions which adjusts the field to the empirical electrodynamic imperceptibility of hydrogen: electrodynamic particle fields with half-integer spin are represented in any point of configuration space as a standing wave with a cancelling counter-wave, so that the particle field vanishes everywhere and always, but everywhere constitutes a positive source of energy, constant in time.

A first preliminary but basic picture of electrodynamic hydrogen is offered by the central transversality (4)–(5). It depicts the hydrogen field as a light wave impinging from the outside – from all directions – on a point, being reflected by itself at that point and running outward again. The incoming and outgoing waves form a standing wave field with a high concentration of energy around the point of reflection, the center of the hydrogen field. The standing concentration of energy at the center is what we perceive as the hydrogen atom. – This picture removes in an unsurpassably simple manner the contradiction of the century of contemporary physics, the wave-particle-dilemma.

On closer inspection it becomes clear that the incoming and outgoing light wave breaks up into a field twin. As it already becomes evident from Ehrenfest's theorem [1] (5)–(17), two light fields of equal amplitude take part in the energy localization at a point, which influence each other, i.e., which mutually refract one another. In our method of solution they are taken into account in such a way that one light field is carried through the calculation as real, the other one as imaginary. In other words: The two centrally transverse twin fields of hydrogen demand calculation with complex components. – Ehrenfest's theorem, in a one-component calculation, shows that the centers of energy of the two twin fields obey Newton's mechanics. The cause of this mechanical interaction are the two twins of that Kepler system which reflects itself in the hydrogen spectrum and lets us perceive the hydrogen atom mechanically.

Finally subjecting the general solution (59), which is still open with respect to the signs of the components according to the statements made after (46), to the central transversality condition (4)–(5), the following must be valid:

$$\boldsymbol{E}\cdot\operatorname{grad}\varepsilon_{\text{int}}^{\text{hyd}}=\boldsymbol{E}\cdot\boldsymbol{r}^0=e^{i(m+1/2)\phi}\;e^{i(\pm\phi/2-\omega t)}$$

$$\cdot\begin{pmatrix}4CR_E^R\;P_l|^m|\cos\theta\,e^{-i\phi}\\ i4CR_E^R\;P_l|^m|\cos\theta\,e^{-i\phi}\\ 4CR_E^R\;P_l|^m|\sin\theta\end{pmatrix}\cdot\begin{pmatrix}\sin\theta\cos\phi\\ \sin\theta\sin\phi\\ \cos\theta\end{pmatrix}=0. \tag{62}$$

The condition (62), first of all, is fulfilled trivially because of the cancelling functions of the vector (59). Beyond that it is fulfilled if we omit the cancelling functions and attribute to the first two components a positive and to the third one a negative sign.

For the electromagnetic energy we get from (59) and (61)

$$U^{\text{hyd}}=\boldsymbol{E}_{\pm}^{\text{hyd}}\cdot\overset{*}{\boldsymbol{E}}{}_{\pm}^{\text{hyd}}+\boldsymbol{H}_{\pm}^{\text{hyd}}\,\overset{*}{\boldsymbol{H}}{}_{\pm}^{\text{hyd}}\neq 0\,, \tag{63}$$

where the asterisk denotes complex conjugation, as should be expected.

[1] H. Sallhofer, Z. Naturforsch. **43a,** 1039 (1988).
[2] H. Sallhofer, Z. Naturforsch. **44a,** 504 (1989).
[3] H. Sallhofer, Z. Naturforsch. **41a,** 468 (1986); and corrigenda in H. Sallhofer, Z. Naturforsch. **41a,** 1087 (1986).
[4] H. Sallhofer, Z. Naturforsch. **44a,** 715 (1989).
[5] H. Sallhofer, Z. Naturforsch. **45a,** 1038 (1990).
[6] M. v. Laue, Handbuch der Radiologie, Bd. VI, 1, pp. 102ff., Leipzig 1933.
[7] H. Sallhofer, Z. Naturforsch. **44a,** 167 (1989).

Index

Abdera 6
Abraham, Max 115
absolute temperature 189
absorption of light 73
acoustics 143
action-angle variables 176
Adams Prize 20
adiabatic invariance principle 175
alkali elements 104
alpha particles 38
Alsos 103
Ampère, A. M. 17, 29
Anaximander 5, 9, 215
Anderson, Carl D. 204
Ångström, Anders J. vii
angular frequency 178
angular momentum 79, 106, 107, 111, 128, 162, 176
  generalized definition of 108
  orbital 92
  spin 92
  quantization of 104, 179
Anschluss 147
apeiron 9
Aston, F. A. 19
atmospheric electricity 142
atom, nuclear 37
  Thomson's 17
atomic spectra 39, 104
atomicity 21
atomists 6
Auschwitz 103

Balmer formula, relativistic generalization of 92
Balmer, Johann J. vii, 54
Balmer series 28, 59, 68, 90
Barón, Máximo 15, 30
Becquerel, Antoine-Henri 54
being 6
Beltrami fields 215
beryllium 69
Bethe, Hans A. 94, 201, 207
big bang 9
binding energy 64
Biraben, F. viii
Biswas, Tarun 101, 116
Blackett, Patrick M. S. 204
Bode's Law 41
Bohr formula 161, 176, 179, 184
Bohr, Niels vii, 29, 40, 51, 80, 85, 146, 178, 217
  quantization 175
  quantization rules 91
  theory 104, 181
Bohr's assumption 63
  correspondence principle 180
Boltzmann constant 189
Born, M. 131, 136, 145, 165, 179, 217, 220
Bose–Einstein statistics 193
boundary condition 133
bound states 129, 130, 152
bra 109
bracket notation 110
Brackett series 59, 90
Bragg, William L. 94
Brillouin, Leon 94
Brownian motion 143
Bunsen, Robert 54

c-numbers 185, 193
Cagniac, B. viii
capillarity 143
cathode rays 23

Cavendish, Henry 19
Cavendish Laboratory 18, 19
celestial mechanics 181
Čerenkov radiation 90
Chadwick, James 46
Chandrasekhar, S. 215
characteristic equation 153, 154
Chernobyl 221
Clarke, Arthur C. 46, 47
classical electrodynamics 156
classical mechanics 175
commutation 160
commutativity, lack of 180
commutator bracket 111
commutator, Heisenberg's 181
complex conjugate 188
Compton effect 190, 195
Compton-scattered radiation 190
Condon, Edward U. 94
conic section 162
conjugate momentum 120
conjugate transpose hermitian adjoint 109
conservation of momentum 123
conservation laws 222
constant of the motion 123, 125
convergence limit 59
Copenhagen interpretation 211, 217
Coulomb potential 121, 179
Coulomb, C. A. 17, 29
Coulomb's law 63
Crookes, W. 18, 21, 30
Crookes tube 18
cross product 122
Cunningham, Ebenezer 172
curator 87
current term, Ritz's 69
Cuvier, Georges 51
cyclotrons 215

Dalton, J. 21, 31
Darwin, Charles 52, 174, 207
de Broglie, Louis 7, 129, 144, 165, 187, 206, 216, 229
de Broglie wave 194
de Lange, O. L. 141
de Sitter 131
de Unamuno, Miguel 15
Debye, Peter J. W. 94, 144
degrees of freedom 106, 109
delta function, Dirac 190
  Kronecker 160
Democritus 6, 53, 215
density matrix 156
Des Coudres, Th. 23, 31
dielectric properties 142
differential equations 126
dipole–dipole interaction 104
dipole moment, magnetic 105
Dirac equation 93, 113, 155, 201
    negative energy solutions 202
    positive energy solutions 202
  hole 204
  spinors 109
Dirac, Paul A. M. viii, 7, 108, 115, 145, 165, 171, 206
Dirac's breakthrough 181
discharge tube 24
discovery of radioactivity 54
divectors 9
Doppler principle 175
dot product 110
Duality, Ionian-Eleatic 6

East–West tension 224
Eckart, Carl 94
Eddington, Arthur S. 173
Edmonds, Jr., James D. 119, 136
Ehrenfest, Paul 7, 219, 229
eigenstates 110

eigenvalue 110
    spectrum 153
Einstein, Albert 36, 55, 85, 115, 144, 178, 217
Einstein's objection 41
    letter 225
elbower society 227
electric discharge 59
electricity, atmospheric 142
electrochemistry 17
electrodynamics, classical 63, 156
electrolysis 17
electromagnetic field 177
electron 18, 25, 56
    dynamics 90
    orbital 39
    quantum theory of 195
    relativistic equation of 171
    ring 77
    spin 101, 113, 132, 171
    spinless 129
    theory 17
    wave 187
elements, transmutation of 46
Eliot, T. S. 103
elliptical orbit 76
energy 120
    relativistic corrections to 92
energy-frequency relation of Planck 149
energy-mass relation of Einstein 149
engineering science 87
ether 7, 18, 23
Everett, E. 20
Ewald, Peter P. 94
exclusion principle 105, 113, 203

factorization method 159
Faraday, M. 17, 30
Fermi–Dirac statistics 188
Fermi, Enrico 189, 207

Fermi statistics 89
Fermilab, 226
Feynman, Richard P. 93, 130, 136
field, concept of 17
fine structure, addition 91
fine structure constant 89, 92, 152, 171, 199
fire 10
first Laplace vector 164
Fleischmann, Martin 46
Fluorescent lighting 21
focus 162
force, centrifugal 39
  divectorial 10
  electromagnetic 44
  gravitational 44
  inertial 39
Force-free magnetic fields 215
Foster, John Steward 216
Fourier coefficients 180
Fowler, Ralph H. 172
Franklin, Benjamin 17
frequency, mechanical 64
Fresnel, Augustine 54
friction 87

Geiger, Hans 38, 46
general relativity 143
German ultranationalism 89
ghost fields 220
Goldstein, E. 18, 21, 30
Gordon, Walter 195, 199, 207
Goudsmit, advice for young physicists 114
Goudsmit, Samuel ix, 92, 196
Greek metaphysics 5
ground state 178
group theory 108
gyromagnetic ratio 107

Haas, Arthur 60

Hahn 221
Hamilton vector 104
Hamilton–Jacobi theory 175, 176
Hamilton, William R. 216, 228
Hamiltonian 120, 121, 176, 193
  classical 129
  equations 176
  mechanics 181
  non-relativistic 200
  operator 155, 158
Hansen, Hans 66
harmonic oscillator 158, 180, 188
Hauksbee 21
Hayles, W. H. 20
Heisenberg commutator 181, 195
Heisenberg, Werner x, 10, 85, 94, 130, 131, 145, 165, 179, 213, 217, 229
Heisenberg's basic guess 132
  model 213
  operators 133
  product rule 182
Heitler, Walter 94
Helmholtz, H. 18, 30
Henderson, W. Craig 19
Heraclitus 5, 9, 215
Hertz, Heinrich R. 18, 22, 30, 177
Herzfeld Karl 94
Higgins, William vii
Hilbert, David 86, 153
Hilbert space 191
Hindu philosophers 53
Hittorf, J. W. 18, 21, 30
holes 203
Houston, William 94
Hurwitz, Adolf 86
Huygens, Christian 54
hydrogen atom 121, 128
  radius of 65
  spectrum of 59, 113, 152

hydrogen molecule 77, 171

illusions of the senses 5
inertia, law of 8
Infeld L. 23, 166
inner product 110
interaction energy 177
interpretation controversy 216
ionisation potential 65
ion 17
Isomorphism, Maxwell–Dirac 7
isotopes 39
Itzykson, C. xi

Jacobian identity 182, 192
Jordan, Pascual 88, 131, 145, 165, 181, 217
Julien, L. viii

Kelvin, Lord 19, 31
Kepler problem 154
  structure 212, 220
  system 42
ket 109
kinetic energy 73
Kirchhoff, Gustav 54
Klein, Felix 87
Klein–Gordon equation 155, 195, 197
Klein, Oscar 195
Kohlrausch, R. 17, 30
Kramers, Hendrik A. 179, 185
Kronecker delta function 160
Krönig, Ralph 114
Kuhn, Heinrich 201, 207

Lagrangian 120, 123
Lakhtakia, Akhlesh vii, 211, 228
Lamb shift 93, 201
Lamb, Jr., Willis E. 93, 201, 207
Lanczos, Cornelius 191, 207

Landé, Alfred 94, 105, 145
Langevin, P. 19
Laplace vector, first 164
  second 162
Laplace–Runge–Lenz vector 162
Laplacian 151
Laporte, Otto 94
lattice dynamics 142
law, normative 9
laws, conservation 9, 222
Lederman, Leon 229
Lemaître, Georges E. 173
Lemonick, Michael D. 47
Lenard, P. E. A. 18, 22, 30
Leucippus 6, 53
Lie, Sophus 192
Lincoln, F. J. 19
Lindemann, Ferdinand 86, 146
linear operator 110
linear momentum 108, 176, 178
Lipson, H. S. 30
lithium 69
Lodge, Oliver 22
logarithms 180
logos 9
London, Fritz 94
Lorentz, Hendrik A. 88, 145, 217
Lorentz transformation 198
lowest energy state 178
Lyman series 59, 68, 90

Mach, Ernst 42, 212
Macháček, Martin 42, 46
magic numbers 216
magnetic field 120
  force-free 215
magnetic moment 100, 107
Marsden, Ernest 38, 46
Marshall Plan 223

master prototype system 136
matrix algebra 133
  mechanics 145, 162, 181, 185
  multiplication 110
matter waves 150
Maxwell–Dirac isomorphism 7, 211, 212
Maxwell, James Clerk 19, 30, 54, 177
Maxwell's equations 176, 177
McClelland, A. J. A. 19
mechanical frequency 64, 71
mechanics, celestial 181
  classical 111, 175
  theoretical 87
Medawar, Sir Peter viii
Melissus 5
Mendeleeff, Dmitri 56
metals, electron theory of 89
Meyer, Lothar 56
Michelson, Albert A. 91, 200
microscopic scale 109
Millikan, Robert A. 104, 115
model, Bohr's 51, 73, 88, 151, 217
  Heisenberg's, absence of 130
  Kelvin's 26
  nuclear 38
  plum-pudding 28, 37
  quantum mechanical 41
  raisin-muffin 28
  Rutherford's 35, 57
  Sallhofer's 212
  Saturnian 57, 74
  Schrödinger's 41, 153
  Sommerfeld's 85
  Thomson's 26, 57
model is not the thing 35
molecular hydrogen 77, 174
molecule of electricity 18

momentum 72, 120, 162
angular: *see* angular momentum
conjugate 120, 124
conservation of 123
linear 108, 176, 178
ordinary 123
monad 53
Morse, Philip M. 94
Mossotti, Ottavio F. 19
Muller, Wilhelm 89

Nagaoka, Hantora 28, 31, 37, 46, 57, 80
Natural Philosophy 18
Nazis 89, 147
neutron 39
Newton, Isaac 79
Newton's corpuscles 55
laws 72
second law 121
Nicholson, John William 37, 46, 60, 80,
Nietzsche, threefold vision of truth 8
non-commutative algebra 192
nuclear reactions 46
nucleus, positive 21
nucleus, tiny 39

Oersted, H. C. 17, 30
old quantum theory 152
operator 108
method 141
Oppenheimer, John R. 194
optical dispersion 179
orbit, circular 91
orbit, elliptical 91
orbit, hyperbolic 153
orbit, Keplerian 132
orbital electrons 63, 72, 217
orbiting electrons 40
oscillations 87

oscillator 178

Paneth, Fritz 46
parabolic coordinates 157
Parmenides 5, 215
Paschen, Friedrich 67
Paschen series 59, 90
Pauli–Darwin theory 199
Pauli operator 162
  spin matrices 112, 196
  spinors 109, 111, 113
  wave equation 196
Pauli, Wolfgang ix, 94, 108, 115, 132, 162, 179
Pauling, Linus C. 94
perception 6
periodic motion 176
Periodic Table 27, 28, 55, 62, 105
Perrin, J. 18, 22, 30
perturbation theory 157
Peters, Kurt 46
Pfund series 59, 90
phase wavefunction 151
phase waves 149, 150
phosphorescence 21
photoelectric effect 55, 178
photon 36
physics and poetry 194
Plücker, J. 18, 21, 30
plagiarism 114
Planck–Dirac constant $\hbar = h/2\pi$ 178
Planck, Max ix, 51, 55, 88, 115, 145, 177, 213, 217
Planck's constant 55, 79, 92, 103
  concept of light emission 61
  hypothesis 177
plane wave 202
planetary orbit laws 62
planetary system 39
Plimmer, M. D. viii
Pons, B. Stanley 46

Poisson bracket 111, 119, 124, 127, 134, 181
identity with Heisenberg's commutator 185
Poisson, Simeon D. 181
polarity 39
polynomial method 141, 153, 154
position coordinate 176
eigenstate 111
observable 110
postulate, Bohr's second 91
Poynting, J. H. 19
probability 110
negative 197
wave 221
proton 39
core 57, 69, 75
Ptolemy's epicycle theory 42
Pye, W. G. 19

q-numbers 185
quadratic integrability 159
quanta 55, 61, 103
quantization, rules for 90
Bohr–Sommerfeld 91
general 91
quantum electrodynamics 192
quantum mechanics 43, 72, 171, 177
new 88, 145
old 88, 145, 152
principles of 109
three formulations of 191
quantum number, azimuthal 199
principal 159, 200
radial 154
second 91
quantum numbers, four 105
quantum physics, Schrödinger's 145
quantum-mechanical scattering 195
quantum state 133
quantum transition 129

quarks 201
quaternions 8

Röntgen, Wilhelm 18, 22, 30, 88
radioactivity, discovery of 52
Rajopadhye, Vilas Y. 171
Ranjitsinhji x
Rayleigh, Lord 19
relativistic precessional effect 106
relativity, general 143
resonance phenomena 87
Retherford, Robert 201
Retherford, Richard C. 207
Richardson, O. W. 19
ring 74
Ritz equation 70
Ritz, Walter 69
Robertson, Daniel Stewart 51
Rolph, J. M. 20
Rose, Dennis 5
Roosevelt, Franklin 225
rotation operator 112
Rubinowicz, Adalbert 92
Ruth, Babe x
Rutherford, Ernest ix, 19, 28, 46, 80
Rutherford's model 178
Rydberg constant 55, 59, 67, 179
  equation 70
Rydberg, Johannes R. vii, 55

Sömmerfeld, Arnold ix
Salam, Abdus 206
Sallhofer, Hans x, 7, 41, 42, 46, 211, 228
Saturn's rings 26
scalar potential 150, 176
scattering 38
  angles 38
Schopenhauer 143
Schrödinger, Erwin x, 85, 116, 164, 187, 206, 217

Schrödinger's wave equation 113, 132, 144, 145, 191, 195
time-dependent 156
time-independent 156
Schweber, Silvan S. 42, 47
Schwinger, Julian 93
Searle, G. F. C. 19
second Laplace vector 162
second Pauli principle 204
senses, illusion of 5
Segre, Emilio 204
Series, George W. x, 201, 207
shift operations 158
shift operator method 153
shift operators 162
shirker society 227
Slater, John C. 173, 185
solar system, miniature 38
Solvay Conference 60
Sommerfeld, Arnold 7, 72, 85, 145, 165, 173
Sommerfeld, white Jew of science 89
special relativity 89, 205
species, evolution of 52
extinction of 52
specific heats 143
spectral lines 59
Spectroscopy 55
spectrum, atomic 39, 104
spectrum, hydrogen 90
spherical harmonic 152
spherical polar coordinates 152
spin 93, 106, 101, 110, 196
angular momentum 108, 199
energy 108
half-integer 7
states 111, 203
spin-half case 110
spinors 7, 109, 218
Standard Model 219, 225
Stark effect 157

state 109
stationary state 65, 156
statistical thermodynamics 148
Stern, Otto 78
stimulated emission 189
Stoney, G. J. 25, 30
Strategic Defense Initiative 224
Stueckelberg, Ernest C. G. 94
sub-atom hypothesis 25
Superconducting Super-Collider 113, 218
synchrotron radiation 40

Teller, Edward 94
Thales of Miletus 5
theory of relativity 149
Thomas, Llewellyn H. 115, 174
Thomas precession 113
Thomson, George Paget 16, 22, 29
Thomson, Joseph John ix, 15, 30, 53, 80, 177
    atom 17, 28
    experiment 24
Three Mile Island 221
Tilton, Homer Benjamin 35
time dilation 149
Tomonaga, Sin-Itiro 93
total derivatives 122
Townsend, J. S. E. 19
trajectory 109, 110, 162
transformation matrix 190
travelling wave 150
two-electron system 188

Uhlenbeck, George E. 92, 102, 115, 196
unified field theory 9, 148
unit of charge 17
unity of opposites 9
uranium atom 75

vacuum fluctuations 201
pumps 21, 24
tubes 89
van Heerden, Pieter J. 42, 43, 47
vector potential 119, 176
quantized 193
Vlasov, A. D. 166
Volta, A. 17, 29
voltage 120
von Bayer, H. C. 30
von Laue, Max 94, 188, 217
von Liebig, Justus 25

water 8
wave, compressible 23
de Broglie 40
electromagnetic 87
gravitational 40
longitudinal 23
matter 40
sound 23
transverse 23
wave equation, free particle 135
wave mechanics 141, 145, 149, 187, 212
wavefunction 145, 151, 187, 202, 216
electrodynamic interpretation of 156
momentum-space 161
radial 153
wavelength 178
Weber, W. E. 17, 30
Weinberg, Steven 226, 229
Wentzel, Gregor 94, 186
Weyl 184
White, William B. 85
Wien, Max 144
Wien, Wilhelm 54, 145
Wigner, Eugene P. 108, 115, 192, 206, 207
Wilson, C. T. R. 19, 20
Wilson, H. A. 19

Wilson, Robert 226
Wood, Robert 73

X-rays 18, 20, 22
  diffraction of 87, 142

Young, Thomas 54

Z-number 58
Zeeman effect 105
Zeeman spectra 179
Zeleny, J. 19
Zeta-Puppis 68
Zhirinovsky 226
Zuber, J. xi